The World's Beef Business

THE WORLD'S BEEF BUSINESS

JAMES R. SIMPSON
Associate Professor
Food and Resources Economics Department
University of Florida, Gainesville

DONALD E. FARRIS
Professor
Agricultural Economics Department
Texas A & M University, College Station

IOWA STATE UNIVERSITY PRESS / Ames, Iowa, U.S.A.

Composed and printed by The Iowa State University Press, Ames, Iowa 50010.

First edition, 1982

Library of Congress Cataloging in Publication Data

Simpson, James R.
The world's beef business.

Includes index.
1. Beef industry. 2. Cattle trade.
I. Farris, Donald E. II. title.
HD9433.A2S53 338.1′76213 81-20770
ISBN 0-8138-0960-6 **AACR2**
ISBN 0-8138-1924-5 (pbk.)

To **FLOYD R. SIMPSON,**
a fine agricultural economist,
and to **VIOLA FARRIS,**
and in memory of
WILLIAM FARRIS,
good farmers

A NOTE ON THE CENTER FOR TROPICAL AGRICULTURE AT THE UNIVERSITY OF FLORIDA. This book is published by the Iowa State University Press in conjunction with the Center for Tropical Agriculture at the University of Florida. The Center, directed by Dr. Hugh Popenoe, is a component of the Institute of Food and Agricultural Sciences (IFAS). It draws upon more than 1,000 scientists in the statewide complex of agricultural teaching, research, and extension. The Center's activities include involvement in both graduate and undergraduate training, foreign student coordination, and technical assistance and cooperation with the numerous internationally oriented organizations and centers at the university.

CONTENTS

PREFACE

The World's Beef Business is written to provide an understanding of and a source of information on the world's beef business. It is for undergraduate and graduate students, managers of cattle and meat operations, professionals, and investors interested in the business, economics, marketing, and trade of the cattle and beef industries.

In much of the world, beef is a supplemental enterprise and/or a by-product of the dairy industry. On the global scene as well as within most countries, it is a major component of the "balance wheel" of feed and food supplies because cattle are widely adaptable and flexible in their use of resources and in their product output. As the world population presses the available food supply, beef will probably have to make the greatest adjustments of all the major food industries. By describing, analyzing, and comparing the diverse beef businesses of the world, we hope to clarify the key factors that shape demand, markets, prices, and costs under different and changing conditions. This book examines low-cost and high-cost systems, and efficient and inefficient systems. It identifies bottlenecks, practices, and policies that impede or enhance productivity in this vital source of human nutrition.

A special effort is made to present the subject as an interrelated set of functions and to point out that each one is a vital, necessary component. It is by dissecting the whole into its parts, by comparing them, and by tying the various aspects back into a whole through policy considerations that the reader obtains perspective. Thus the book is a synthesis: It brings together scattered and voluminous material into one organized source and transforms a complex subject into a relatively clear and understandable set of functions. In addition, a special effort has been made to provide facts, figures, examples, and a conceptual basis for decisions, which makes the text a useful manual as well as a reading experience for those who would like to know about one of the world's largest and most exciting businesses.

The chapters run the gamut from production to retailing and make forays into the marketing of live cattle, trade, diseases, demand, and cattle breeds. Emphasis is given to representative geographic areas that are most important in terms of production, trade, and distribution to provide the reader with a sense of where, why, when, and how the industry operates. In effect, a systems approach has been adopted. Special care is taken to explain the way in which systems evolve and consequently considerable attention is given to the developing countries, few of which have had the benefit of modern management and technology. This orientation also has the advantage of opening the subject to allied interests serving the beef industry and to nonindustry groups such as anthropologists, development specialists, and policymakers.

Numerous people deserve credit for assisting in the preparation of this book. John Hopkin, Head of the Department of Agricultural Economics at Texas A & M University, and Leo Polopolus, Chairman of the Food and Resource Economics Department at the University of Florida, provided encouragement and understanding and allowed flexibility in reorientation of job responsibilities to permit completion of the manuscript. Special thanks are due reviewers who suggested many improvements: James McGrann, Bill Boggess, Joe Conrad, Jack Loosli, Tim Olsen, Mike Burridge, Phil Martin, A. J. Dye, Gerald Mott, and Norman Beaton.

Credit goes to Lorri Tomlinson and Sharleen Simpson for typing and reviewing many of the drafts, and to the staff editors at Iowa State University Press for encouragement and skill in guiding the manuscript through the editorial and production process.

We also want to thank Hugh Popenoe, Head of International Programs and the Center for Tropical Agriculture, Institute for Food and Agricultural Sciences, University of Florida, for generously providing partial financial assistance for publication of the book as well as travel funds used in preparation of the manuscript. In addition, we are particularly grateful to the Rockefeller Foundation for granting us a place as scholars-in-residence at their Villa Serbelloni facility on Lake Como near the town of Bellagio, Italy. It was in this magnificent and peaceful setting that a major part of the final draft was completed.

This book, which was nearly a decade in preparation, is truly an international study. It grew out of our work at Texas A & M University, and it was there that the first draft was written. Substantial modifications were made in Chile and Costa Rica while Jim Simpson was there on long-term assignments. Additional modifications were made by Don Farris while on short-term assignments in Syria and Tanzania. Changes were made during numerous international trips and as part of local, regional, and national extension and research projects in the United States.

Our thanks go to Itsuko Simpson for careful review of galley and page proofs and for endless patience and to Billye Farris for continual encouragement and fortitude.

We conclude that while the beef industry is a versatile, complex, multiproduct industry providing humankind with products for both survival and the good life, a vast amount of unnecessary waste exists in some economies and much improvement can be made everywhere. We are proud to be associated with people the world over who are working to bring about the needed changes that will help producers, middlemen, and consumers in the developing countries and developed areas of the globe.

All errors or misinterpretations in this book are the responsibility of the authors.

The World's Beef Business

1
OVERVIEW

The world's beef business is a heterogeneous mixture of production and marketing systems that range from that of the nomadic East African herdsmen, some of whom still use a spear to protect their cattle from lions and only occasionally sell an animal for beef in order to buy staples, to the intensive grain-finishing systems of Europe and North America, which produce beef as an efficient business enterprise. In between, much of the world's beef supply is a by-product of draft cattle or a secondary product of the dairy industry. In Central and South America, Australia, and Africa, beef is produced from cattle raised on forage that other animals generally would not use. In most areas where beef cattle are raised, beef production offers the most efficient use of resources; otherwise, the beef enterprise could not compete with sheep, goats, dairy cattle, hogs, or poultry for land, capital, or feedstuffs.

The simple economic relationships just described are not devoid of waste and inefficiency. In fact, most of the world's beef production and marketing systems, like many sectors of agriculture and other economic activities, are woefully wanting in production and management skills. We write this book with the hope that it may contribute toward a better understanding of the beef business, toward improving the efficiency of beef production and marketing, and thereby toward raising the nutrition and well-being of humankind.

INTRODUCTION

The world's beef business is huge by any set of measures, and it is growing rapidly. World beef cattle numbered approximately 1.2 billion head in 1979 compared with 920 million head in 1960 and 691 million head just before World War II (Table 1.1). Production of beef, another measure of size, is equally impressive. World production was 17 million metric tons in the late 1930s, 28 million metric tons in 1960, and 47 million metric tons by 1979 (Table 1.2). The vastness these figures represent can be grasped by considering the volume of production they signify. Since one metric ton of bone-in beef carcasses occupies ap-

James E. Vance Photo/Fort Worth *Star Telegram*

Table 1.1. World cattle inventory by region, late 1930s to 1979

Region	Late 1930s	Year[a] 1950	1960	1970	1979	Share of world cattle 1960	1979	Increase 1950 to 1979	Increase 1960 to 1979
			(1,000 head)			*(%)*		*(%)*	*(%)*
Africa	80,000	88,000	114,000	156,766	170,110	12	14	93	49
North and Central America (other than USA)	28,971	31,679	46,664	56,170	63,521	5	5	101	36
	66,029	77,321	96,236	112,303	110,864	10	9	43	15
South America	107,000	133,000	160,000	198,350	216,119	17	18	63	35
Asia	228,000	215,000	292,990	351,731	366,579	32	30	70	25
Europe	103,000	96,000	113,400	124,020	134,535	12	11	40	19
Oceania	18,000	19,000	22,800	31,414	36,203	2	3	91	59
USSR	59,700	53,300	74,115	95,161	114,086	8	10	114	54
Total world	690,700	713,300	920,300	1,125,915	1,212,017	100	100	70	32
Developed world					425,019	. . .	35		
Developing world					786,997	. . .	65		
(Centrally planned economies)[b]					(217,751)	(. . .)	(18)		

Source: Compiled from FAO *Production Yearbook:* 1979 from 1979 issue, 1970 from 1972 issue, 1960 from 1962 issue, late 1930s and 1950 from 1954 issue.

Note: Does not include buffaloes.

[a]Totals contain an estimate of missing countries. Thus columns may not add to totals.

[b]Includes parts of developed and developing world.

Table 1.2. World beef and buffalo meat production by region, late 1930s to 1979

Region	Late 1930s	Year[a] 1950	1960	1970	1979	Share of production 1960	1979	Increase 1950 to 1979	Increase 1960 to 1979
			(1,000 head)			(%)		(%)	(%)
Africa	1,210	1,520	1,800	2,490	2,852	6	6	88	58
North and Central America (other than USA)	953	1,056	1,497	1,925	2,333	5	5	121	56
USA	3,617	4,844	7,183	10,006	9,704	26	21	100	35
South America	3,380	4,020	4,360	6,041	6,865	16	15	71	57
Asia	1,900	1,750	1,330	4,048	5,016	5	11	187	277
Europe	4,800	3,900	6,320	8,896	10,508	23	22	169	66
Oceania	720	800	890	1,458	2,525	3	5	216	184
USSR[a]	...	...	...	5,381	6,966	16	15	...	...
Total world	16,580	17,890	28,000	40,245	46,769	100	100	...	67
Developed world				20,090	31,530	...	67		
Developing world				10,899	15,239	...	33		
(Centrally planned economies)[b]				(9,256)	(11,685)	(...)	(25)		

Source: Compiled from FAO *Production Yearbook:* 1979 from 1979 issue, 1970 from 1972 issue, 1960 from 1962 issue, late 1930s and 1950 from 1954 issue.

Note: Indigenous production only; does not include imported animals.

[a]USSR and Mainland China excluded until 1970, but world totals from 1960 contain an estimate of missing countries. Thus columns may not add to totals.

[b]Includes parts of developed and developing world.

proximately 1.78 m^3, 1979 world production would cover about 83 km^2 of beef piled 1 m high. The cash value of world production is equally impressive. On the basis of an international price of $1,600 per ton, the total value of 1979 production was $75 billion.

A few facts about the United States meat industry will also help place the industry in better perspective. In 1979 receipts from sales of cattle and calves totaled $34.8 billion, or about 21 percent of cash receipts from farming. There are about 6,300 federally inspected slaughter plants with over 170,000 employees. Their average hourly earnings at $6.44 in 1977 were about $0.80 higher than the average wage for all manufacturing (American Meat Institute 1978).

World cattle inventory increased about 32 percent from 1960 to 1979 (Table 1.1), primarily in Oceania (59 percent) and the USSR (54 percent). The United States registered the lowest rate of increase, 15 percent, because inventory numbers were drastically reduced owing to the cattle cycle. Despite these growth differences, there was little change in relative position of the different regions. In 1979 the greatest cattle concentrations were in Asia (30 percent of total world cattle numbers excluding buffalos) and in South America (18 percent of the total). The United States had 9 percent, while Europe accounted for 11 percent of world cattle numbers. The USSR had about 10 percent of the total. About 35 percent of all cattle are in the developing nations, while 65 percent are in the developed world. About 18 percent of the world's cattle are in centrally planned economies. The great discrepancies found in the share of cattle compared with the share of beef production between the developed and the developing world and the centrally planned economies are shown in Table 1.3. With 35 percent of the cattle, the developed countries produced 67 percent of the beef.

Population densities of the major cattle producing countries vary considerably (see Appendix, Table A.1). For example, the United Kingdom has 229 inhabitants per square kilometer whereas Australia has only 2 and the United States 24. Cattle per person for these same countries varies from 0.24 in the United Kingdom, to 1.89 in Australia; in the United States the figure is 0.50 compared with 0.90 in South America.

Table 1.3. Share of cattle and beef production between world sectors

Sector	Cattle numbers	Beef and buffalo meat production
	(percent of world total)	
Developed world	35	67
Developing world	65	33
Total	100	100
Centrally planned economies	18	25

Source: Tables 1.1 and 1.2.

The average for the whole world is 0.28. The developed countries have 0.37 cattle per person, whereas the developing countries have 0.25 head (Table 1.4). (Cattle inventory and other related statistics are given in Appendix Table A.1, for most countries in the world, which are compared by major regions.)

The great differences in the ratio of cattle to people among countries are also reflected in the variety of beef products around the world. At the retail level, a myriad of fresh and frozen beef cuts as well as a multitude of canned products containing various amounts of beef can be found in different nations. Corned beef, for instance, has been sufficiently cooked so that 1 kg of that product is equivalent to 1.43 kg of chilled, deboned beef.

The type of beef product in greatest demand also varies with the country. Hamburger is a national favorite in the United States, and of course the hot dog (which contains varying amounts of beef) has become identified with this country. Chileans, on the other hand, hardly ever eat hamburger and they rarely eat steak as North Americans know it outside the capital city of Santiago. Nearly all their meat is parted at the major muscles or, where applicable, it is deboned. In some areas of the world parts of the entrails are considered a delicacy and command high prices in relation to meat. On the wholesale side, diversity is also a significant factor. For instance, one meat packer organization in South America, Camara Argentina de la Industria Frigorifica (CADIF), lists more than eighty different subprimal boneless specialty cuts that require only minimal cutting at the retail level.

Complexity and diversity also characterize the world's beef cattle production systems. Although the advanced systems of the world are efficient and achieve high output per beef cow, not many areas have adopted advanced production, management, or marketing technology, largely because their herds are generally small (ten head or less) and because reproduction periods are long (gestation is 9 months in a cow, and 14–36 months are required for fattening) in comparison with other types of livestock or poultry. Also, many decision makers have believed that the beef industries are too traditional and complex to improve much, or that cattle production and marketing is only for "the big boys." International trade negotiations have done little to foster improved trade relations, as the protectionism in the 1970s attests.

Efficiency and other improvements introduced by advanced technology depend on more than knowledge and skill of the producer, processor, and others in the beef chain. There must be political and economic stability and a legal and economic framework to foster long-term investment and stable international trade (Aziz, 1975). For example, some of the world's best beef producing resources have been underutilized in

Table 1.4. Livestock and related statistics by region, 1979

Region	Total area	Permanent pastures	Population	Population density	In agriculture	Total cattle	Cattle per person	Total world cattle	Permanent pasture per head
	(1,000 ha)		*(1,000)*	*(mi²)*	*(%)*	*(1,000 head)*	*(head)*	*(%)*	*(ha)*
Africa	3,031,299	806,336	455,873	15	66.1	170,110	0.37	14	4.74
North and South America	2,241,492	354,693	363,981	16	11.8	174,385	0.48	15	2.03
South America	1,781,851	444,704	238,912	13	32.4	216,119	0.90	18	2.06
Asia	2,757,442	623,041	2,509,010	91	58.6	366,579	0.15	30	1.70
Europe	487,055	87,016	481,726	99	15.8	134,535	0.28	11	0.65
Oceania	850,956	460,772	22,318	3	21.8	36,203	1.62	3	12.73
USSR	2,240,220	374,300	263,500	12	17.3	114,086	0.43	9	3.28
Total world	13,390,315	3,150,862	4,335,310	32	45.9	1,212,017	0.28	100	2.60
Developed world	5,618,166	1,269,936	1,154,780	21	13.1	425,019	0.37	35	2.99
Developing world	7,772,149	1,880,926	3,180,530	41	60.0	786,997	0.25	65	2.39
(Centrally planned economies)[a]	(3,521,422)	(739,809)	(1,398,920)	(40)	(49.4)	(217,751)	(0.16)	(18)	(3.40)

Source: Adapted from FAO *Production Yearbook*, 1979.
[a]Includes part of developed and developing world.

parts of South America and Africa because of the investors' lack of confidence, inadequate disease control, and violent fluctuations in available export markets. Dietary levels, costs, prices, and stability of world food supplies cannot improve unless these impediments are corrected.

GRAIN FATTENING OF CATTLE: AN INTRODUCTION TO ANALYSIS

The practice of feeding grain to cattle being fattened for slaughter while many people are starving around the world has become such an emotional issue that it provides a convenient point of departure for a central objective of this book, which is to analyze policy in the world's beef business. The principal target of allegations about morally wrong practices has been the United States, where beef is king among meats. Per capita consumption of beef in the United States exceeded pork for the first time around World War II, when readily available grasslands made it cheaper to mass produce fresh beef than pork. The advent of feedlots in the 1950s in response to increased demand for beef resulting from higher incomes and cheap grain resulting from great technological advances in agriculture meant that a "way of eating" was born.

By the early 1970s, there was a sudden recognition that the world's population was increasing at a geometric rate. Concomitantly, a series of natural and man-made events led to extreme grain shortages throughout the world. Although these shortages were short-lived, a major ramification was considerable publicity suggesting that beef cattle are relatively poor converters of grain to meat (Svedberg 1978). Let us analyze this criticism to determine whether feeding grain to cattle is "wrong." This is an important topic since grain feeding is the main method of cattle fattening in the United States and most of Europe, and since more grain feeding is expected in the lesser developed countries (LDCs) (FAO 1979*b*).

To begin with, beef cattle are versatile utilizers of crop residue or grain in fields where yields are so poor that they are uneconomical to harvest. Much of the expanded grain production has been destined for the cattle finishing industry. If legislation were adopted that prevented grain from being fed to cattle in feedlots, initially there would be substantial oversupplies of grain and much lower prices in the absence of an export market to absorb the differences in production. In the absence of effective demands there is no reason to believe grain exports would increase to any degree, or that they would be high enough to offset surplus production.

Without doubt, many hungry people around the world would like to eat U.S. grain, and the leaders of their respective countries probably would like to receive the grain if it were free or inexpensive. Thus we can say that there is a demand but that "effective demand," that is, purchasing power, is insufficient. In order for U.S. grain to reach the hungry people, someone must pay for it. It is unlikely that farmers will give away their produce, and the record indicates that the U.S. taxpayers' altruism manifests itself only in crisis situations and favors only a limited amount of "food for peace." In fact, it is not clear whether surplus grain shipments are a stimulus or deterrent to increased food production in the recipient countries.

Proponents of controlling grain feeding will point out that beef cattle are less efficient in converting grain to meat than are hogs or poultry, and that in an age when energy conservation is of wide concern, something should be done to reduce grain feeding. On the other hand, although beef cattle are less efficient on an individual basis in converting grain to meat than are hogs or poultry, the situation changes on a product basis. About 8 kg of grain are required per 1 kg of beef on a liveweight basis, whereas the comparable ratio for hogs is 4:1 and for poultry 3:1. In the case of beef cattle, the dam and sire of an animal that will eventually be fed for slaughter probably receive little or no grain, and the animal itself will be on full grain feed for only about four to six months out of its fourteen- to sixteen-month life. All hogs and poultry, on the other hand, are maintained and fed almost exclusively on grain. The result is that the ratios of grain per kilogram of product are actually higher for pork and poultry than for beef (Table 1.5).

Cattle are grain fattened in the United States and most of Europe because the input/output price ratios are favorable to this mix (Table 1.6). Also, the demand for beef (examined in more detail in Chapter 2) is sufficiently high that people are willing to pay premium prices over close substitutes. In developing countries beef prices are frequently set at artificially low prices with respect to grain so that grain feeding is generally not economically viable.

Table 1.5. Grain utilization rates for meat production in OECD-Oceanic countries in kilograms of grain fed per kilogram of product

Product	1962	1975	1985[a]
		(kg of grain fed per kg of product)	
Pork	5.80	5.40	5.00
Poultry	4.00	3.50	3.30
Beef and veal	2.40	2.90	3.10
Milk	0.14	0.16	0.16
Eggs	3.60	3.40	3.30

Source: Regier, 1978, pp. 64–66.
[a]Projected.

Table 1.6. Grain utilization rates for meat production in selected developed countries, 1975

Product	United States	Canada	Japan	Australia	Other OECD countries
	(kg of grain fed per kg of product)				
Pork	8.30	7.80	6.60	3.00	3.50
Poultry	4.20	2.20	2.30	2.20	3.00
Beef and veal	3.60	4.00	1.20	0.40	2.30
Milk	0.30	0.30	0.20	0.10	0.11
Eggs	4.00	4.40	2.30	2.50	3.10

Source: Regier, 1978, p. 65.

The possibility of fattening a higher percentage of cattle on forages rather than on grain rations is a recurring proposal in the United States. In 1974–1975 it was in response to high grain prices, international shortages of grain, a grave concern about humankind's ability to produce sufficient foodstuffs, and heavy long-term losses by cattle feeders. In 1978–1980 an interest in producing leaner beef, combined with reduced supplies of cow beef as beef cycle number eight was completed and as the industry embarked upon cycle number nine, led to speculation about the economics of finishing cattle on forage to provide increased supplies of beef for the ground beef trade. This is a major consideration since about 45 percent of all beef in the United States is consumed in ground form. A 1979 study of this possibility (Connor and Rogers) clearly indicated, however, that "grow-out" programs to furnish a source of relatively lean beef for the ground beef trade is not generally feasible. In addition, we should note that inflation, a major concern in the late 1970s and early 1980s, favors shorter feeding periods because of the high rates of interest being paid on borrowed capital, or because of the opportunity cost of capital invested in cattle, feed, and operating expenses. In other words, the proportion of cattle fed grain is controlled by economics, and the industry adjusts quite rapidly to changes in economic criteria.

In summary, although it appears intuitively obvious and is emotionally satisfying to pressure for less feeding of grain to cattle, in reality there is little logical justification for "rich countries" to adopt policies that would encourage "simpler" food consumption habits as they would do little, if anything, to relieve the long-run food shortage in poor countries. Furthermore, as Svedberg (1979) points out, paradoxically a major effect of such policies would be a lowering of world grain prices that would cause a decline in prices of other agricultural products, primarily through shifts in production. This result would hurt developing countries rather than help them, as they are net exporters of agricultural products. The possibility of serious consequences elsewhere, even though such policies might be of importance to the United States, demonstrates the need for careful economic analysis in the evaluation of different situations in the beef business. In fact, our thrust in this book is to stress the importance of highlighting financial and economic aspects and of integrating them to the extent possible with technical production and marketing relationships.

WORLDWIDE PRODUCTION AND CONSUMPTION OF BEEF

Let us now examine beef production and consumption data more closely. Referring back to Tables 1.1 and 1.2, we find that several points stand out. The United States produces 21 percent of the world's beef despite having only 9 percent of the world's cattle. South America, with 18 percent of world cattle, produces 15 percent of the beef. Europe has 11 percent of the world cattle and 22 percent of the production. Remarkable changes have taken place in some countries while production in others has barely increased (see Appendix Table A.2). In many cases the largest percentage increases have occurred in countries with low output. South Korea emerges as the country with the greatest increase (410 percent from 1960 to 1979), with Taiwan second and Nicaragua third. The United States increased production 38 percent over the nineteen-year period 1960–1979, while the USSR shot up by 115 percent. The nine European Economic Community (EEC) countries increased production modestly, about 50 percent on the average.

Per capita consumption of beef for the selected countries for which the production data are given is presented in the Appendix Table A.3. The differences are startling, to say the least. Whereas per capita consumption in Greece increased 311 percent from 1961–1979, it declined 39 percent in Peru, 19 percent in Honduras, and 15 percent in Uruguay. Most of this change may be attributed to government policy. Wide variations occur among all the countries. In the EEC, for example, the changes in per capita consumption of beef ranged from an 8 percent decline for Denmark to a 64 percent increase in Italy. The United States registered a 16 percent growth, while the increase was 23 percent in Australia and 206 percent in Japan.

The United States went from 42.8 kg per capita in 1961 to 49.8 kg in 1979. Over this same period, the Japanese increased their consumption from 1.6 to 4.9 kg. Australians eat about 50 kg annually, while Argentines top the list with a whopping 90 kg. European countries vary from 11 to 33 kg, most being slightly over 20 kg.

Figures on total beef consumption for fifty-three selected countries appear in the Appendix Table A.4. The United States was the leader by far, consuming 11.0 million metric tons in 1979 compared with 6.6

million metric tons consumed in the USSR, about 2.3 million in Argentina, and 2.1 million in Brazil. The French, West Germans, and Canadians consumed 1.7, 1.5, and 1.0 million metric tons, respectively. Increases in total consumption over 1961–1979 follow approximately the ordering given under per capita consumption. The Republic of Korea is the largest gainer (612 percent), followed by Taiwan (559 percent), Greece (353 percent), and Japan (275 percent).

About 26 percent of the world's beef is consumed in North America (Canada and the United States), while 22 percent is eaten in Europe, and 12 percent in South America (Table 1.7). The Central American countries, the Caribbean countries, and Mexico account for about 3 percent of the world's total. The largest regional gain in total consumption from 1961 to 1979 was in Central America (174 percent), followed by the Soviet Union (150 percent) and Asia (115 percent).

The United States imports about a million metric tons of beef annually, making it the world's largest beef importer (Figure 1.1). The quantity is relatively stable as there is a formula tied to production that sets the amount of beef that can enter the country each year. Imports expanded during 1973 because quotas were relaxed in a government effort to fight rapidly rising beef prices. The second largest beef consuming country,

Table 1.7. Consumption of beef and veal in fifty-three selected countries by region, 1961 and 1979; and beef, veal, and buffalo meat in the world, 1979

Region	Consumption by year		Change	Total, 1979	
	1961	1979	1961–1979	Selected countries	Estimated world
	(1,000 metric tons)		*(%)*	*(%)*	
North America	8,519	11,986	41	31	26
Central America, Caribbean, and Mexico	487	1,332	174	3	3
South America	4,007	5,845	46	15	12
Europe					
EEC	5,219	6,806	30	17	15
Western	825	1,465	78	4	3
Eastern	1,085	1,952	80	5	4
Total	7,129	10,223	43	26	22
USSR	2,652	6,636	150	17	14
Oceania	547	981	80	3	2
Other	1,023	1,969	92	5	4
Total selected	24,364	38,972	60	100	83
Other	...	7,797	...	...	17
World	...	46,769	...	...	100

Source: Estimates for the selected regions and countries are given in the Appendix, Table A.4.

Note: Estimates for the world derived by assuming that world production equals consumption. See Table 1.2. Also, that the statistics from Table A.4 (U.S. Department of Agriculture) can be used with FAO statistics (Table 1.2).

the USSR, began importing beef in 1974 and has continued to bring in small quantities ever since (Figure 1.1). Production and consumption of beef for two major exporting countries, Argentina and Australia, are presented in Figure 1.2. Cattle in both countries are raised and fattened on pastures and to some extent on postharvest grain fields, so that weather as well as price and world grain or beef conditions plays an important part in production. Political considerations are of major importance, especially in Argentina, which reacts quickly to changes in world demand for its product. The erratic pattern displayed in Figure 1.2 clearly indicates that we need to consider the multitude of relationships at both business and government levels when identifying the components of the world's beef business.

CONCEPTUALIZING THE WORLD'S BEEF BUSINESS

Traditional marketing theory provides at least three different ways in which to break down marketing systems into their components according to: (1) behavioral systems, (2) institutions, and (3) functions. These classification schemes can also be used at the pro-

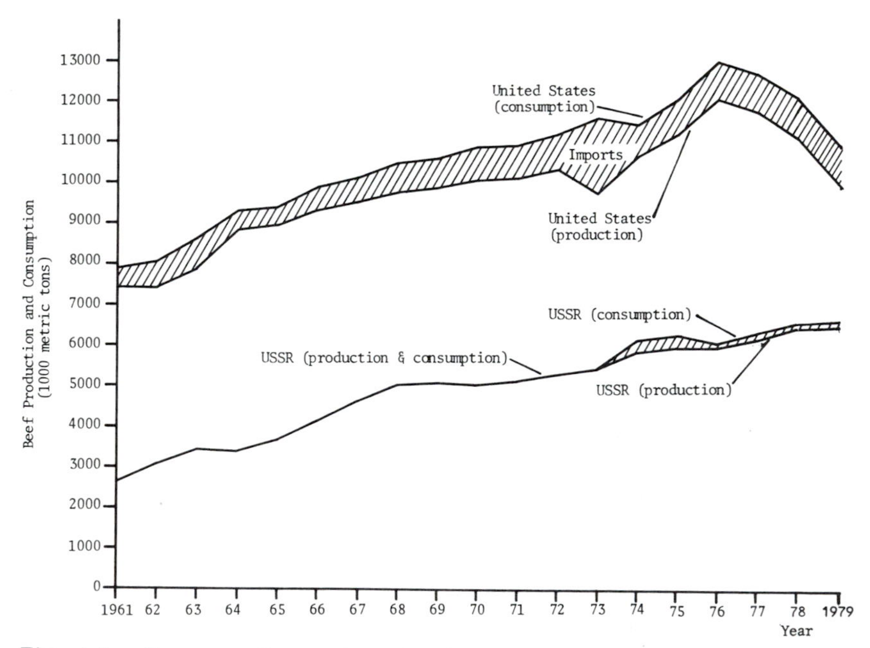

Fig. 1.1. Consumption and production of beef in the United States and USSR, 1961–1979.

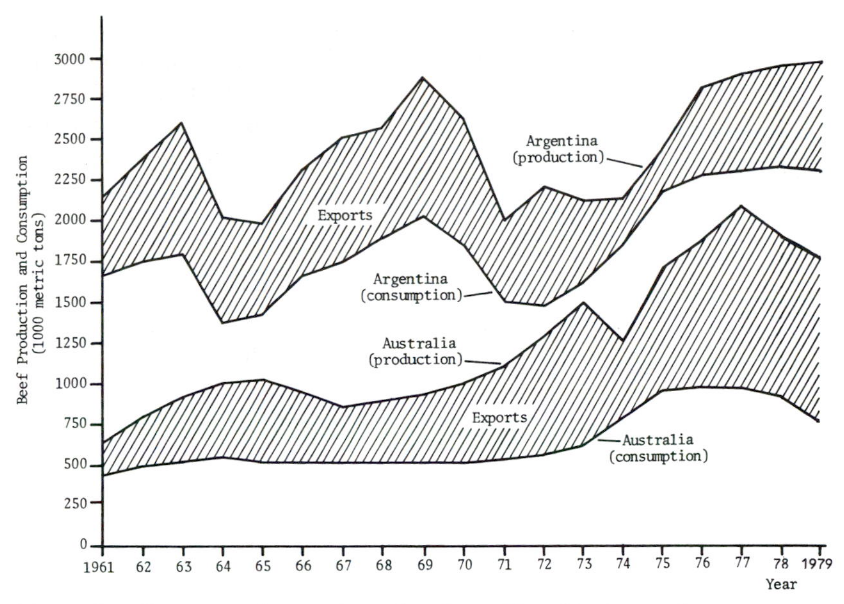

Fig. 1.2. Consumption and production of beef in Argentina and Australia, 1961–1979.

duction level or, as we do with this book, for the beef business as a whole. Looking at systems from the behavioral aspect involves studying types of problems encountered in daily activities. This approach is used mainly for problem solving, for example, by means of efficiency studies of markets, firms, or industries (Kohls and Downey 1972).

The institutional approach involves the study of various agencies and business structures that perform, or are related to, the marketing and production processes. It focuses on the "who" in the question "Who does what?" by considering the nature and character of the various middlemen and related agencies, as well as the arrangement and organization of the marketing and input supply machinery. Examples of firms, various middlemen, and agencies in the beef business are presented in Table 1.8, which shows that the number of organizations is as extensive as the jobs they perform. Each one is a necessary link in an efficient system, whether made up of brokers, truckers, health inspectors, statisticians, order buyers, or others. Paradoxically, the most efficient systems have the greatest amount of specialization. In other words, there is no ex ante reason to believe that cost to consumers would decline if the number of middleman specializations was reduced.

The common approach to marketing and production systems is to examine by function—in effect, to consider the major specialized ac-

Table 1.8. Two approaches to conceptualizing the world's beef business

Functional	Institutional
Retail	Supermarkets or meat markets
	Government weights and standards
International trade	Government statistics
	Brokers
Wholesale/distribution	Freight forwarders
	Insurance companies
Slaughter/processing	Shipping companies
	Government meat inspectors
Feeding/finishing	Butcher shops
	Centralized meat processors
Stocker/backgrounding	Commission men
	Jobbers
Cow/calf	Warehouses
	Meat truckers
	Packing plants
	Government graders
	Futures markets
	Order buyers
	Feedlot owners
	Speculators/investors
	Auction markets
	Farmers/ranchers
	Breed associations
	Livestock associations
	Lenders or financial institutions

tivities performed. The functional approach attempts to answer the "what" in the question "Who does what?" It considers the job to be done, not the agency. Several types of activities are considered within each function, such as exchange (buying, assembling, selling), physical activity (storage, transportation, production, processing), and facilitating activity (standardization, financing, risk bearing, market intelligence). Each function has a number of institutions related to it. For example, meat storage requires interaction with insurance agencies, health inspectors, brokers, and owners of the facility and finance agencies, to mention just a few.

Our description of the world's beef business is based on the functional approach, according to which we will identify the various steps (functions) that take place from the retail counter down to production of the calf. Seven stages or levels are identified (Table 1.7). The first function listed, and the point of departure in this book, is retailing because this is the point at which supply interacts with demand to set price. The subject of Chapters 3, 4, and 5 is an analysis of the various production systems around the world. Then, economic aspects in relation to production are discussed, followed by a description of marketing live cattle and meat. After focusing on world trade in beef, the final chapter concentrates on projections and policy considerations.

The picture painted thus far merely hints at the multitude of interrelationships. This is a good way to begin, for the world's beef business is complicated. In order to explain it, we will separate the various parts from the whole, just as a butcher does when he cuts up a beef carcass. By analyzing beef production, processing, trade, and other aspects in terms of systems, it will be possible to determine why certain facets of the industry operate as they do, why the production of beef has increased at the rate it has, why prices change over time, and why they vary between regions or countries. Not all points will be covered in detail as the purpose of this book is to explain the big picture, the why of the world's beef business, and what changes will be required to increase production sufficiently to meet future needs.

The year 2000 is less than two decades away. Certainly there will be major changes in beef production, marketing, and consumption patterns. Some countries will make great adjustments while others will remain relatively traditional. It is important, for example, to recognize that livestock production systems in the developing world will have to undergo changes more radical than crop production systems if expectations from this subsector are to be fulfilled. Output will have to rise about 4.7 percent annually compared with 3.6 percent for crops if potential demand is to be met.

There is a great challenge to improve breeds and breeding, to adjust land tenure patterns, and to develop improved range management practices. As experts at the Food and Agriculture Organization of the United Nations (FAO) have recently written, "While production will continue to be based primarily on grasslands and by-products, increasing use will have to be made of feedgrains and concentrates" (FAO 1979*a*, p. 92). How does this view square with cries from many quarters for less use of grains for cattle feeding? Greater offtake rates (calculated as the number of animals slaughtered in a year divided by the total herd and multiplied by 100) will require a substantial commitment from both national and international agencies. Will the necessary policies, which demand a good understanding of the stimuli for increased production, be forthcoming? The amount of annual investment required for livestock is staggering. About $3.5 billion was invested in 1980 for the ninety developing countries, but by 2000 that figure, in constant 1980 dollars, should increase to about $10.9 billion, an annual compounded growth rate of 5.9 percent (FAO 1979*a*, p. 102). Will lending agencies provide this capital?

We are facing a challenge to make the world's beef business more efficient and more responsive to the ever-changing desires of final consumers. Sound analysis and a good economic basis for policy are needed but can only be derived from a good understanding of the component parts and the interrelationships existing in this complexity known as the world's beef business.

REFERENCES

American Meat Institute. 1978, 1980. *Meatfacts, 1978,* and *Meatfacts, 1980,* Washington, D.C.

Aziz, S., ed. 1975. *Hunger, Politics and Markets.* New York: New York University Press.

Connor, J. Richard, and Rogers, Robert W. 1979. Ground Beef: Implications for the Southeastern U.S. Beef Industry. *J. Agric. Econ.* 11(2):21-26.

Kohls, Richard L., and Downey, W. David. 1972. *Marketing of Agricultural Products,* 4th ed. New York: Macmillan.

Regier, Donald W. 1978. *Livestock and Derived Feed Demand in the World GOL Model.* USDA, FAER no. 152.

Studemann, John A., Huffman, Dale L., Purcell, J. C., and Walker, Odell L., eds. 1977. *Forage-Fed Beef: Production and Marketing Alternatives in the South.* Georgia Agric. Exp. Sta. SCS Bull. 220. Athens: University of Georgia.

Svedberg, Peter. 1978. World Food Sufficiency and Meat Consumption. *Am. J. Agric. Econ.* 60:661-66.

United Nations, Food and Agriculture Organization. 1979a. *Agriculture: Toward 2000.* Rome.

______. 1979b. Committee on Commodity Problems: Intergovernmental Group on Meat. *Utilization of Grains in the Livestock Sector: Trends, Factors and Development Issues.* Prepared for the Eighth Session, May 7-11, 1979 (CCP:ME 79/6 February). Rome.

______. *Production Yearbook.* Rome. Various issues.

U.S. Department of Agriculture. *Foreign Agricultural Circular: Livestock and Meat.* Periodical.

______. 1973. World Meat Economy in Perspective. *Livestock and Meat Situation.* USDA, LMS00:27-31.

2
DEMAND FOR BEEF

The driving force behind the beef business, as for all food products, is consumer demand. In other words, what will consumers pay at different levels of market supply, and what market premiums and discounts do they set for different qualities? For a given level of demand at a specific time, the consumer depends on the supply. Over a specific time period, on the other hand, the demand for beef is determined largely by consumer income and price of the product. Prices of other foods and nonfood items also influence the demand for beef, but normally these effects are so slight that often they are not measurable with current statistical techniques, even in the higher income countries.

DEMAND AND PRICES IN A COMPETITIVE ECONOMY

Beef prices are determined by three principal means. One, used primarily in centrally planned economies, is strong government intervention in the market through price setting at some or all levels, from the farm gate to the retail counter. A second possibility is some government intervention. In Costa Rica, for example, retail prices are set for certain cuts that have wide appeal among lower income people. Costa Rica also uses a two-price system for meat from cattle destined for export, which is priced according to a formula tied to the *Yellow Sheet* (a U.S. trade publication), and for domestic cattle, the price of which is governed by a combination of free and fixed retail prices of the various cuts (Armstrong 1968). A third possibility is one in which prices are set in the open market by competition. The price determination mechanism is the subject of this chapter.

In a market economy the pricing machinery is expected to transmit the orders concerning where products will be allocated, when, and how much. The fluctuating competitive prices guide and regulate production, consumption, and distribution of goods over time and place. The price setting mechanism begins with demand, which is a schedule of different quantities of a commodity that buyers will purchase at different prices at a given time and place. Demand is peculiar in that it is not visible, even in a market, but it exists and influences prices as much as supply does. Demand should not be confused with consumption; rather, it is the willingness of consumers to purchase different quantities at various prices. Consumption is determined at that point on the demand curve where consumers and suppliers agree to the price and quantity of the trade.

The law of demand states that with lower prices more will be purchased and, conversely, with higher prices, less will be purchased. The relationship between prices charged and the amount sold is affected by several factors: as buyers take more of the product, their utility (that is, satisfaction) drops; likes and dislikes differ among consumers; and there are wide variations in the distribution of incomes. However, for a given level of price and income in a society, beef consumption can be predicted with reasonable accuracy.

A schedule or statistical table showing the relationship between quantities and prices can be translated into a demand curve such as the one shown in Figure 2.1. The demand curve does not represent the price and quantity, only the effect of different prices on the quantity purchased. A shift in demand, due to factors such as changes in tastes and preferences or price changes in close substitutes, will mean a shift to a new curve, which would indicate either less or more of the product taken at the same price. A shift in quantity demanded is a movement along the demand curve.

Supply schedules and curves are defined in the same way as demand schedules and curves, except that supply is used in place of demand. The law of supply states that as prices increase, more is offered for sale than at lower prices. The intersection of the supply and demand curve reveals the price and quantity sold and purchased at a particular time and place. Prices are continuously moving toward or around an equilibrium price, but an equilibrium is not readily found or maintained owing to a continuing stream of new market information. Consider, for example, the

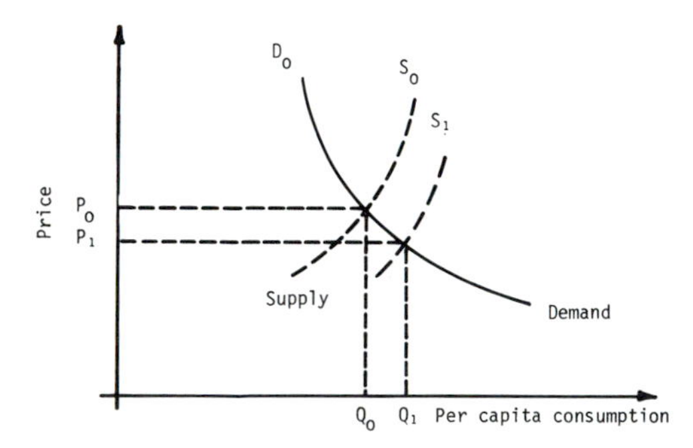

Fig. 2.1. Demand, supply, and equilibrium price and quantity.

variation in prices for the same type and quantity of cattle in just one morning's trading at an auction market or on the futures exchanges. A shift in supply, such as that from S_0 to S_1 in Figure 2.1, takes place in the short run because of changes in maintenance costs (for example, feed grains for cattle), in a seller's need for cash, or in general expectations about the futures situation. Changes in costs of production, the cattle cycle, government policy, or technological advances are some of the reasons for longer-run supply shifts.

The aggregate demand in a specific market or for a country as a whole represents the consolidation of numerous variables and forces, most of which can be measured mathematically. To understand the aggregate demand for beef, we must first identify the components and specify exactly the product and market conditions under which demand is being examined. For example, we sometimes use demand to mean the average annual demand for beef products at retail for the entire U.S. market, and at other times to mean the demand for USDA Choice yield grade 2–4 500 kg slaughter steers at a certain location and time, such as Omaha, Nebraska, on November 1, 1980.

Demand is first expressed for particular beef products at retail on the basis of consumer consideration of beef prices relative to prices of other food products, as well as the amount consumers can afford to spend for each item. Demand for carcass beef at wholesale, demand for live slaughter cattle, and demand for feeder cattle are all derived from this retail demand (Figure 2.2). The derived demand relationship may not be entirely measurable at any given time and place in the short run, but for longer periods it can be identified. The marketing margin for beef in the United States during the decade 1969–1979 for example, was generally 35–40 percent despite rather wide price fluctuations. In other words, when supply and demand determine the average retail price, one can estimate the net farm value of live slaughter cattle by simply subtracting the marketing cost from the composite retail price of a carcass.

MARKETING MARGINS AND PRICE SPREADS

Prior to embarking on a detailed analysis of the demand for beef, we should look at a few more definitions. Two widely used terms are "marketing margins" and "price spreads." Care must be taken to avoid misinterpreting these terms for although they are related, they denote different measurements.

The farm-retail *price spread* for a food item is the difference at one point in time between the weighted average retail store price and the average value of a corresponding amount of raw food material as sold at

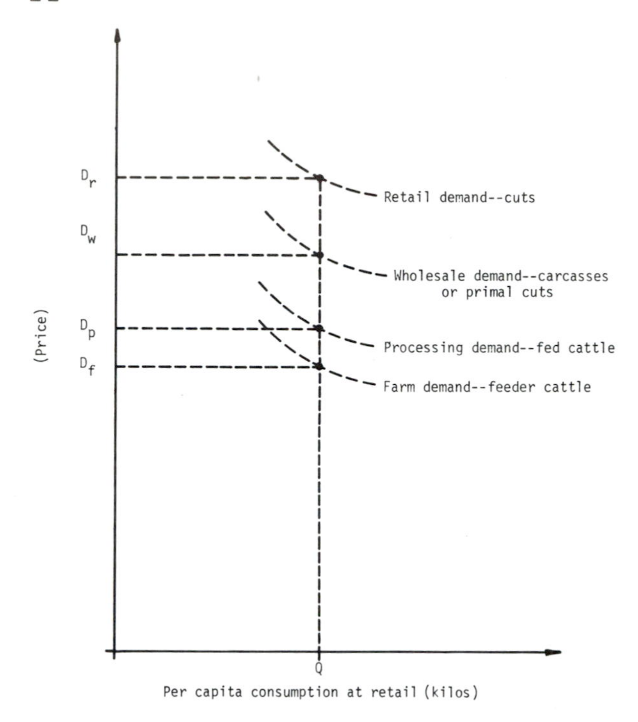

Fig. 2.2. Demand for beef and beef cattle from the production unit to the retail counter.

prices at or near the farm. In terms of beef, the farm-to-retail price spread is the difference between the weighted average retail price per kilogram of a certain grade of beef (Choice in the case of the United States) and the farm value of the quantity of Choice grade live animals that are equivalent to one kilogram of retail cuts, less the estimated value of by-products. Thus the farm-to-retail price spread normally includes all costs incurred, as well as profits of the agencies involved in marketing that kilogram of beef eventually sold at retail. These include charges for assembling, processing, transporting, packaging, and distributing products from the farm gate to the consumer.

An example of two methods used for computing the spreads is shown in Table 2.1. In both cases the farmer's share is calculated at 58 percent of the retail price. The percentage that the farmer receives depends on the level of beef prices, with higher prices yielding a larger percentage of the final price. The highest percentage since 1965 was 66 percent in 1968, while the lowest was 57 percent in 1976 (USDA 1978).

The farmer's share of the price spread for beef in Lima, Peru, in 1972 was 66 percent (Table 2.2). The marketing cost was \$0.46 (\$1.36 –

Table 2.1 Farm-to-retail price spreads for beef in the United States calculated by two methods, 1977

	Distribution of retail price according to marketing function, Choice beef, all types, 1977[a]			Distribution of retail price according to carcass values, Choice beef, all types, 1977[b]	
Item	Cost, value, or price	Percent of retail price	Item	Value or price	Percent of retail price
	($/kg)			*($/kg)*	
Farm value (farmer's share)	1.76	58	Farm value (farmer's share)	1.76	58
Assembly and procurement	0.04	1	Farm by-product allowance	0.24	8
Processing	0.18	6	Gross farm value	2.01	66
Intercity transportation	0.03	1	Net carcass price	1.92	63
Wholesaling	0.18	6	Carcass by-product allowance	0.03	1
Retailing	0.85	28	Gross carcass value	1.98	65
Retail price	3.04	100	Retail price	3.04	100

[a]Adapted from USDA, 1978.
[b]Adapted from Ball 1979.

$0.90) compared with $1.28 ($3.04 − $1.76 in Table 2.1) in the United States five years later. The key here is that given the circumstances, an efficient marketing system is one that moves goods from producers to consumers at the lowest possible cost, while providing the services that both consumers and producers desire and are willing to pay for. This is why, even though the spreads given above are interesting, they are difficult to interpret across countries. It would be more useful to compare price spreads within a country in order to determine whether, without a reduction in services provided, there would be a gain in efficiency. These changes are difficult to identify and evaluate precisely when a number of functions are involved. For this reason another measure, gross margins, has been developed.

Table 2.2. Farm-to-retail price spreads by marketing function in Lima, Peru, 1972

Item	Distribution of retail price according to marketing function, Prime grade, 1972[a]	
	Cost, value, or price	Percent of retail price
	($/kg)	
Farm value (farmer's share)	0.90	66
Assembly and procurement	0.08	6
Processing	0.04	3
Intercity transportation and wholesaling	...	...
Retailing	0.34	25
Retail price	1.36	100

[a]Adapted from Fenn, 1977, Table 9.

Gross margins—the difference between dollars paid and dollars received—of packers and retailers do not take into account all marketing functions. Rather, they represent the portion for a packer's or retailer's labor cost, packaging, overhead, other costs, and any net profit. They exclude some costs included in the spread, such as charges for transportation and services performed by businesses other than meatpackers or retailers. The concept of gross margins is very important when countries embark on price fixing policies or when they set wage and price controls, as the United States did in the early 1970s (Simpson 1978). Price spreads are more useful in the continual monitoring that takes place on conduct and performance in the industry than are gross margins, however.

THE NATURE OF DEMAND FOR BEEF

Total demand for beef is a summation of separate demands for individual cuts at the retail store or in restaurants (Nix 1978). The concept of substantially different demands is illustrated in Figure 2.3, which shows individual demand curves for steaks, roasts, and ground beef. The graph is just another means of representing the beef display counter at a modern supermarket where each item has its own demand configuration, related to the price of that and other items in the display case. The individual demand curves can be aggregated to form the demand for beef as a whole. In an orderly and reasonably stable market, these relationships are close to equilibrium, that is, the point at which supply just equals demand at a certain price. In this situation the real difference in demand is explained by the product's price and consumer's income (Polopolus 1978).

Figure 2.3 shows that for a given price, say price P_r, if there were an unlimited amount of beef, more steak would be demanded than roast (level Q_{s_1}) and more roast (Q_r) than ground beef (Q_{gb_1}). However, supply must be taken into account in reaching an equilibrium. The result is that the limited supplies interact with demand to maintain the price of steaks at levels higher than the other products, and this interaction limits the consumption of steak. As a matter of fact, over a certain range of prices, demand for individual items may actually converge and consumers will be somewhat indifferent toward the individual items. In the United States, for example, when ground beef was at record high prices in 1978 and 1979, it was common practice for primal cuts of chuck to be ground and sold as ground beef. This is one factor that has led to the research mentioned in Chapter 1 concerning the possibility of forage feeding cat-

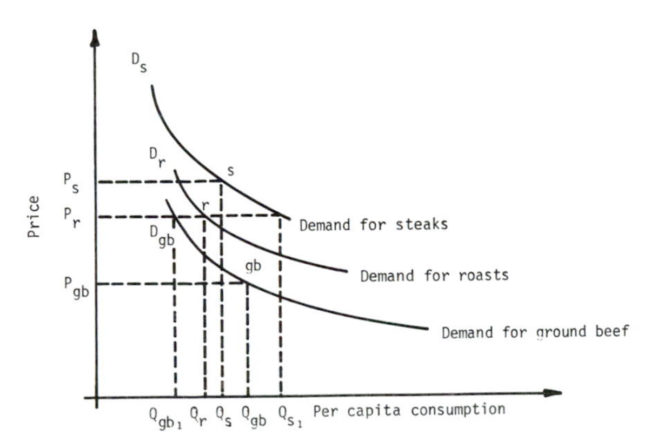

Fig. 2.3. Demand for different cuts of beef at retail.

Demand for beef can be enhanced by the form and the atmosphere in which it is presented. This American steak grill in West Germany is an example of the many ways to introduce customers to the versatility and good quality of beef.

tle to produce lean beef. Similarly, when the price of steak is high relative to roasts, a butcher can stretch profits by cutting a lower proportion of roasts from a carcass. If the demand for ground beef were greater than for roasts or steak, virtually all the carcass could be sold as ground beef. Since this does not happen, the demand curve for ground beef lies to the left and below the demand curves for roasts and steaks.

Prices of retail cuts change in relation to one another, depending on supply as well as demand conditions. This changing pattern of relative prices is used to argue against government interference and setting of retail prices. In January 1978 the average price in the United States for porterhouse steak, boneless rump roast, and ground beef was $2.45, $1.81, and $0.87 per pound, respectively. By December 1978 cow slaughter had declined sharply, so that the supply of ground beef was reduced and its price had increased by 37 cents per pound to $1.24. Roasts and steak also increased in price by a greater absolute amount but a smaller percentage. When prices for different products, whether they be beef or other commodities, are not allowed to seek their equilibrium prices in relation to close substitutes, the result may be a black market. This has been the outcome in many countries having fixed prices.

Demand in the short run is influenced mainly by events in the market, the weather, holidays, and so forth, while the main elements of demand change little. These longer-run main elements are level of personal income, price of the product and of competing products, and consumer tastes and habits. In mathematical terms, the average demand for beef in a particular country or area can be formally specified as:

$$Q_B = f(P_B, P_P, P_C, P_F, P_I, I) \qquad \textbf{(2.1)}$$

where

Q_B = quantity demanded per capita
f = a function of
P_B = average price of beef
P_P = average price of pork
P_C = average price of chicken
P_F = index of other food prices
P_I = price index of nonfood
I = after-tax income per capita

The foregoing factors have been shown to account for over 90 percent of the variation in consumer demand for beef in the long run because many of the variables remain constant over time (Breimyer 1961). As a result, for a given market the price of beef is often the only measurable adjuster for quantity of beef demanded within a year. Price varies because of rather frequent changes in the available supply.

THE ROLE OF INCOME

Income is the key variable in explaining the differences in demand among markets, among groups within the market, and among countries. In general, increases in income explain changes in demand over the long run, as well as the difference in demand among groups. For our purposes, then, demand for beef at retail can be reduced from the above formula to the following one:

$$Q_B = f(P_B, I) \qquad \textbf{(2.2)}$$

that is, the quantity of beef demanded per capita is a function of the average price of beef and after-tax income per capita. For periods of relatively stable income it can be reduced even further to:

$$Q_B = f(P_B) \tag{2.3}$$

The surprisingly stable relationship $Q_B = f(P_B)$ has been verified many times although explanation of all factors influencing the demand for beef requires an even more complicated formula than the first one given above. Attempts to measure the influence of separate variables on demand have not been consistently successful, however, except for the simple relationship of $Q_B = f(P_B, I)$, where just two variables, price and income, are used. The form of the statistical equation most often used is:

$$Q_B = aP_B^{b_1} I^{b_2} e \tag{2.4}$$

where

$a =$ a constant
$b_1 =$ price elasticity of demand
$b_2 =$ income elasticity of demand
$e =$ random error

This simple equation has been used to explain over 75 percent of the variation in the per capita consumption among leading beef consuming countries of the world (Regier 1978). It can explain even more of the variation in consumption of a given country, and it can explain how per capita consumption varies by income group.

In our judgment and experience, a useful rule of thumb for estimating world demand for beef is:

$$Q_B = aP_B^{-1.0} I^{1.0} \tag{2.5}$$

This formula states that a 1 percent change in price will change the quantity demanded 1 percent in the opposite direction, and a 1 percent change in income will change quantity demanded 1 percent in the same direction. These values approach –0.5 and 0.5 as the level of beef prices decline substantially or as the level of income increases substantially. They increase above –1.0 and 1.0 as the price of income levels move in opposite directions. For example, the elasticities are –0.7 and 0.5 for the United States and –1.20 and 1.20 for Japan (Table 2.3).

The relationship of price and income to quantity of beef demanded is given in Figure 2.4, which depicts two levels of income, moderate and

Table 2.3. Demand and income elasticities for red meats in selected countries and regions, 1970

Country or region and commodity	Price elasticity					Income elasticity
	Beef					
	Finished	Other	Pork	Poultry	Mutton	
United States						
Beef, finished	−0.7	0.20	0.10	...	...	0.50
Beef, other	0.4	−0.80	0.10	0.10	...	0.35
Pork, other	0.4	...	−0.80	0.10	...	0.25
Poultry	0.3	...	0.20	−1.00	...	0.90
Canada						
Beef	...	−0.60	0.30	0.15	...	0.70
Pork	...	0.40	−0.70	0.15	...	0.15
Poultry	...	0.30	0.20	−0.80	...	0.90
EC-6						
Beef	...	−0.70	0.30	0.10	...	0.60
Pork	...	0.50	−0.80	0.12	...	0.50
Poultry	...	0.38	0.50	−1.07	...	1.00
Mutton	...	0.15	0.15	...	−0.25	...
Other Western Europe						
Beef	...	−0.60	0.20	0.10	...	0.70
Pork	...	0.20	−0.70	0.20	...	0.60
Poultry	...	0.10	0.20	−0.80	...	0.90
Mutton	...	0.15	0.15	...	−0.25	...
Japan						
Beef	...	−1.20	0.26	0.35	...	1.20
Pork	...	0.20	−0.90	0.11	...	0.90
Poultry	...	0.50	0.17	−1.10	...	0.60
Mutton	...	−0.40	0.20	0.30	−0.40	0.50
Oceania						
Beef	...	−0.50	...	...	0.20	...
Pork	...	0.20	−0.40	...	...	0.10
Poultry	...	...	...	...	...	...
Mutton	...	0.40	...	...	−0.80	...
Mexico and Central America						
Beef	...	−0.40	0.10	...	...	0.70
Pork	...	0.10	−0.30	...	...	0.60
Poultry	...	...	...	...	...	...
Argentina						
Beef	...	−0.40	...	...	...	0.30
Pork	...	0.20	−0.40	...	...	...
Mutton	...	0.20	...	...	−0.40	...
Brazil						
Beef	...	−0.60	0.30	...	...	0.40
Pork	...	0.20	−0.60	...	...	0.40

Source: Regier, 1978, vol. 2.

high, in a relationship that measures price per kilogram on the vertical axis and per capita consumption on the horizontal axis. The demand curves slope down and to the right and thus indicate, according to the law of demand, that a greater quantity is taken at lower prices. This figure also shows that a greater quantity is taken at a higher price by the

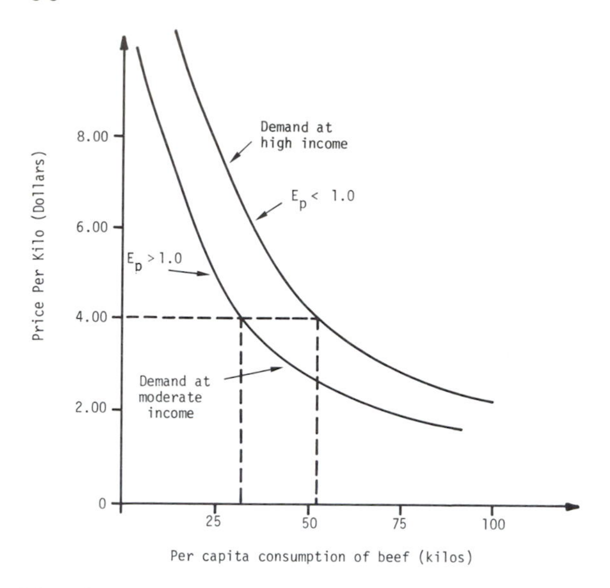

Fig. 2.4. Demand for beef at two levels of income.

high income group than is taken by the moderate income group. Furthermore, even though the shape of the curve for the high income group is similar to the moderate one, the elasticity—that is, the percentage change in quantity with a 1 percent change in price—is less than one for the high income group and greater than one for the moderate income level.

INTERNATIONAL PRICE AND INCOME ELASTICITIES

Specific estimates of beef demand vary depending on the country, income group, estimating procedure, and time period under study. An attempt has been made by Regier (1978) to bring the results together and to construct an interactive econometric model showing the mathematical relationships at the world level between feed grains and livestock. The results of the analyses show that price elasticities for beef are about –0.60 to –0.70 for most developed countries, and about –0.40 for most developing countries (Table 2.3). In other words, as income levels increase, the demand becomes more elastic. This means that in high income countries a greater quantity of beef will be taken for a given change in price than will be taken in low income countries. Mainly because of government import quotas, Japan, where per capita income is relatively high but per capita beef consumption is very low, has a very high price elasticity, –1.20. This means that a 1 percent decrease in price will bring about a 1.2 percent increase in quantity.

Cross price elasticities with other meats are also given in Table 2.3. For example, the value of 0.10 in the column labeled pork means that for both finished and other beef in the United States a 1 percent increase in the price of pork will increase the quantity of beef purchased by about 0.1 percent. As the table shows, the cross price elasticities are positive, whereas the direct or own price elasticities are negative.

Income elasticities for the various types of livestock products are shown in the last column. Argentina has a much higher per capita income than Mexico or the Central American countries and thus has a much lower income elasticity for beef (0.30 versus 0.70). The 0.70 may be taken to mean that a 1 percent increase in income will increase the quantity demanded by 0.7 percent. It is our belief that the estimates are probably low for the late 1970s, considering the rapid increase in prices due to inflation. We should also take into account the tendency to underestimate elasticities from time series data, and the sources of these estimates.

An important aspect of the differences in income elasticities, apart from income, is the amount of the product consumed. Argentina consumes about 8 to 10 times as much beef as the other countries do. In other words, the level of consumption is also significant. An illustration of the relationship between price and quantity taken, assuming a constant (not equal, but constant) income level, is depicted in Figure 2.5. The Japanese take a smaller quantity at a higher price than do the

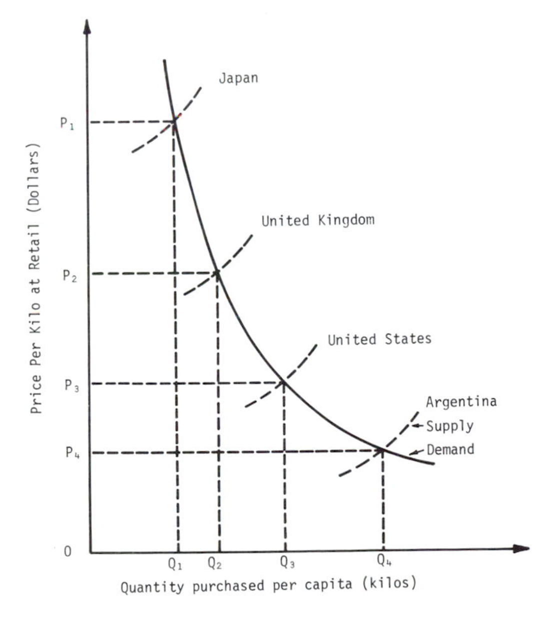

Fig. 2.5. Illustration of demand and equilibrium price for steak, assuming a constant income level.

citizens of the United Kingdom. The Argentines take quite high levels, both because of taste and preference and because of low price.

The elasticities for beef depend not only on the country but also on the type of cut and grade. Coleman (1966) has estimated that the elasticity for all beef in the United States is –0.85; for low grade and processing beef it was calculated at –0.56, and for high grade beef at –1.24. Thus a meaningful economic distinction can be made between the two classes of beef. There are important political ramifications, especially with respect to beef import policy, as the results show that for the United States, where virtually all beef imports are a low quality product, prices can change considerably with little more of the product demanded (Ball 1979; George and King 1971). Recall that the demand elasticity for beef in Japan is –1.20, about the same as the U.S. high quality beef elasticity—and most of the beef that Japan imports is a high quality product.

INTERNATIONAL BEEF PRICES AT RETAIL

Retail meat prices vary considerably around the world. On May 6, 1980, sirloin steak was $34.48 per kilogram in Tokyo, $16.16 in Stockholm, $6.81 in Washington, and just $2.71 in Brasilia (Table 2.4). The prices for pork and broilers vary considerably less because hogs and broilers are raised almost entirely in confinement and are fed on grain, the price of which varies only according to transporta-

Table 2.4. Retail meat prices in selected world capitals, May 6, 1980

City	Steak, sirloin, boneless	Roast, chuck, boneless	Pork chops	Roast, pork, boneless	Broilers, whole
			(U.S. $/kg)		
Tokyo	34.48	24.39	7.30	7.83	3.56
Bern	18.71	8.75	8.45	12.98	3.14
Stockholm	16.16	7.04	19.68	14.00	4.68
Brussels	13.20	7.31	5.72	6.06	3.36
London	11.66	6.04	5.22	4.25	2.36
Rome	11.26	10.07	6.52	7.11	3.08
Bonn	11.06	8.02	5.24	6.35	2.82
Paris	10.37	10.61	5.94	6.82	3.82
Canberra	8.96	5.33	6.13	4.93	2.55
Madrid	8.53	6.42	4.50	7.17	1.60
Ottawa	6.98	4.53	3.76	3.15	1.95
Washington, D.C.	6.81	4.72	3.73	4.39	0.93
Buenos Aires	6.16	5.60	5.60	7.84	3.36
Mexico City	4.08	4.04	3.66	4.58	2.35
Brasilia	2.71	2.50	3.49	5.08	1.22
Median	10.37	6.42	5.72	6.35	2.82

Source: Compiled from USDA, 1980, p. 20.

tion, availability, and import duties. Since cattle can be both raised and fattened on grass, their value is considerably lower in countries such as Mexico or Brazil, where they are raised in an extensive type industry, than in the European countries, where cattle raising is an intensive enterprise.

Although the absolute level of prices given in Table 2.4 is a useful comparative device, it fails to take into account differences in purchasing power. There are many statistical deflators, one of the best for our purposes being wage rates. By calculating the work time required to purchase various meat items in selected world capitals it can be shown that the ranking of countries according to real cost changes somewhat. In Washington, where about 59 kg of beef were consumed in mid-1978, an average of only 51 minutes of work time was required to purchase 1 kg of beef (Table 2.5). In Argentina, where one hour and two minutes of time was required by production workers to buy 1 kg of beef, it is still a good buy relative to other countries.

The data in Table 2.5 for sirloin and chuck roast are plotted in Figure 2.6, with hours representing price on the vertical axis and per capita consumption of all beef as a proxy for consumption of the two items on the horizontal axis. Japan, just as in Figure 2.5, is in the upper

Table 2.5. Work time required to purchase selected meat items in world capitals, mid-1978, as an indicator of real cost

Capital	Approx. rank	Per capita[a] consumption, 1977 Beef	All meat	Sirloin steak	Chuck roast	Pork chops	Roast pork	Canned ham	Whole broilers
		(kg)				*(hr and min/kg)*			
Washington	1	58.8	112.7	:51	0:33	0:44	0:41	0:67	0:14
Buenos Aires	2	88.2	107.9	1:02	0:35	1:28	...	...	1:02
Ottawa	3	52.8	100.5	1:02	0:37	0:46	0:36	0:53	0:35
Canberra	4	70.6	119.8	1:24	0:46	0:59	0:58	1:28	0:32
Copenhagen	5	16.2	68.4	1:46	0:46	0:56	0:58	0:46	0:18
Brussels	6	28.2	83.6	1:47	0:56	0:48	0:48	1:11	0:25
The Hague	7	20.6	63.1	2:01	1:10	1:05	1:18	1:04	0:24
Bonn	8	23.3	79.4	2:08	1:17	0:55	1:49	...	0:20
Stockholm	9	18.7	58.1	2:06	1:17	1:01	1:47	1:14	0:35
Paris	10	31.3	84.7	2:43	1:32	1:55	1:53	2:49	0:57
London	12	24.2	70.5	3:20	1:39	1:27	1:14	1:12	0:35
Rome	11	23.9	61.0	2:32	2:13	1:16	1:16	1:15	1:16
Mexico City	13	15.6	28.4	3:55	3:47	3:48	5:08	...	2:34
Brasilia	14	19.9	33.4	3:46	3:17	4:28	8:31	9:36	2:01
Tokyo	15	4.1	26.8	6:11	4:14	2:18	2:19	2:50	0:42

Source: Compiled from USDA, 1980, pp. 6–7.

Note: Mid-1978 national average for production workers calculated in local currencies.

[a]National averages.

Beef is more expensive in Japan than most other countries. Import restrictions limited total per capita supply of beef to less than 5 kilos during the 1970s.

left side while Argentina is in the lower right side. The interesting aspect is that the close relationship delineated demonstrates, at the international level, the value of the earlier stated relationship $Q_B = f(P_B, I)$.

PRICE DISCOVERY AND DETERMINATION

Now that we have discussed the tools for relating beef price changes to quantities, we are in a better position to discuss how prices are actually set. The key factor is that in an open, competitive market prices are set by the interaction of supply and demand forces, subject to the impact of supply and demand shifters. The intangible quality of this interaction makes most people feel uneasy. Price discovery, then, is inexact, for it is concerned with the interaction of buyers and sellers and with their expectations about current and future

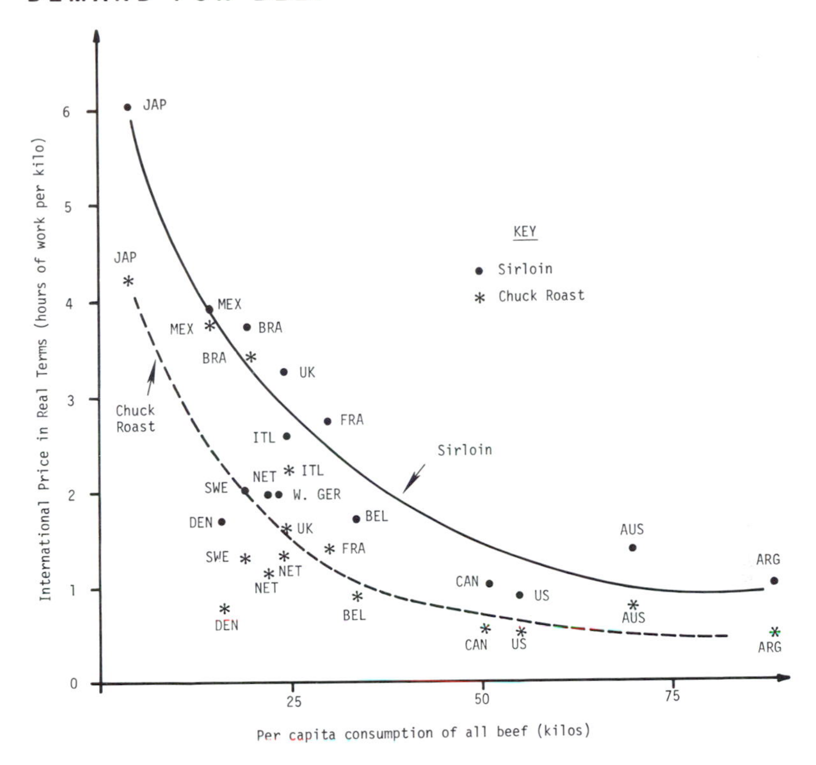

Fig. 2.6. International demand for sirloin steak and chuck roast, mid-May 1978.

events. As a result, price projections or forecasts, and the methods for making them, differ radically depending on the time period involved. In the very short run, one day up to 3 or 4 days, the market fluctuates primarily because of current beef supplies and factors that might affect supplies—for example, inclement weather, threat of a transportation or packinghouse strike, or sudden adverse regulatory changes such as the temporary removal of a beef import quota.

The next length of time is the short term, 1 month or so. Here forecasts will be influenced by new market information about cattle on feed, a possible drought, seasonal factors such as holidays, or delayed reactions to very short-term events. In other words, during these two time periods the emphasis is mainly on the supply side.

The intermediate run can be considered to cover up to 6 months. More emphasis will now be placed on expectations about the number of cattle that will be coming out of feedlots or off pastures, seasonal effects, world supply conditions, and consumer reactions to price changes. During the period known as the long run, which covers up to 1 or 2 years, attention focuses on the interaction of demand shifters, for example, prices of competing products or information that might affect tastes and preferences, such as successful marketing of ground beef blended with soy analogs. On the supply side, credit policies affecting national herd expansion, weather, the influence of competing products, and expectations about the long-run profitability of the beef business are the key factors.

The very long run, up to 10 years or so, is much more important in the cattle business than in other types of livestock operations or agriculture because of the biological nature of beef cattle, which leads to fairly predictable cattle cycles, a phenomenon that is discussed in detail in Chapter 9. First, however, we should examine beef production systems around the world and the applications of economic principles to the beef business.

REFERENCES

Armstrong, Jack H. 1968. *Cattle and Beef: Buying, Selling and Pricing Handbook.* Lafayette, Ind.: Purdue University Extension Service.

Ball, Eldon. 1979. Elasticities and Price Flexibilities for Food Items. *Livestock and Meat Situation.* USDA, LMSOO:22–25.

Breimyer, Harold F. 1961. *Demand and Prices for Meat: Factors Influencing Their Historical Development.* USDA Tech. Bull. no. 1253, ERS.

Coleman, D. R. 1966. Elasticity of Demand for Two Grades of Beef. *Illinois Agric. Econ.* 6(2):11–18. Urbana: University of Illinois.

Fenn, M. G. 1977. *Marketing Livestock and Meat,* 2nd ed. Rome: United Nations Food and Agriculture Organization.

George, P. S., and King, G. A. 1971. *Consumer Demand for Food Commodities in the United States with Projections for 1980.* Giannini Foundation Monogr. no. 26. Berkeley: University of California.

Nix, James E. 1978. *Retail Meat Prices in Perspective.* USDA, ESCS-23.

Polopolus, Leo. 1978. *Factors Affecting World Food Consumption.* Food and Resour. Econ. Dept., Staff Paper 99. Gainesville: University of Florida.

Regier, Donald W. 1978. *Livestock and Derived Feed Demand in the World GOL Model.* USDA For. Agr. Econ. Rep. no. 152, ESCS.

Rojko, Anthony. 1978. *Alternative Futures for World Food in 1985,* vol. 2. USDA World GOL Model, Supply-Distribution and Related Tables, ESCS.

Simpson, James R. 1978. *An Analysis of Wage and Price Controls and Cattle Supplies: A Primer for Meat Packers and Processors.* Food and Resour. Econ. Dept., Staff Paper 105. Gainesville: University of Florida.

U.S. Department of Agriculture. 1978. *Developments in Marketing Spreads for Food Products in 1977.* AER no. 398.

______. 1980. More Price Rises Than Declines in Fifteen Capitals. *For. Agric.* 18(6):20.

3

BEEF PRODUCTION IN THE UNITED STATES

The United States beef system differs from the systems described in Chapters 4 and 5 as it is almost completely separated from the dairy business, and the vast majority of slaughter cattle are fattened in confinement feedlots rather than on pasture or crop residue. In addition, there is considerably more specialization of production and marketing functions than is found in most other areas of the world. One major exception is Canada, and thus much of the basic system described in this chapter also fits that country.

REGIONAL PRODUCTION

The United States has a population of about 225 million, which, combined with high per capita incomes, makes it the largest beef consuming country in the world. However, only 2.3 percent of the population are engaged in agriculture. There are 0.50 cattle per person compared with 0.37 head average for all the developed countries and 0.25 head for the developing countries. The neighboring countries of Canada and Mexico average 0.52 and 0.44 head per capita, respectively.

The inventory of beef cows 2 years and older in the United States has grown from 16 million head just after World War II to 36 million head in 1980, in effect more than doubling in the postwar years. During that period of time, milk cows decreased in number from 24 to 11 million, with the net result that the all-cow inventory has increased from 40 million to 47 million. Pasture land, however, has declined from about 250 million ha. to about 243 million as a result of technological changes such as increased use of feedlots.

As we shall see in Chapter 11, the U.S. beef trade is a classic example of interdependence in a world economy. Although the United States has the capacity to easily produce enough beef for domestic consumption, it is the largest beef importer in the world because other countries have a comparative advantage in production of the low quality, nonfed beef that makes up a large part of the U.S. consumer's diet. Ground beef

The Cattleman Magazine Photo

If it were not for cattle, this probably would be idle land. These cows must do a lot of walking to get feed and water.

and beef for processing account for about 40 percent of beef consumption. These beef imports (about 8 percent of U.S. production) are, however, more or less offset by exports of fed beef, hides, tallow, and other by-products. The net result was a positive trade balance of $63 million for bovine products in 1978, but a deficit of $183 million in 1979 (Appendix Table A.6).

Cattle production and marketing systems in the United States, as in other areas of the world, are continuously evolving and changing. A century ago, for example, central and west Texas ranchers were simultaneously involved in cow-calf and fattening operations. Steers were held until they matured (3–5 years of age), at which time they were trailed by the owners to railheads. The early 1900s witnessed expansion of the railroads and introduction of better breeding stock, which stimulated ranchers to improve their production practices. As time passed, communications systems continued to improve and transportation rates became cheaper relative to the animals' value. Specialization of operations soon became common and ranchers on better quality land finished out steers on improved pastures while operators on more marginal land (marginal in the sense of low productivity and restricted agricultural possibilities) specialized in raising calves.

Between World Wars I and II, corn belt farmers firmly established themselves as feeders, and most of their cattle supply came from the northern Great Plains. By the 1950s considerable expansion had been made by grain farmers in the Panhandle area of northern Texas. An abundance of feedstuffs led entrepreneurs to begin feeding cattle on a large scale with purchased as well as home raised grain. Substantial improvements in grain sorghum, rapid increase of feedlot size, progress in disease control and management techniques, and improved refrigeration and marketing methods, among other factors, caused confinement feedlot systems to virtually replace grass finishing of slaughter cattle.

The development of feedlots, plus improved transportation, has led central and east Texas ranchers to specialize as cow-calf operators. Except for hauling small numbers of cattle to auction markets, these ranchers now have little or nothing to do with transportation. The net result, however, is that the basis of the cattle raising system is now a commercial type herd which produces feeder calves that are grazed as stockers and that afterward enter feedlots. Some are moved directly at weaning age from the ranch to feedlots (Williams and Farris 1975).

Beef cattle raising constitutes a relatively important agricultural enterprise in most regions of the United States. It occurs throughout various topographic, climatic, and soil conditions favorable for raising feeder cattle destined for feedlot placement or direct slaughter. On many operations cattle raising is the sole or primary enterprise, while on others it is a secondary source of income. Because more than 70 percent of beef production is from cattle fed-out on a ration of concentrates in confinement feedlots (called "fed" cattle), most farms and ranches with breeding cattle are geared to supply animals for these feeding operations.

There are relatively few operations in which the owner both raises calves and fattens them to slaughter weight.

The Southeast, Southwest, and North Central regions of the United States, each with nearly one-fourth of the nation's beef cows, are the most important regions for feeder cattle raising (Figure 3.1). The Southeast is a region that has great flexibility in shifting from crop agriculture to beef cattle operations because much of the land is of poorer quality than in other major crop producing areas such as the corn belt in the Midwest. Although the western United States is usually thought of as a major cattle producing area, in reality it has only about one-tenth of the national breeding herd. The largest cow herds are found in the Pacific West and Great Plains, where 7.3 percent and 8.2 percent, respectively, of the farms and ranches carry an inventory of 200 or more cows. Less than 1 percent of the farms in the North Central states are in this category (Table 3.1).

The five regions delineated in Figure 3.1 have been segmented according to different cattle raising systems. For example, the Southeast accounts for 20.1 percent of the cow-calf units in the United States that specialize in producing feeder calves for sale at weaning or as yearlings (Table 3.2). The West, on the other hand, accounts for only about 7.5 percent of the cow-calf feeder operations in the United States. About 88 percent of all the units purchasing feeders for fattening (excluding large commercial feedlots) are found in the North Central region.

Fig. 3.1. Cattle raising regions of the United States (Boykin et al. 1980).

Table 3.1. Structure of U.S. beef cattle raising, 1974

Region	Farms and ranches		Cows and heifers that have calved		Distribution of farms and ranches with beef cattle by cow herd size					
	Total	Proportion	Total	Proportion	1–19	20–49	50–99	100–199	200 and more	Total
		(%)	*(1,000)*	*(%)*			*(%)*			
Southeast	294,444	30.5	9,268	23.1	54.7	30.5	9.8	3.4	1.6	100.0
Southwest	173,104	17.9	9,199	23.0	40.3	33.7	15.0	6.7	4.3	100.0
West	59,704	6.2	3,903	9.7	50.6	21.6	12.2	8.3	7.3	100.0
Great Plains	107,829	11.2	8,395	20.9	23.8	30.3	23.5	14.2	8.2	100.0
North Central	329,435	34.2	9,325	23.3	51.4	33.3	11.5	3.1	0.7	100.0
All regions	964,516	100.0	40,090	100.0	47.3	31.4	13.0	5.4	2.9	100.0

Source: Boykin, Gilliam, and Gustafson, 1980, pp. 44–45.

Table 3.2. United States beef cattle raising systems by region, 1976

Region	Cow/calf feeder	Cow/calf slaughter	Feeder purchase slaughter sale[a]	Feeder purchase stocker sale	Mixed	All systems
		System as a percent of all regions				
Southeast	20.1	12.2	2.8	4.2	6.6	16.2
Southwest	24.8	3.5	1.7	37.4	22.0	21.1
West	7.5	3.9	1.1	4.2	6.0	6.4
Great Plains	36.0	25.5	6.3	39.7	34.7	32.6
North Central	11.6	54.9	88.1	14.5	30.7	23.7
All regions	100.0	100.0	100.0	100.0	100.0	100.0
		System as a percent of an individual region				
Southeast	87.0	6.9	1.4	0.9	3.8	100.0
Southwest	82.1	1.5	0.7	6.2	9.5	100.0
West	82.1	5.6	1.4	2.3	8.6	100.0
Great Plains	77.2	7.2	1.6	4.2	9.8	100.0
North Central	34.2	21.3	30.5	2.1	11.9	100.0
All regions	69.9	9.2	8.2	3.5	9.2	100.0

Source: Boykin, Gilliam, and Gustafson, 1980, p. 14.
Note: Based on the number of farms and ranches.
[a]Commercial feedlots excluded.

Another way of looking at the U.S. structure is to examine the importance of each system within a region. About 87 percent of the units in the Southeast, for example, are cow-calf operations producing feeder cattle, whereas only 34.2 percent of the North Central operations are oriented toward the production of feeder calves owing to the predominance of crop farmers, a significant percentage of whom purchase feeder calves for feeding in confinement lots. The purchase of weaned feeder cattle for grazing as stockers with sale to feedlots for finishing is a relatively minor part of any region's system. The Southwest has the highest proportion of stockers, 6.2 percent, and most of these are located near large commercial feedlots in the grain growing areas.

Cattle of the West and Southwest regions graze on a combination of private, state and federal rangelands (Van Arsdall and Skold 1973). In these areas public ownership of forage resources adds another dimension to cattle raising systems for, depending on a host of factors, some ranges may be grazed only in the summer, or perhaps the spring and fall if they are used for recreation purposes as well. Operators must adhere to agency specifications concerning seasonal use and stocking rates and most of them must have base property that provides the remaining spatial and nutritional requirements for the cattle herd when public lands cannot be used. The use of these public lands for grazing has become a major policy issue since the growing population, especially in the West, is favoring the use of public lands for recreational purposes.

CATTLE BREEDS AND BREEDING SYSTEMS

The United States has a long history of importing cattle to expand and upgrade herds. Shorthorns, introduced in 1783, were the first European breed to arrive. By 1800 significant numbers of Devons were imported. Herefords, introduced in 1817, were quickly adopted from as far north as Canada, where they were found to be cold tolerant, to Texas, where they were mated with Texas Longhorns. During the middle to late 1800s, other English breeds such as Aberdeen Angus, Lincoln Red, and Sussex were brought over, but the dominant breed until the early 1900s remained the Shorthorn. The leading position has now been taken by Herefords (Mason 1969).

By the mid-1960s cattle breeders in the United States recognized that few advances could be made in upgrading the English breeds. Crossbreeding with Zebu cattle had been practiced for many years in the southern United States, but northern breeders did not yet appreciate the benefits of crossbreeding. At the same time, preferences began to develop for larger cattle and leaner meat. The result was that Charolais cattle, which were an established breed in the United States since before World War II, suddenly became popular. More important, the so-called "exotic" breeds of Europe such as Simmental, Limousin, and Chianina were "discovered" and demanded by cattlemen.

Sanitary regulations did not permit direct importation of live cattle or semen into the United States from Europe because hoof-and-mouth disease was endemic in most of the countries from which "exotics," that is, breeds not heretofore available, could be imported. Strict import controls in the 1960s led Canadian breeders to take the lead in developing exotic breeds. U.S. breeders purchased heavily from them so that by mid-1970 European exotics were common in the United States. Regulations were changed about 1970, however, to permit importation of semen provided that bulls were quarantined at the point of origin. It should be observed that Argentina was testing the Italian breeds in the early 1960s, but little interest was shown by researchers and breeders because "exotics" had not yet become fashionable.

The development of new breeds was begun by cattlemen in the warm climate areas of the United States and can be traced back to the introduction of *Bos indicus* cattle in 1849, when two Mysore (Longhorned Zebu) bulls were imported into South Carolina. Only a few head of Zebu cattle were introduced into the United States during 1850–1890, but then the wave of imports grew rapidly, so that crossbreeding with Zebu cattle became fairly common in Texas just after the turn of the century. The Brahman breed, which was developed from crosses of a number of *Bos*

Brahman cross cows are well adapted to the heat and humidity of the U.S. Gulf Coast. Their calves are typically one-half to three-fourths British breeds.

indicus breeds, developed to such an extent that by 1942 the American Brahman Breeders Association was organized.

Definite attempts in the 1920s and 1930s to fix new breeds led to the establishment of the foundation Beefmaster herd in 1931. The International Brangus Breeders Association was founded in 1949, the Red Brangus breed in 1946, and the American Charbray Breeders Association in 1949. Santa Gertrudis cattle were well established in the late 1930s, but the Santa Gertrudis Breeders International was not organized until 1951. Recently developed breeds and crossbred types are given in the Appendix, Table A.11.[1]

At the same time that cattlemen in the United States were experimenting with Zebu crosses, the Brazilians were performing similar experiments. The first Zebu arrived in Rio de Janeiro in 1875 and, by the end of the century, large-scale introductions were being made. The demand and major importations continued except for the period during World War I. By the 1920s Brazilians were exporting Zebu breeding stock, some of which entered the United States via Mexico. The Brazilian government stopped issuing import permits in 1930 owing to fear of rinderpest, but such vast improvements have been made that breeders of Zebu type cattle in the United States are still interested in importing cat-

1. Much of the information in this section has been drawn from Rouse 1970 and 1973.

tle from Brazil. Future importation of foreign cattle into the United States will be greatly facilitated by the Fleming Key (Florida) quarantine station, which opened in 1979. This is the first facility in the United States for long-term quarantine of live cattle. It should play an important role in improving cattle breeds in the United States.

Because the beef business in the United States is so closely tied to genetic improvement and development of new breeds, our discussion now turns to some principles of selection and crossbreeding.

SELECTION AND CROSSBREEDING

The process of choosing certain animals of one generation to be parents of the next generation and of determining the number of progeny they will be allowed to have is known as selection. It is the real foundation for cattle improvement. Although the beneficial results are well demonstrated, negative selection is still practiced in some countries. Pastoral herdsmen, for example, often sell the largest and fattest animal for slaughter, keeping the least thrifty for reproductive purposes. Similarly, in societies that use oxen for draft, the cattle raisers tend to castrate the best males. Furthermore, in most pastoral situations, all animals are herded together and selection proceeds on a natural rather than a managed basis.

But selection on the basis of individual merit is widely practiced for selective improvement in cattle, and most progress to date can be credited to individual animal selection (Warwick and Legates 1979). This method, which is strictly a phenotypic process, allows traits such as conformation or growth rate to be evaluated directly from individual animal performance, especially if records are kept. Several important traits such as maternal ability in brood cows are expressed only by females, however, so that selection of breeding males cannot be based strictly on their performance. The relationship between beef cattle traits and improvement factors given in Table 3.3 shows that reproduction—i.e., weaning rate—has a high economic value but that heritability is low. High improvement comes from crossbreeding, however. Conformation has a very high heritability, but the economic value of selection for this trait is usually low, especially where cattle are sold by the head, as in most of Africa, Asia, and parts of Latin America.

Whereas selection on the basis of individual merit is strictly phenotypic, progeny testing is a method of evaluating the genotype of an animal on the basis of its progeny's performance. Since each parent contributes a sample half of its genes to each offspring, evaluating an individual on the basis of one or a few offspring can be misleading. The trouble is that while progeny tests can provide extremely accurate ap-

Table 3.3. Relationship between beef cattle traits and improvement factors

	Improvement factors			
Trait	Heritability (h^2)	Inbreeding depression	Improvement from cross-breeding	Economic value
Reproduction (weaning rate)	Usually low	High	High	High
Production (weaning weight, average daily gain, milk production, etc.)	Medium	Medium	Medium	Medium
Product (carcass traits)	High	Low	Low	Low
Conformation	Very high	Practically zero	Practically zero	Low

Source: Compiled by Tim Olson, University of Florida.

praisals of an individual's breeding value, the time and expense involved in obtaining the information can become prohibitive. Also, effective progeny testing programs require large populations.

The extent to which quantitative traits can be improved depends on heritability of the trait and the amount of selection practiced. Individual merit of the animals is the basis for selection, although information on ancestors, collateral relatives, and progeny are valuable guides.

Selection should be directed only toward the traits of real importance, as selection for more than one trait reduces the selection pressure on any single trait. Selection for economically unimportant traits such as ear size, details of color marking, or conformation is fallaciously reasoned to do no harm. In reality, such selection is indirectly harmful because it reduces the selection intensity for more important traits. Selection of individuals is a difficult and long-term process. One way to complement individual selection is to introduce crossbreeding, a genetic technique which the U.S. beef cattle industry uses extensively.

Although fertility and viability are low in heritability, they are strongly affected by heterosis, also known as hybrid vigor. Heterosis is reflected in increased productivity among crossbreeding or crosslining individuals compared to that, on the average, of the parental types. Growth, for example, is reasonably heterotic. Inbreeding, a system of mating animals more closely related than average individuals of the population to which they belong, depresses performance in most beef traits, especially those related to fertility and livability. On the other hand, outbreeding, a system of mating animals that are less related than average individuals of the population being intermated, improves performance. Crossbreeding, a mating system that combines hereditary material from two or more pure breeds, will maximize heterosis (Koger, Cunhua, and Warmick 1973). Crossbreeding is also useful for incorporating desired characteristics of two or more breeds in complementary combinations. Crossbreeding has thus been used to overcome an environmental difficulty such as heat stress in European breeds, to obtain heterosis (hybrid vigor), or both (Koger, Peacock, Kirk, and Crockett 1975). In addition, crossbreeding can help breeders to upgrade native stock or develop a new breed.

The cowman has two possibilities in a crossbreeding program. The original breeds such as English and Zebu types can be used continuously to reap the benefits of heterosis, but this approach is generally not a satisfactory one when environmental stress is a major factor. The environmental problem can be overcome by using dams and sires with varying degrees of Zebu breeding in a rotation program. Another possibility is to use a composite or synthetic breed developed from a crossbred foundation. Composite-bred cattle offer considerable advantages where it is necessary to maintain a certain percentage of a particular type of breeding, for example, in tropical and semitropical areas a certain degree of Zebu breeding is usually necessary for efficient beef production.

One advantage of using Zebu breeds, of which the Brahman is a leading representative, is that they combine well with all the European breeds with which they have been crossed, and have been responsible for high levels of heterosis. In fact, Brahman bulls have been used so successfully in crossbreeding programs in the southern United States that six new composite or synthetic breeds based on Brahmans have become well established: Santa Gertrudis, Beefmaster, Brangus, Braford, Barzona, and Charbray (Putnam and Warwick 1975). New crosses such as the Simbrah (Simmental and Brahman), Brahmanstein (Brahman and Holstein), Limousin-Brahman, and others have recently received the attention of the industry.

The basic objective of any beef cattle crossbreeding program is to maximize the sum of the additive genetic values and heterosis levels for the three major traits that govern performance in commercial beef cattle herds: (1) weaning rate, that is, the calf crop; (2) maternal ability, which includes milking and motherly attention; and (3) growth rate of the calf (Koger 1979). Selection for these traits is more difficult if only one breed is used, as the principal method of herd improvement is by sire selection. In crossbreeding, on the other hand, progress can be accelerated because additive genetic values are improved through selection both within and between herds.

Several mating systems are commonly used to arrive at crossbred cattle. One is the *two-breed terminal cross,* which is the simplest of all plans and will fit any size herd. Bulls of breed "A" are bred to cows of breed "B" and the calves sold. This system offers 100 percent of the possible hybrid vigor in the calves but not in the cows (maternal heterosis), where it is most important. Another difficulty with this

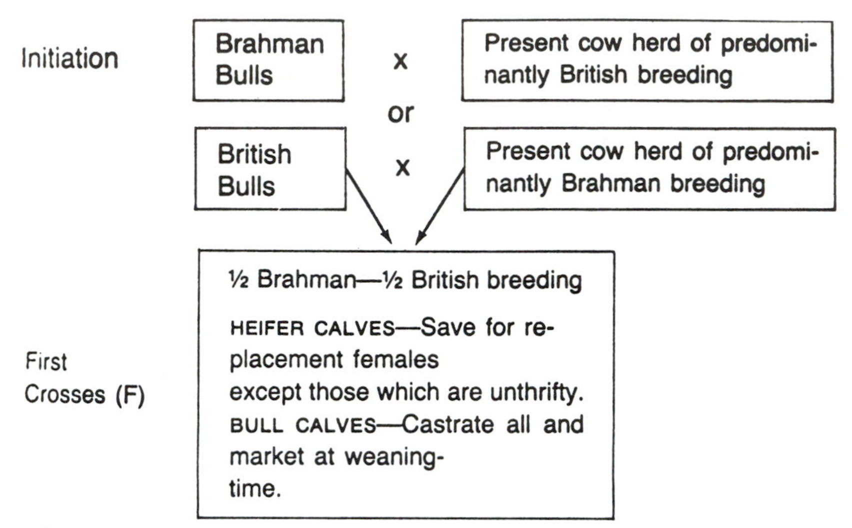

First-cross cows can be backcrossed to Brahman or British bulls. As soon as numbers are large enough, first-cross cows are divided into two herds. One herd uses only Brahman bulls and the other only British bulls of the parent breed. These two herds remain intact during breeding with no change in mating from year to year.

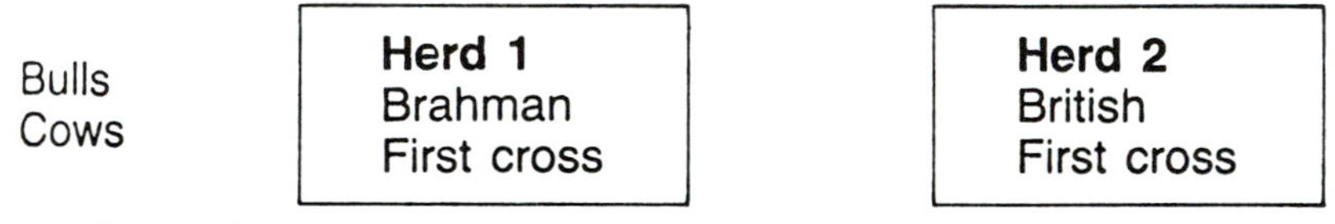

Heifers produced in Herd 1 are selected and placed in Herd 2 for replacements; heifers in Herd 2 go into Herd 1. Bulls are always mated to cows which are most distantly related. Here's how the program will look eventually.

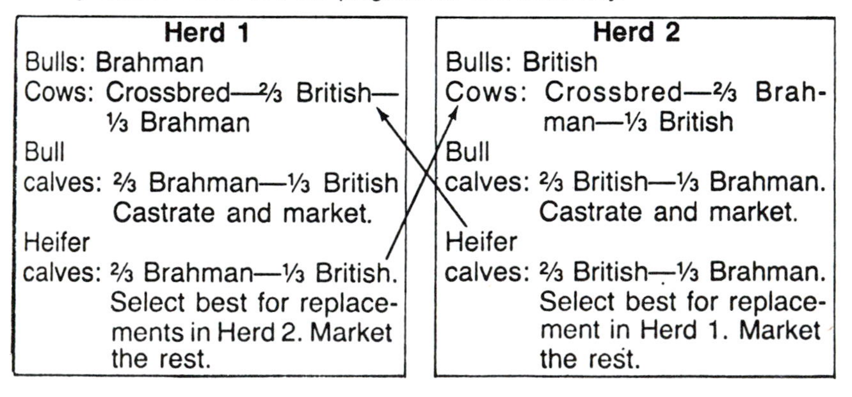

Fig. 3.2. Example of two-breed rotational cross (From *Progressive Farmer® Beef Cattle Book,* by Tom Curl. Copyright © 1978 Oxmoor House, Inc. Reprinted by permission.)

system is that cows and bulls must regularly be replaced from outside the herd.

The second type of mating system is called the *two-breed rotational cross* or *criss-cross.* Figure 3.2 shows that bulls of two different breeds are used each breeding season. Bulls of breed "A" are mated with cows sired by bulls of breed "B," and vice versa. Female replacements can be produced, but cows and calves have only about two-thirds of the maximum potential hybrid vigor. Although herds of more than fifty cows and at least two pastures are required, this is the most commonly used systematic program of crossbreeding (Crockett, Koger, and Franke 1978).

Three-breed rotational crosses are a third type of mating system. The major advantage is that replacements can come from within the herd and selection can be practiced on them. Hybrid vigor is about 86 percent of the maximum for both cows and their calves. Major disadvantages are that at least three breeding pastures and seventy-five cows are required, and record keeping is more difficult.

The *three-breed terminal cross* is a system in which breed "A" bulls are mated to cows that are crosses of breeds "B" and "C." Cows and calves have 100 percent of their hybrid vigor potential, but all breeding herd requirements must be obtained from outside the herd. This plan works with any size herd and requires only one breeding pasture, but replacements are a problem.

A *two-breed rotational cross with terminal sire breed* is theoretically possible but unlikely in practice as it would require at least 150 cows produced by the two-breed rotational cross system.

Composite breeds are not extensively used, but they do offer numerous advantages. One is that a relatively constant percentage of Brahman breeding can be maintained in the cow herd and in the calves to be sold. Another benefit is that operators whose herds are too small to allow them to effectively use a rotational crossbreeding program can take advantage of the greater fertility and increased weaning weights that result from heterosis. A third advantage is that a composite allows the cattleman to select replacement heifers and bulls from within the herd rather than constantly purchasing them from outside sources. This selection process, which also holds for the rotations, permits a rapid rate of genetic improvement and at the same time reduces the possibility of introducing disease into the herd. A final advantage, especially for warm climates, is that composites permit development of a cattle type that performs best under local conditions.

There are several ways of introducing composites into a crossbreeding program. One way is to simply purchase a composite breed. Another approach is to build a composite breed by starting with individual breeds. A cattleman might develop a herd, for example, by first crossing a Brahman with an Angus and a Hereford with a Simmental. The F_1 crosses would then be mated to produce the first generation of the composite, which would be one-quarter of each original breed. Selection of the most productive of the second and later generation females should provide for high levels of fertility and good weaning weights if care is taken to avoid inbreeding. If replacement herd bulls come from within the herd, at least 100 cows, and preferably 200, are needed to prevent the mating of closely related animals. After several generations of selection, the original percentages of each initial breed will probably change because the producer selects the animals that do best under the local environmental conditions (Olson 1979).

Another way in which a composite system could be used, even though few cattlemen follow the practice, would be to cross two unrelated composites or an animal of a composite of conventional breeds in a rotational crossbreeding system involving a three-breed rotation such as Angus × Braford, Angus × Beefmaster, and Hereford × Brangus. In this system a high degree of heterosis is maintained by these types of rotations because the more breeds that are included, the higher the percentage of heterosis. Although several breeds are a minimum requirement, the actual level of performance depends on the extent to which each breed performs both as a breed and in combination with other breeds. Crosses of Brahman and Angus, for example, give higher levels of heterosis than crosses of Hereford and Angus.

A successful crossbreeding program has four essential features (Koger 1979). First, the matings must be planned and executed systematically to maintain high levels of heterozygosity. Failure to plan the crosses will cause the cattle to degenerate into merely mixed breeding. A second rule is that only simple crossbreeding systems should be attempted. Complicated systems such as the two-breed rotation or terminal crosses generally fail to achieve the systematic matings necessary to maintain high levels of heterosis. Third, an effective breed combination should be chosen for particular climatic conditions, terrain, feed supply, market, and desirable growth characteristics. A last rule is that rigid selection and culling procedures must be followed. In Koger's opinion the only three systems that merit close attention in warm climates are (1) two-breed rotational crossbreeding where one of the sires is Brahman, Brahman derivative, or Brahman crossbred, (2) three-breed terminal crossbreeding, and (3) selection from one of the composite breeds.

The development and adoption of new breeds continues with increasing force and determination. New techniques, such as embryo transplants, have potential in speeding up cattle breeding. In this particular procedure the egg of a purebred or superior cow is fertilized, usually artificially, by a purebred bull. Approximately 1 week later the fertilized egg, or embryo, is extracted and transferred to the uterus of a common, commercial cow that carries it to term, whereupon the cow gives birth to a purebred calf. The donor cow can keep ovulating each month as her reproductive cycle is uninterrupted by a 9-month pregnancy. Although this technique is expensive, it does permit establishment of a genetically superior breeding herd in just a few years, whereas 20 or more years are now required. As of the late 1970s transfer companies in the United States charged about $450 to as much as $1,800 for each pregnancy. Despite the apparently high cost, about ten companies were specializing in the procedure (*Wall Street Journal* 1979).

FEEDLOTS AND CATTLE FEEDING

The total number of cattle feedlots in the twenty-three major cattle feeding states has shown a steady decline although the number of cattle marketed has increased (USDA 1978). In 1975 about 137 thousand lots marketed 20.5 million head of cattle (Table 3.4). By 1978 the number of lots had dropped to about 127 thousand while marketings increased to 26.6 million head. The decline in number of lots took place almost entirely in those with 1,000-head or less one-time capacity. However, the number of cattle fed in these smaller lots actually increased.

The net result in feedlot structure change is that in 1978 the 1,000-head or less capacity lots accounted for 98.5 percent of all feedlots, but they only marketed 32.0 percent of the cattle. In the meantime, the number of 32,000-head or more lots remained almost constant (66 versus 65, respectively). Their marketings increased, however, from 15.9 to 19.3 percent of the total. Overall, there is a definite move toward greater concentration, although the large total number of lots indicates that cattle feeding is still in the hands of many widely dispersed operators.

Who owns the lots? A 1979 study showed that the largest feedlots are generally operated under names unrelated to their corporate ownership. Of special significance is the move by big business into feedlots. For example, Cargill, the world's largest grain-trading house and owner of the second largest meatpacking firm in the United States (MBPXL) owns the greatest amount of feedlot capacity in the United States (Table 3.5). The integration advantage of grain, cattle feeding, and meat packing is an appealing concept and some companies take it one step further as seven of the top twenty also have stock or commodity brokerage house ties. Despite this advantage, the large grain companies are inclined to use brokers and traders other than their own to avoid revealing futures market positions (Richardson 1979).

Table 3.4. Cattle feedlots and marketings, twenty-three states

Item	1973		1974		1975		1976		1977		1978	
	(no.)	*(%)*	*(no.)*	*(%)*	*(no.)*	*(%)*	*(no.)*	*(%)*	*(no.)*	*(%)*	*(no.)*	*(%)*
Total												
Lots	146,220	100.00	137,737	100.00	136,696	100.00	132,535	100.00	131,904	100.00	127,425	100.00
Marketings (000 head)	25,304	100.00	23,330	100.00	20,500	100.00	24,170	100.00	24,853	100.00	26,645	100.00
1,000 head												
Lots	144,180	98.61	135,815	98.61	134,919	98.70	130,739	98.70	130,018	98.57	125,523	98.51
Marketings (000 head)	8,941	35.33	8,261	35.41	7,246	35.35	7,926	32.90	7,917	31.85	8,542	32.06
1,000–1,999 head												
Lots	865	0.59	747	0.54	647	0.47	664	0.48	824	0.62	845	0.66
Marketings (000 head)	1,138	4.50	981	4.20	813	3.97	935	3.87	1,177	4.73	1,374	5.16
2,000–3,999 head												
Lots	471	0.32	484	0.35	440	0.32	446	0.32	401	0.30	412	0.32
Marketings (000 head)	1,287	5.08	1,065	4.57	953	4.65	1,158	4.79	1,186	4.77	1,300	4.88
4,000–7,999 head												
Lots	280	0.19	258	0.19	262	0.19	267	0.20	239	0.18	230	0.18
Marketings (000 head)	1,811	7.16	1,541	6.61	1,386	6.76	1,781	7.40	1,654	6.66	1,568	5.88
8,000–15,999 head												
Lots	218	0.15	212	0.15	211	0.16	209	0.15	221	0.17	217	0.17
Marketings (000 head)	3,170	12.53	2,854	12.23	2,620	12.78	3,087	12.65	3,583	14.42	3,626	13.61
16,000–31,999 head												
Lots	137	0.09	148	0.11	151	0.11	149	0.11	140	0.11	133	0.11
Marketings (000 head)	4,124	16.30	4,172	17.88	4,216	20.56	4,911	20.56	4,846	19.50	5,081	19.07
32,000+ head												
Lots	69	0.54	73	0.05	66	0.05	61	0.04	61	0.05	65	0.05
Marketings (000 head)	4,833	19.10	4,456	19.10	3,266	15.93	4,372	17.83	4,490	18.07	5,154	19.34

Source: USDA, *Livestock and Meat Situation,* various issues.

Table 3.5. Top twenty cattle feedlots in the United States, 1979

Rank	Company	Number of lots	One time capacity	Grain company tie	Packer tie	Brokerage house ties
1.	Caprock Industries	3 Texas 2 Kansas	216,000	Cargill Nutrina Feeds	MBPXL	Cargill Investors
2.	Monfort of Colorado	2 Colorado	200,000		Monfort Packing	
3.	AZL Resources	4 Texas 1 Nebraska	186,500			Bromagen & Hertz
4.	Northwest Feeders Inc.	6 Washington	183,000		Iowa Beef Processors	
5.	Hitch Enterprises	2 Oklahoma 1 Kansas	177,000		Booker Custom Packing	
6.	Barrett-Crofoot	2 Texas	177,000			
7.	Catus Feeders	2 Texas 1 Oklahoma	153,000			One major stock-holder with REFCO
8.	Western Beef, Inc.	2 California 2 Texas 1 New Mexico	124,000	Western Beef Grain Co.		One director is RB&H broker
9.	Friona Industries	3 Texas	112,000		Village Packing	
10.	Foxley & Co.	2 Nebraska 1 Texas	110,000			Agent for RB&H
11.	Allied Mills	2 Texas	105,000	Continental Grain Wayne Foods		Conti-Commodities
12.	Miller Feedlots	2 Wyoming 1 Texas 1 Colorado	104,000			
13.	Harris Feedlot	1 California	100,000		Diamond Meat San Jose Meat	
14.	Dekalb Ag. Research Inc.	1 Texas 1 Arizona 1 Kansas 1 Mississippi	96,000	Arizona Feeds		Heinold
15.	Hi-Plains Feedyard	1 Texas	90,000			
16.	Red River Feedyard	1 Arizona	85,000			
17.	Valley View Cattle Co.	2 Texas	81,000			
18.	Fat City Feedlots	1 California	80,000			
19.	Wilhelm Co.	2 Colorado	80,000			
20.	Monson & Son Cattle—Van Degraff Feedlot	1 Washington	75,000		Washington Beef Processors	

Source: Richardson, 1981.

The top twenty feedlots have a one-time capacity to feed over 2.4 million head, with nearly all of that located in the seven major cattle feeding states. Numbers of cattle on feed fluctuate substantially, but using an average of eight million head, the top twenty control about 30 percent of all cattle on feed. One advantage that large multicommodity firms have in feeding cattle is their financial ability to spread risk and cope with the prolonged periods of losses that characterize the business.

One reason that cattle feeding has become a big business is the relatively high investment involved (Richards and Karzan 1975). One study shows, for example, that a 500-head capacity (on a one-time basis), traditional dirt open-lot system would have cost about a quarter of a million dollars in 1979 (Simpson, Baldwin, and Baker 1980). A 5,000-head facility would have cost about $1.1 million and a 10,000-head unit nearly $2 million (Table 3.6). On a per head of capacity basis, this translates to $532, $227, and $198, respectively. A total confinement, all-concrete, "flume floor" cattle feeding facility would be slightly less expensive because a smaller area per head would be required. Consequently, many of the major cost items such as feed bunks, fence, and land would be reduced. "Flume floor" is the name given to a system having flush gutters at 2.4- to 4.6-m intervals across pens that are 9.1 to 12.2 m wide. Utilizing animal trampling and a natural inclination for cattle to stand uphill, the floor slopes to the gutter from each side to move manure downslope into the gutter. Water is circulated through the gutters almost

Table 3.6. Summary of total and per unit investment costs for flume and open-lot systems in Florida, 1979

	Flume lot			Dirt open-lot		
Item	500	5,000	10,000	500	5,000	10,000
			($)			
			Total investment			
Feeding facilities	63,578	451,165	872,191	108,848	475,175	918,008
Supporting facilities	122,843	403,243	661,142	134,206	507,632	804,430
Waste management	27,490	130,500	247,200	12,880	59,140	105,300
Subtotal	213,911	984,908	1,780,533	255,934	1,041,947	1,827,736
Construction management	10,000	90,000	150,000	10,000	90,000	150,000
Total	223,911	1,074,908	1,930,533	265,934	1,131,947	1,977,738
			Investment per head of capacity			
Feeding facilities	127	90	87	218	95	92
Supporting facilities	246	81	66	268	102	80
Waste management	55	26	25	26	12	11
Subtotal	428	197	178	512	209	183
Construction management	20	18	15	20	18	15
Total	448	215	193	532	227	198

Source: Simpson, Baldwin, and Baker, 1980.

Cattle feeding facilities can be low cost, especially in low rainfall areas as in the southern Plains of the U.S. where no shelter is provided.

continuously. There are various types of all-concrete lots, depending on the waste removal system. The investment costs and operating expenses given in Table 3.6 are for Florida feedlots, but costs are typical of those found in other areas of the United States (Scofield 1974).

The trend toward larger feedlots can also be attributed to their greater efficiency, which means reduced operating expenses as well as lower fixed costs (Edwards 1973). Furthermore, larger lots can afford to employ skilled managers who are specialized in buying cattle and other inputs, and who have the knowledge and ability to use modern cattle marketing techniques (Dyer and O'Mary 1972). A most important consideration is that larger lots result in economies of size since they feed more cattle (Gee, Van Arsdall, and Gustafson 1979). The 1980 study by Simpson, Baldwin, and Baker shows, for example, that the nonfeed operating costs per head fed for a traditional open-lot system operating at 90 percent of capacity are $49, $18, and $17 for the 500-, 5,000- and 10,000-head capacity units, respectively (Table 3.7). Operation of the lots at 60 percent of capacity causes per head costs to increase to $67, $23, and $22, respectively, as few of the expenses are reduced in direct proportion to the number of head fed.

The initial investment costs have been translated to a per head fed basis in Table 3.6 assuming an annual turnaround of 2.73 lots of cattle.

Table 3.7. Total annual investment, operating and replacement costs, and cost per head fed, of three sizes of flume and dirt open-lot systems at 90 and 60 percent of capacity, 1979

Feedlot capacity and type of lot	Total annual cost				Cost per head fed			
	Initial investment[a]	Equipment replacement cost	Operating expenses	Total	Initial investment	Equipment replacement cost	Operating expenses	Total
				($)				
				90% of capacity				
500 Head								
Flume	29,438	14,064	60,591	104,093	24	11	49	84
Dirt open-lot	34,962	13,533	59,796	108,291	28	11	49	88
5,000 Head								
Flume	141,318	49,696	191,299	382,313	12	4	15	31
Dirt open-lot	148,817	53,147	222,006	423,970	12	4	18	34
10,000 Head								
Flume	253,807	97,072	355,964	706,843	10	4	15	29
Dirt open-lot	260,013	106,093	416,790	782,896	11	4	17	32
				60% of capacity				
500 Head								
Flume	29,438	14,064	56,584	100,086	36	17	69	122
Dirt open-lot	34,962	13,533	55,106	103,601	43	17	67	127
5,000 Head								
Flume	141,318	49,696	172,869	363,883	17	6	21	44
Dirt open-lot	148,817	53,147	186,763	388,727	18	6	23	47
10,000 Head								
Flume	253,807	97,072	319,583	670,462	16	6	19	41
Dirt open-lot	260,013	106,093	365,286	731,392	16	7	22	45

Source: Simpson, Baldwin, and Baker, 1980.
[a]Amortized over fifteen years at 10 percent.

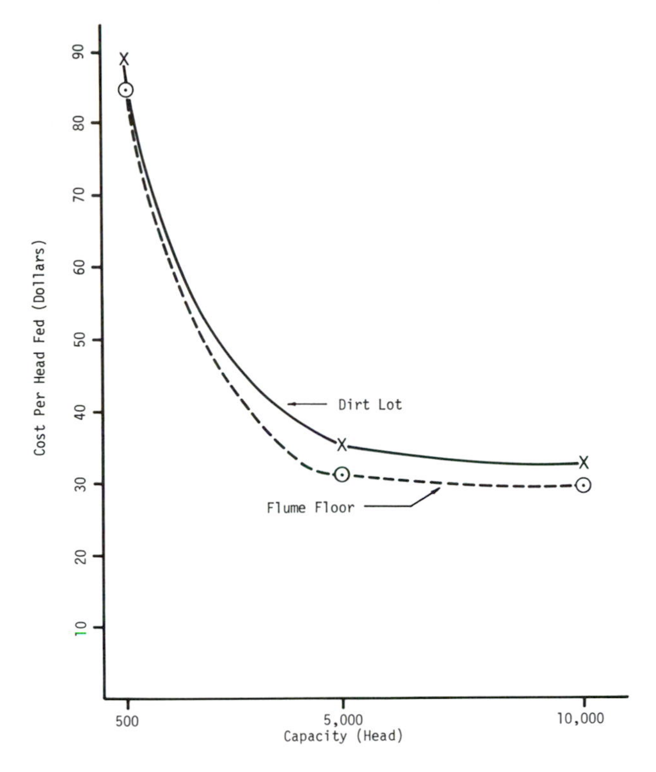

Fig. 3.3. Total feedlot cost per head fed, flume floor, and dirt open lots at 90 percent of capacity, 1979 (Simpson et al. 1980).

Many feedlots keep cattle on feed for 6 months, so that their turnaround would be less, say 1.9 to 2.0 lots, and consequently the cost per head would be higher than that shown. In addition to the initial investment and operating expenses there is equipment replacement cost. The total cost per head is $88, $34, and $32 for the traditional dirt open-lot system operating at 90 percent of capacity, and about $3 per head less for the flume floor system. Figure 3.3 shows the great economies of size available.

Cattle feeding involves high risks because of variable production factors such as weather, disease, labor problems, and input price changes. Risk is most related, however, to the great variability in prices of feeder and fed cattle. An example of the monthly margin movement is detailed in Table 3.8, which presents complete budgets for Great Plains custom cattle feedlots. These budgets, which are published several times a year by the U.S. Department of Agriculture in *Livestock and Meat Situation,* are designed to show both how and why the margins change.

Table 3.8. Great Plains custom cattle feeding

Purchased during / Marketed during	July / Jan. 79	Aug. / Feb.	Sept. / Mar.	Oct. / Apr.	Nov. / May	Dec. / June	Jan. 79 / July	Feb. / Aug.	Mar. / Sept.	Apr. / Oct.	May / Nov.	June / Dec.	July / Jan. 80	Aug. / Feb.	Sept. / Mar.	Oct. / Apr.
								($/head)								
Expenses																
600 lb feeder steer	358.02	359.52	380.00	370.50	384.90	404.34	448.44	481.38	528.66	541.56	515.40	454.44	474.00	456.78	485.28	470.58
Transportation to feedlot (300 mi)	3.96	3.96	3.96	3.96	3.96	3.96	3.69	3.96	3.96	3.96	3.96	3.96	3.96	3.96	3.96	3.96
Commission	3.00	3.00	3.00	3.00	3.00	3.00	3.00	3.00	3.00	3.00	3.00	3.00	3.00	3.00	3.00	3.00
Feed:																
Milo (1,500 lb)	62.55	59.10	58.65	62.55	61.20	58.65	60.75	60.45	60.60	62.25	64.95	73.95	80.10	71.85	71.25	70.50
Corn (1,500 lb)	67.65	66.75	63.75	68.85	69.45	66.90	71.70	72.30	72.15	75.45	79.65	86.55	90.75	84.75	81.75	80.55
Cottonseed meal (400 lb)	39.60	36.80	38.40	40.00	43.20	43.20	44.40	44.00	44.00	42.40	42.40	41.60	43.20	44.00	44.40	45.20
Alfalfa hay (800 lb)	37.20	38.40	39.00	40.00	40.00	41.00	43.00	42.20	43.20	44.20	41.00	41.00	40.40	39.80	39.60	40.80
Total feed cost	207.00	201.05	199.80	211.40	213.85	209.75	219.85	218.95	219.95	224.30	228.00	243.10	254.45	240.40	237.00	237.05
Feed handling and management charge	21.00	21.00	21.00	21.00	21.00	21.00	21.00	21.00	21.00	21.00	21.00	21.00	21.00	21.00	21.00	21.00
Vet medicine	3.00	3.00	3.00	3.00	3.00	3.00	3.00	3.00	3.00	3.00	3.00	3.00	3.00	3.00	3.00	3.00
Interest on feeder and ½ feed	23.08	23.00	24.04	23.81	25.82	26.73	29.31	31.02	33.53	34.32	33.04	30.24	31.56	31.73	35.47	37.56
Death loss (1.5% of purchase)	5.37	5.39	5.72	5.56	5.77	6.07	6.72	7.22	7.93	8.12	7.73	6.82	7.10	6.85	7.23	7.06
Marketing[a]	F.O.B.	F.O.B.	F.O.B.	F.O.B.	F.O.B.	F.O.B.	F.O.B.	F.O.B.	F.O.B.	F.O.B.	F.O.B.	F.O.B.	F.O.B.	F.O.B.	F.O.B.	F.O.B.
Total	624.43	619.92	641.52	642.23	661.30	677.85	735.28	769.53	821.03	839.26	815.13	765.56	798.07	766.72	795.94	783.21
								($/cwt)								
Selling price required to cover[b]																
Feed and feeder cost (1,055 lb)	53.51	53.08	55.00	55.10	56.70	58.15	63.29	66.32	70.89	72.52	70.40	66.05	68.98	66.02	68.40	67.01
All costs	59.13	58.70	60.75	60.82	62.62	64.19	69.63	72.87	77.75	79.48	77.19	72.50	75.57	72.61	75.37	74.17
Selling price ($/cwt)[c]	61.28	65.14	72.15	75.72	75.73	70.48	69.25	63.50	68.80	65.49						
Net margin/cwt	+2.15	+6.44	+11.40	+14.90	+13.11	+6.29	−0.38	−9.37	−8.95	−13.99						
Cost/100 lb gain:																
Variable costs less interest	47.27	46.09	45.90	48.19	48.72	47.96	50.11	50.03	50.38	51.28	51.95	54.78	57.11	54.25	53.65	53.62
Feed costs	41.40	40.21	39.96	42.28	42.77	41.95	43.97	43.79	43.99	44.86	45.60	48.62	50.89	48.08	47.40	47.41
Unit prices																
Choice feeder steer 600–700 lb Amarillo ($/cwt)	59.67	59.92	63.50	61.75	64.15	67.39	74.74	80.23	88.11	90.26	85.90	75.74	79.00	76.13	80.88	78.43
Transportation rate ($/cwt/100 miles)[d]	0.22	0.22	0.22	0.22	0.22	0.22	0.22	0.22	0.22	0.22	0.22	0.22	0.22	0.22	0.22	0.22
Commission fee ($/cwt)	0.50	0.50	0.50	0.50	0.50	0.50	0.50	0.50	0.50	0.50	0.50	0.50	0.50	0.50	0.50	0.50
Milo ($/cwt)[e]	4.71	3.94	3.91	4.17	4.08	3.91	4.05	4.03	4.04	4.15	4.33	4.93	5.34	4.79	4.75	4.70
Corn ($/cwt)[e]	4.51	4.45	4.25	4.59	4.63	4.46	4.78	4.82	4.81	5.03	5.31	5.77	6.05	5.65	5.45	5.37
Cottonseed meal ($/cwt)[f]	9.90	9.20	9.60	10.00	10.80	10.80	11.10	11.00	11.00	10.60	10.60	10.40	10.80	11.00	11.10	11.30
Alfalfa hay ($/ton)[g]	93.00	96.00	97.50	100.00	100.00	102.50	107.50	105.50	108.00	110.50	102.50	102.50	101.00	99.50	99.00	102.00
Feed handling and management charges ($/ton)	10.00	10.00	10.00	10.00	10.00	10.00	10.00	10.00	10.00	10.00	10.00	10.00	10.00	10.00	10.00	10.00
Interest, annual rate	10.00	10.00	10.00	10.00	10.50	10.50	10.50	10.50	10.50	10.50	10.50	10.50	10.50	11.00	11.75	12.75

Source: USDA, 1979, p. 13.

Note: Represents only what expenses would be if all selected items were paid for during the period indicated. The feed ration and expense items do not necessarily coincide with experience of individual feedlots. For individual use, adjust expenses and prices for management, production level, and locality of operation. Steers are assumed to gain 500 lb in 180 days at 2.8 lb per day with feed conversion of 8.4 lb per pound gain.

[a]Most cattle sold F.O.B. the feedlot with 4 percent shrink.

[b]Sale weight 1,056 lb (1,100 lb less 4 percent shrink.)

[c]Choice slaughter steers, 900–1,100 lb, Texas-New Mexico direct.

[d]Converted from cents per mile for a 44,000 lb haul.

[e]Texas Panhandle elevator price plus $0.15/cwt handling and transportation to feed lots.

[f]Average prices paid by farmers in Texas.

[g]Average price received by farmers in Texas plus $20/ton handling and transportation to feedlots.

Budgets are also published for the corn belt of the North Central United States.

Each column in Table 3.8 signifies the purchase and sale months. Cattle purchased in April 1979, for example, would be sold 6 months later in October 1979. A Choice 600 lb steer at Amarillo was priced at $90.26 per cwt ($1.99 per kg) in April for a total of $541.56 per head. The steers are assumed to gain 500 lb (227 kg) in 180 days at 2.8 lb (1.27 kg) per day with a feed conversion of 8.4 lb (3.78 kg) of feed per pound of gain. Transportation would add $3.96 per head and commission charges another $3.00. Total feed cost would be $224.30 per head, while a feed handling and management charge would be $21.00. The total cost including other expenses would be $839.26.

The selling price in October 1979 averaged $65.49 per cwt ($1.44 kg) so that our average High Plains cattle feeder would have lost $13.99 per cwt ($0.31 per kg). Total selling price of the 1,056 lb (480 kg) animal was $691.57, so that our average cattle feeder would have lost $147.69 per head. The break-even price required to cover all costs would have been $79.48 per cwt ($1.75 per kg). The margins vary between months mainly because of price movements. Cattle purchased in November 1979 and sold in May 1979 gave a net margin of $13.11 per cwt ($0.29 per kg). However, cattle feeders purchasing cattle just a few months later, in February 1979, and selling them in August of that year, would have lost $9.37 per cwt ($0.21 per kg).

Net margins per kg from Great Plains cattle feeding are shown on a monthly basis in Table 3.9. Cattle feeders had a positive margin in only 23 of the 70 months from January 1974 to October 1979. If the feeders had placed the same number of cattle on feed each month, they would have suffered an average loss of $1.82 per cwt ($0.04 per kg) over that period.

Table 3.9. Net margin per hundredweight by month of sale from Great Plains custom cattle feeding, 1974-1979

Month sold	Year					
	1974	1975	1976	1977	1978	1979
				($)		
January	1.15	−5.04	−4.26	−11.42	−1.35	2.15
February	−5.92	−8.35	−7.63	−9.61	0.82	6.44
March	−8.19	−5.66	−11.22	−7.21	5.28	11.40
April	−3.67	0.29	−0.48	−3.10	9.10	14.90
May	−5.51	7.55	−4.93	−0.05	12.59	13.11
June	−9.94	11.81	−6.15	−3.41	8.58	6.29
July	−6.11	10.21	−8.23	−3.73	5.64	−0.38
August	−3.46	8.21	−10.61	−6.20	0.63	−9.37
September	−6.46	9.11	−10.06	−5.94	−1.12	−8.95
October	−6.09	5.19	−12.21	−6.33	−2.02	−13.99
November	−4.23	2.02	−9.96	−5.40	−5.91	
December	−0.19	1.61	−9.23	−1.40	−1.05	

Source: USDA, *Livestock and Meat Situation*, various issues.

CAPITAL FLOWS IN CATTLE FEEDING

The longer periods of losses by cattle producers naturally lead one to ask why entrepreneurs would continue to place cattle on feed despite long periods of sustained losses; where the equity comes from; and what equity requirements are. We briefly discuss these questions in this section.

Capital requirements associated with beef production in feedlots are high in relation to many other agricultural production subsectors. For example, operating costs of feedlots in the major twenty-three feeding states for feeder calves, the cost of feed, and other operating expenses ranged between $7.4 billion and $12.6 billion annually over the period 1970–1976 (Hottel and Martin 1976). A high proportion of the total costs are normally financed from outside sources and thus represent short-term, rapid-turnover debt funds. Over this same period, minimum equity requirements increased from an estimated $0.7 billion in 1970 to a high of $2.1 billion in mid-1974. By early 1975 they were down to about $1.5 billion. A major reason for the fluctuation is that equity capital requirements are directly related to cattle and feed prices (Martin and Dietrich 1977).

The work by Hottel and Martin aptly illustrates how equity reserves are drawn down during periods of prolonged losses. They calculate, for example, that equity reserves, i.e., funds in excess of those required to support outstanding debts, would have increased from $2.04 million in the fourth quarter of 1969 to $4.08 million by the fourth quarter of 1973 if there had been no net income withdrawals. Because of consistent heavy losses over the 18-month period from the fourth quarter of 1973 through the first quarter of 1975, the same industry-wide reserves would have declined to $3.99 million. If there had been a 25 percent net income withdrawal, the equity reserves for all feeders in the twenty-three states would have increased to a high of $3.64 billion in the second quarter of 1973, and then would have declined to a negative $673 million in the first quarter of 1973.

The impact of the losses sustained by the industry during the disastrous 1973–1975 period varied, of course, among individual cattle feeders. While many cattle owners had sustained substantial losses, few feedlots actually liquidated their holdings under foreclosure. Firms that raised equity capital through the sale of limited partnership interests were able to shift some of those losses to their investors. As a result of those losses, much of the equity capital provided by limited partnerships was lost or withdrawn from the sector during 1974, but had returned by late 1975.

Use of limited partnerships, commonly called "funds," came on the cattle feeding scene in the late 1960s. Under this arrangement, there are one or more general partners and one or more limited partners. By 1973, less than 5 years after the funds were pioneered, as much as $350 million of the tax-induced equity capital was channeled into cattle feeding (Meisner and Rhodes 1974). This amount represented about 20 percent of the capital in the fed cattle industry.

The major advantage to the limited partner has been a reduction in income taxes for the short term, although the expectation of economic return has been an important incentive. Another incentive has been the "Schmaltz" effect, which is a desire of investors to associate themselves with what they believe is a glamorous or exciting industry. The trade-off is between glamour with risk and a tax shelter. The principal advantage to the general partners, who may be cattle feeders or simply promoters, is the availability of large amounts of capital without restrictions on management. This equity can be leveraged two or three times for purchase of cattle or feed. The general partner may take a percentage of the profits or charge a fixed fee.

A major reason for success of the funds is that agriculture has been taxed on a cash rather than an accrual basis. The advantage to the "big boys" with their "funny money" has been receipt of credit from the Internal Revenue Service as soon as the money is invested, and deduction of prepaid expenses such as feed and interest, which gives them considerable maneuverability for timing gains and losses. As a result of controversy about unfair advantage for nonagricultural interests and investors with large amounts of capital, Congress, in the Tax Reform Act of 1976 required accrual accounting and tax deductions on items not consumed in the year purchased by limited partnerships. The result is that although nonagricultural investors have continued to invest in funds, the impact of funds from this source has declined substantially.

Limited partnerships have contributed substantially to the formation and growth of larger cattle feeding firms in the United States. Indirect benefits have been the attraction of related industries such as slaughter plants to feedlot areas, greater use of capacity, and advantages from economies of size. Previous market relationships have broken down with the advent of funds as feedlots have taken on more of the character of cost-plus operations rather than risk-taking speculations. In fact, it is now useful to carefully distinguish between ownership of a feedlot and ownership of the cattle being fed.

There is, of course, deep long-run concern in the United States over the shift to bigness in cattle feeding, and philosophical questions about size will continue to plague the feeding industry despite the highly tax-oriented money returning to the more traditional tax shelters on Broadway and in the oil industry. This experience may be instructive to the developing countries as they seek methods for attracting risk capital to the livestock industry. The point is, nonagricultural interests will invest large amounts of capital in rather risky ventures if they can be guaranteed some certain benefit, and if the activity is properly promoted. Just as legislation was required to limit nonagricultural capital in this case, it is also true that agricultural development can be "pulled" through the legislative process rather than always being "pushed" via government action programs—and it may be more effective.

FUTURES TRADING

Both producers of calves and cattle feeders can reduce risk by using the commodity futures markets to lock in prices for inputs and fed cattle. Many commodities are traded on futures exchanges, for example, gold, metals, and agricultural products, but all commodities are traded in basically the same manner. There are two major exchanges, the Chicago Board of Trade and the Chicago Mercantile Exchange. Grains are traded on the former market and cattle on the latter one. Speculators buy and sell commodities with the expectation of making a profit through favorable price changes. A person who owns a commodity and sells a contract for delivery at a future time is a hedger. Buyers of a commodity for future use can also be hedgers. Speculators are necessary in that someone must purchase the contracts and they add volume and competition to the market.

Cattle feeders who use futures markets usually hedge cattle for delivery in the month that they are expected to be finished. In most instances it is inconvenient and unprofitable to actually deliver the cattle to the point specified in the contract, so the position is liquidated by the hedger buying back an offsetting contract. Speculators are willing to sell as they do not want to take possession of the cattle. Usually, there is no problem in matching up the number of buyers and sellers as price is set by the interaction of supply and demand. The futures market has an ownership flexibility (liquidity) advantage over regular (nonfutures) contracts in which the intention is to actually deliver the cattle (forward contracting) because positions can be taken or liquidated simply by paying the commission. As in nearly any cattle sale contract, if prices rise more than the hedger anticipated, there is no actual loss, just an opportunity cost by having misjudged price movements. At the other extreme, if the price were to fall below the contract price, the seller would have used the proper strategy by locking in a price.

In addition to hedging fed cattle, a cattle feeder could also lock in grain prices over the feeding periods, as well as feeder cattle price. If all three factors are accounted for—feeder cattle price, feed, and fed cattle price—the individual is fully hedged and a profit (or loss) can be calculated fairly accurately by estimating the basis, death loss, feed conversion, and operating expenses. The basis, which is the spread between futures and local prices, includes transportation, shrink, and other marketing costs, as well as differences in supply and demand factors.

SUMMARY AND CONCLUSIONS

This chapter has provided a brief explanation of feeder cattle production and some economic aspects of fattening them in feedlots. Some concern has been expressed about the desirability of feeding grain to cattle, with the implication that they should be fed on forage. But economic analyses clearly show that with the grain/beef price ratios found in the United States, feedlot fattening will be the lowest cost alternative for years to come (Connor and Rogers 1979). The desirability of feeding grain to cattle does raise serious policy questions about the world's beef business, and it is frequently discussed in Latin America and Australia, two major cattle producing areas that are described in the next chapter.

REFERENCES

Boykin, Calvin C.; Gilliam, Henry C.; and Gustafson, Ronald A. 1980. *Structural Characteristics of Beef Cattle Raising in the United States*. USDA, AER no. 450.

Connor, J. Richard, and Rogers, Robert W. 1979. Ground Beef: Implications for the Southeastern U.S. Beef Industry. *Southern J. Agric. Econ.* 11(2):21–26.

Crockett, J. R.; Koger, M.; and Franke, D. E. 1978. Rotational Crossbreeding of Beef Cattle: Preweaning Traits by Generation. *J. Anim. Sci.* 46:1170–79.

Curl, Tom. 1978. *Beef Cattle Book: A Working Guide for Cattlemen*. Birmingham, Ala.: Oxmoor House.

Dyer, I. A., and O'Mary, C. C. 1972. *The Feedlot*. Philadelphia: Lea and Febiger.

Edwards, Joseph. 1973. Trends in Beef Cattle Production in the United States. *World Anim. Rev.* 5:11–15.

Erickson, Donald B., and Phar, Philip A. 1970. *Guidelines for Developing Commercial Feedlots in Kansas*. Manhattan: Kansas State University, Cooperative Extension Service.

Gee, C. Kerry; Van Arsdall, Roy N.; and Gustafson, Ronald A. 1979. *U.S. Fed-Beef Production Costs, 1976–77, and Industry Structure*. USDA, AER no. 424.

Hottel, J. Bruce, and Martin, J. Rod. 1976. Capital Flows in the Cattle Feeding Industry. *Livestock and Meat Situation* 209:38–43.

Koger, Marvin. 1979. Effective Crossbreeding Systems Utilizing Zebu Cattle. Dept. Anim. Sci., unpubl. paper. Gainesville: University of Florida.

Koger, Marvin; Cunhua, T. J.; and Warmick, A. C., eds. 1973. *Crossbreeding Beef Cattle.* Ser. 2. Gainesville: University of Florida Press.

Koger, Marvin; Peacock, F. M.; Kirk, W. G.; and Crockett, J. R. 1975. Heterosis Effects on Weaning Performance of Brahman-Shorthorn Calves. *J. Anim. Sci.* 40:826–35.

Martin, Rod, and Dietrich, Raymond. 1977. *Characteristics of Commercial Banks, Borrowers and Loans in the Southern Plains Cattle Feeding Industry.* College Station: Texas A & M University.

Mason, I. L. 1969. *A World Dictionary of Livestock Breeds, Types and Varieties.* CAB Tech. Commun. no. 8 (rev.). Bucks, England: Farnham Royal.

Meisner, Joseph C., and Rhodes, V. James. 1974. *Feedlot Economies of Size Revisited.* Agric. Econ. Paper 1974–33. Columbia: University of Missouri, Dept. Agriculture Economics.

Olson, T. A. 1979. Use of Composite Breeds in Florida Crossbreeding Systems. *The Florida Cattleman* 43(3):66.

Putnam, P. A., and Warwick, E. J. 1975. *Beef Cattle Breeds.* USDA Farm. Bull. no. 2228.

Richards, Jack A., and Karzan, Gerald E. 1974. *Beef Cattle Feedlots in Oregon: A Feasibility Study.* Agric. Exp. Sta. Spec. Rep. 170. Corvallis, Oreg.: Oregon State University.

Richardson, Len. 1981. Profit Squeeze Has Revised the Roster of Your Feedlot Competitors. *FarmFutures,* pp. 42–44.

Rouse, John. 1970, 1973. *World Cattle,* 3 vol. Norman, Okla.: University of Oklahoma Press.

Scofield, William R. 1974. Operating Capital Requirements of the Cattle Feeding Industry. *Livestock and Meat Situation.* USDA, LMS 199.

Simpson, J. R.; Baldwin, L. B.; and Baker, F. S., Jr. 1980. *Investment and Operating Costs for Two Types and Three Sizes of Florida Feedlots.* Bulletin 817. Gainesville: Florida Agricultural Experiment Station.

U.S. Department of Agriculture. *Livestock and Meat Situation.* Various issues.

______. *Livestock and Meat Statistics.* Suppl. for 1978. Stat. Bull. no. 522.

Van Arsdall, Roy N., and Skold, Melvin D. 1973. *Cattle Raising in the United States.* USDA, AER no. 235.

Wall Street Journal. 1979. Embryo Transfers Let Superior Cows Produce Many Calves a Year. Wednesday, May 9, p. 1.

Warwick, Everett J., and Legates, James E. 1979. *Breeding and Improvement of Farm Animals,* 7th ed. New York: McGraw-Hill.

Williams, Ed, and Farris, Donald E. 1975. *Economics of Vertical Coordination in Beef Cattle Systems.* Series on Cattle Feeding Systems Part 3, MP-1205. College Station: Texas A & M University, Texas Agricultural Experiment Station.

4

BEEF PRODUCTION IN LATIN AMERICA AND AUSTRALIA

Systems of raising beef cattle vary greatly, both among countries and within them. In some cases, whole regions are dominated by one production system. In the Argentine Chaco, for example, nearly all cattle operations are based on cow-calf units selling calves at or near weaning time. At the other extreme, various types of systems may be found in one locale, as is common in certain parts of Europe such as West Germany, where the cattle are primarily dual purpose types used for beef and milk production. Immediately adjacent to a primary dairy operation in West Germany will be herds in which the calves are sold at weaning to neighbors or a local market. A nearby farmer may grass-fatten the weaned calves for slaughter at 300 kg, while another neighbor may grain-feed weaned calves to 450 kg in a confinement feedlot. In all cases, a number of factors influence the type of system that develops in a specific area. In the final analysis, economics generally determines the prevailing system.

FACTORS INFLUENCING CATTLE SYSTEMS

Cattle systems evolve as the result of a host of factors, some of which are stimulants while other are constraints. The systems are not static, but change over time in response to technological, political, and economic forces. Examining the factors that stimulate or constrain cattle production systems will help us to understand why particular types of cattle production systems have evolved.

Land characteristics act as a prime determinant of cattle production systems because poorer land will more likely be used for livestock production, whereas highly fertile land will be devoted to crop production. The law of comparative advantage, discussed at some length in the chapter on trade, strongly dictates the type of commodity that can be

Beef cows make the best use of this rough and often dry terrain. If it were not for predators it could also be used for sheep.

produced in a certain area. Japan has a high effective demand for beef, but consumes little of it and produces rice instead because this is the highest and best use of their very limited natural resources. Also, policies placing heavy limits on beef imports provide a good indication of society's priorities. This concept of comparative advantage has influenced the use of land in the deserts of Mexico or northern Africa, which are poorly suited to cropping, with the result that raising livestock is a more economically justifiable alternative. Elsewhere, cow-calf operations may be found on quite good farming land as well. Why?

Production complementarity occurs in many farming areas. In the midwestern United States corn farmers have the option of selling their harvest or marketing it through cattle. Traditionally, many of these entrepreneurs have chosen to feed their corn to cattle as it has provided a source of employment during the slack winter months as well as a means of marketing corn. More recently, some farmer-feeders have joined together to feed cattle on a cooperative basis and thus are able to take advantage of size economies. Many other farmers maintain a small herd of beef cattle to harvest crop residues and to provide calves for their feeding operations. Also, on most farms, there is some land that is not suitable for cropping. In Europe, the dual-purpose cattle have proved to be excellent for complementary milk, beef, and crop farming. These livestock producers also frequently feed-out progeny as a means of marketing farm products and by-products.

Tradition and social conditions such as population density and land resource distribution also strongly affect land use. Because of tradition or lack of viable alternatives, peasants in central Mexico farm very poor, rocky, rough semidesert land that is better suited for raising cattle, goats, or sheep. On a national or international scale the land clearly should be in livestock, but people cannot be moved around like pawns on a chessboard. Thus, if farmers have little alternative use for their family's labor, a higher total net return can be realized from cropping than from cattle raising. The return per hour of time invested will probably be lower from cropping than from cattle raising but the total return will be higher. One objective of development projects is to determine the highest and best use of land on a national scale subject to individual cost/benefit ratios, and then to design the projects accordingly.

The level of a nation's economic development, combined with cultural mores, has a strong influence on livestock systems. Much land in Africa, for example, is inhabited by tribal nomads or seminomads who have little "effective interest" in expanding production. That is to say, interviews with cattle owners may indicate an apparent desire to improve production; but the desire may not be strong enough to be translated into action, or the institution of communal property may not facilitate increasing productivity. In many potential exporting countries this view is prevalent enough to reduce any incentive to maintain a packing plant at international standards, which might prevent these countries from competing in world beef markets.

Location is another factor influencing cattle raising systems. In a number of areas such as Sweden, livestock production must be subsidized if operations are to compete successfully with exporting countries such as Australia or Uruguay, which have a comparative advantage in trade. Apart from savings on foreign exchange, which would have to be used to purchase imported beef, Sweden recognizes that in some regions cattle raising may be preferable to other options, such as timber production.

Another aspect of location with an impact on cattle raising is the prevalence of disease. Much of central Africa at first appears to be well suited to cattle raising, but the presence of the tsetse fly causes many zones to be avoided except under special conditions. Thus many countries known for certain diseases are not able to fully compete in world trade.

Level of effective demand is a regulator of production in many countries. After export demand is removed from the calculations, the important factor is domestic demand. Per capita beef consumption in all of Africa and much of the Near East is very low because most individuals don't have the necessary income, that is, effective demand, to purchase higher per capita quantities of beef than are taken at present. The Central American countries of Costa Rica, Honduras, and Nicaragua, like most Latin American countries, could greatly expand production if a market and attractive prices prevailed (Valdez and Nores 1979).

Price of beef obviously has great influence on cattle raising systems. If beef prices are low, livestock owners cannot afford to invest as much in inputs as they could with high prices. The production systems in Europe and the United States are capital intensive because prices are relatively high. High interest rates, inflation, lack of capital, low beef prices, and low wages in South America make operations there more labor intensive. The tendency there is also toward extractive business because the return to livestock management is low compared with alternative investment opportunities.

Government policies are among the most important factors influencing the evolution of cattle raising systems, as government can offer the industry incentives to grow and improve. Most countries interfere with the free market to influence cattle prices in some manner. Frequently, retail or wholesale prices of some or all cuts are fixed. Some countries that export beef and cattle—for example, Costa Rica—have a two-price system, one for national consumption and the other for exports. The dilemma for these countries is that while relatively high prices will lead to greater returns for cattle raisers and will likely lead to expanded output, some consumers may have to curtail their consumption. Discontent over higher prices is politically disruptive, especially in countries trying to improve the nutritional level of their lower-income populations.

Government policy also affects the stability of cattle production since it may create or preclude a favorable investment climate. Beef cattle production requires substantial investment on which the returns may be deferred for several years because the industry is tied to the biological cycle of its product. The threat of expropriation, substantial government involvement, or radical shifts in policy will discourage investments so that landowners might concentrate their investment in cattle rather than land improvements. Chile, Peru, Zimbabwe, and Uruguay are countries in which government policies or political turmoil have prevented the beef cattle industry from continuing to intensify and evolve more efficiently.

We now discuss two beef systems. In both types calves are raised from cows grazing an open rangeland or improved pastures. We distinguish, however, between systems in which the progeny are fed to slaughter weights in an extensive manner on pasture and those in which they are fed-out intensively on pasture.

LATIN AMERICAN CATTLE INVENTORY AND PRODUCTION

The cattle inventory in Latin America and the Caribbean grew about 1.7 percent annually from 1960 to 1978, while beef production increased 2.7 percent per year over the same period (Table 4.1). Brazil had the largest inventory by far, about 89 million head in 1978, followed by Argentina with approximately 61 million head (Table 4.1). Mexico was next with 11 percent of the cattle, followed by Colombia with 9.5 percent. The corresponding number of cattle per person for these four countries was 0.84, 2.29, 0.46, and 0.97, respectively (Appendix Table A.1). Uruguay had the highest number of cattle per person, 3.59, while Peru only had a coefficient of 0.25. Substantial improvements in production have taken place in countries such as Panama and Guatemala, where output increased 173 and 136 percent, respectively. Other countries such as Cuba have demonstrated almost no change.

A country can expand production by increasing its national inventory, or by increasing efficiency. The four measures of efficiency presented in Table 4.2 show the substantial variations that can take place among countries. In Argentina, for example, the weaning rate—also known as calf crop—was estimated at 60 percent in the early 1970s, compared with 48 percent in Colombia. The extraction rate for Argentina, i.e., the number of cattle in the national inventory divided by the number of cattle produced, stood at 20 percent for Argentina compared with 12 percent for Colombia. Another important measure is the time required for animals to reach slaughter weight. In Argentina, which uses a considerable amount of corn stover and short-term finishing, cattle are about 30 months old at slaughter. In Colombia, on the other hand, where cattle are almost exclusively fed-out on grass, they are normally 50 months old at slaughter weight (430 kg).

Efficiency can also be measured by the amount of production per head of inventory, although some care must be taken in such estimates, as production per head will increase during periods of herd liquidation (since production is essentially equivalent to slaughter) and will decrease when the national inventory is increasing. Nevertheless, efficiency does provide a useful comparison, especially when there is considerable difference between periods, as is the case in Table 4.3. The difference among countries is striking; some, such as Honduras, have made considerable progress while others, such as Uruguay, have less improvement in product efficiency. Overall the greatest improvement in productive ef-

Table 4.1. Cattle inventory and beef and veal production in Latin America, by type of climate and country, 1960 and 1978.

Region and country	Cattle inventory		Beef and veal production[a]		Cattle inventory		Production		Annual rate of growth 1960–1978	
	1960	1978	1960	1978	1960	1978	1960	1978	Inventory	Production
	(1,000 head)		(1,000 metric tons)		(%)		(%)		(%)	
Tropical Latin America										
South America										
Brazil	72,829	89,000	1,359	2,254	36.7	33.4	26.3	27.0	0.6	2.8
Colombia	15,100	25,294	...	495	7.6	9.5	...	5.9	0.8	...
Venezuela	8,600	10,231	116	219	4.3	3.8	2.2	2.6	1.0	3.6
Cuba	5,760	5,700	...	142	2.9	2.1	...	1.7	0.0	...
Paraguay	4,004	5,800	...	121	2.0	2.2	...	1.4	2.1	...
Peru	3,590	4,167	...	82	1.8	1.6	...	1.0	0.8	...
Bolivia	2,500[b]	3,772	...	80	1.3	1.4	...	1.0	2.3	...
Ecuador	1,530	2,874	37	70	0.8	1.1	0.7	0.8	3.5	3.8
Central America and Mexico										
Mexico	21,000	29,333	299	677	10.6	11.0	5.8	8.1	1.9	4.6
Nicaragua	1,425	2,774	...	85	0.7	1.0	...	1.0	3.8	...
Guatemala	1,062	2,417	33	78	0.5	0.9	0.6	0.9	4.7	4.9
Costa Rica	1,057	2,002	...	71	0.5	0.8	...	0.8	3.6	...
Honduras	1,350[b]	1,700	15	51	0.7	0.6	0.3	0.6	1.3	7.0
Panama	666	1,396	19	52	0.3	0.5	0.4	0.6	4.5	5.7
El Salvador	779	1,333	...	34	0.4	0.5	...	0.4	3.0	...
Caribbean										
Dominican Republic	949	2,050	22	38	0.5	0.8	0.4	0.5	4.4	3.1
Puerto Rico	433	562	11	25	0.2	0.2	0.2	0.3	1.4	4.7
Temperate Latin America										
Argentina	44,550	61,280	1,918	3,200	22.4	23.0	37.1	38.3	1.8	2.9
Uruguay	7,505	9,424	249	358	3.8	3.5	4.8	4.3	1.3	2.1
Chile	2,913	3,492	...	183	1.5	1.3	...	2.2	1.0	...
Others[c]	965	1,849	1,093	47	0.5	0.7	21.2	0.6	3.7	...
Total	198,567	266,450	5,171	8,362	100.0	100.0	100.0	100.0	1.7	2.7

Source: Based on data in FAO *Production Yearbook*, 1964 and 1978.
[a]Production from indigenous animals. The total for 1960 includes FAO estimates for countries not shown.
[b]Estimated.
[c]Includes all other countries in South and Central America and the Caribbean. Canada and the United States are excluded.

Table 4.2. Measures of efficiency in five South American countries, 1970–1974

Country	Calf crop (weaning rate)	Extraction rate[a]	Carcass yield[b]	Production time to slaughter weight[c]
	(%)	*(%)*	*(%)*	*(months)*
Argentina	60	20	59	30
Brazil	50	13	52	48
Colombia	48	12	52	50
Paraguay	50	13	53	48
Uruguay	58	18	56	36

Source: Rivas and Nores, 1978, p. 13.
[a]National inventory divided by number of cattle produced.
[b]Steers only.
[c]430 kg.

Table 4.3. Production of beef and veal per head of cattle inventory and production per inhabitant in the Latin American countries, 1960 and 1978

Region and country	Production of beef and veal per head of cattle inventory		
	1960	1978	Change
	(kg)		*(%)*
Tropical Latin America			
South America			
Brazil	18.7	25.3	35
Colombia	...	19.6	...
Venezuela	13.5	21.4	59
Cuba	...	24.9	...
Paraguay	...	20.9	...
Peru	...	19.7	...
Bolivia	23.5	24.4	4
Ecuador	...	21.2	...
Central America and Mexico			
Mexico	14.2	23.1	63
Nicaragua	...	30.6	...
Guatemala	31.1	32.3	4
Costa Rica	...	35.5	...
Honduras	11.1	30.0	170
Panama	28.5	37.2	31
El Salvador	...	25.5	...
Caribbean			
Dominican Republic	23.2	18.5	−20
Puerto Rico	25.4	44.5	75
Temperate Latin America			
Argentina	43.1	52.2	21
Uruguay	33.2	38.0	14
Chile	...	52.4	...
Others	...	25.4	...
Total or average	26.0	31.4	21

Source: Table 4.1.

ficiency has taken place in the tropical and subtropical countries as they have the greatest potential for intensifying their beef cattle production systems. Of these, the Central American countries have made the most significant progress because of their access to the North American market.

RANCHING OPERATIONS IN LATIN AMERICA

Extensive beef cattle operations, that is, operations in which breeding herds are kept on large improved pastures or native ranges, are found from the Mexican border of the United States to Tierra del Fuego on the southern tip of South America and, to some extent, on the larger islands of the Caribbean. Extensive systems are found in temperate and warm climates and at low, medium, and high elevations. Their size is largely a function of the density of human population and location relative to urban centers, as the extent of operation is not related to per capita income in a country. El Salvador, for example, is one of the poorest countries in Latin America, yet cattle production is quite intensive and many cattle are used for draft purposes. On the other hand, Argentina has one of the highest per capita incomes in Latin America, yet cattle raising is extensive; Paraguay has an extensive type system, but per capita income is very low (Rivas and Nores 1978).

The typical extensive system in Latin America is based on a cow-calf operation in which the offspring are fed-out on the ranch and raised to 400–500 kg. A major variant of the system involves sale of cattle as calves or yearlings to ranches or farms in fattening areas where they are finished on high-quality pasture or crop residue. Only limited use is made of feedlots as most countries do not have adequate supplies of feed grains, and the beef/grain price ratios are prohibitive. Extensive systems may also vary in grazing practices. In much of Uruguay and parts of Argentina, both cattle and sheep are grazed on the same pastures in rotation. This practice, which is also followed in the temperate areas of Australia and New Zealand, helps producers to avert price risk and to increase pasture carrying capacity. Extensive operations in Australia described later in this chapter are similar to those in Latin America (Morgan 1973).

Many Latin American ranches, like those in the western United States, have tended to be large holdings. In Paraguay in the 1960s, for example, two companies—IPC and Liebegs—owned more than 200,000 head of cattle. The King Ranch had over 80,000 head in Brazil, and operations with 20,000 to 25,000 head of cattle are still common. Since World War II the size of holdings has declined because public sentiment has forced dissolution of large holdings by foreign multinational corporations. This decline is also related to agrarian reforms and the division of land among heirs of original cattle barons.

One major advantage of *intensive operations* is their flexibility in rapidly and effectively increasing input when cattle prices are high, and reducing input when prices are low. A critical factor in this profit-maximizing process is fertilizer, as forage yields depend to a large extent on the amount of fertilizer used. Intensive operations contrast sharply with the extensive systems common in the dry areas of western Costa Rica, Argentina, or Australia, where lower initial investment costs are offset by fewer management alternatives. Apart from varying supplemental feed to a certain extent, backgrounding calves, or varying inventories through selective culling, entrepreneurs on extensive operations can do little to increase profit through input adjustment. There is considerably more potential to exercise innovative management in the wet-tropical and semitropical areas than in dry or temperate zones.

Some form of intensive beef production systems can be found in almost every country of the world, but tropical and subtropical areas with high rainfall have the greatest potential for intensive beef production because an abundant amount of lush forage can be produced throughout most of the year. The advantage of such areas over dryer warm areas, or temperate climates, is that the combination of production possibilities is greatly expanded since the cattle raiser potentially has more control over production and thus additional latitude in determining both how and what to produce.

Production specialists generally agree that inadequate nutrition is a major inhibitor of higher output of livestock products, especially in warm climates, even though warm wet areas have the potential to produce abundant forages (Maynard and Loosli 1969). One difficulty, however, is that poor soils and irregular rainfall cause forages to be generally low in protein and digestible energy compared with grasses in more temperate regions. If the forages are not adequately used, or if the animals lose weight or make no gains from grazing, then forage as a resource is not fulfilling its productive role.

The feeding of grain concentrates on a supplementary basis is often proposed as a complement to warm-climate grasses. The drawback of this approach is that concentrates (feedstuffs other than forages) in warm areas are generally too costly to use as animal feed because beef prices are quite low in relation to grain prices. As a result, various nontraditional feedstuffs have to be evaluated, for example, rice bran, copra, sesame, peanuts, molasses, sugarcane stalks, sugarcane tops, corn products, cocoa pod meal, coffee meal, coffee pulp and hulls, copra cake, fruit by-products, bananas, citrus pulp, pineapple by-products, and vegetable derivatives (Preston and Long 1979; Ranjhan 1979). One exotic commodity in Latin America that has been receiving considerable attention over the past decade is cassava (yucca). A coordinating facility has been established at the Center for Tropical Agriculture in Cali, Colombia, for further research on this crop.

Some other warm-climate products that have potential for livestock feeding in the warm climates are sweet potatoes and yams, cottonseed cake and meal, leguminous and nonleguminous leaf meals, seaweed and unicellular algae, yeast grown from hydrocarbons on petroleum products, fish soluble flour and ethamol,[1] high lysine corn, and whey (which is the residual milk from cheesemaking). More recently, cow manure and poultry litter have received attention as potential supplements, especially in the temperate areas. Urea, one of the nonprotein nitrogen compounds, has been acclaimed as an excellent source of nitrogen, as it enables livestock producers to add protein equivalent without providing additional energy. Urea can be effectively incorporated by adding it to molasses or mixing it in a low-protein grain feed.

It has been established that natural grasslands in warm climates generally provide poor grazing for ruminants, even though there are more than 3,000 grasses and at least 1,000 legumes native to the warm climates which might prove useful as forage, either in their present or improved forms. Furthermore, areas of warm climate may have other problems—for example, about 24 percent of the land area 30° north and 30° south of the equator experiences extreme moisture and temperature conditions. Only about nine percent of this land area has no climate restrictions (McDowell 1972).

There is no consensus on the potential for improving forage production in warm climates, and the views range from very pessimistic to optimistic. The pessimists point to the questionable economic feasibility of improving production on open areas characterized by low rainfall and low carrying capacities because of the need for additional brush control, seeding, fencing, improved water supplies, and intensive management. They note that the semiarid areas in some parts of Australia, Africa, Mexico, Colombia, and Brazil are of doubtful potential since profitable improvements generally require multiple use of land. Fertilizers have been tried on some of the better sites, but poor soil and moisture limitations, as well as high costs, usually limit their profitability. Production specialists therefore believe that productivity improvements on the semiarid grasslands should center around slowing land deterioration and reversing some of the traditional views on management, rather than around widespread adoption of intensive production.

Researchers are more optimistic about the potential for increasing

1. Ethamol is a urea-molasses compound.

Steers are fattened on high quality forage to achieve the fastest, most economical gains and to yield the most profit in export markets.

livestock production in the wet areas, especially if improved forages are considered. They point out that many improved perennial forages are now common, or relatively common, in warm climates. When evaluating the forages, many specialists have found that yield is not necessarily a good indication of the advisability of planting different forages, as heavy applications of fertilizer on varieties such as Pangola grass are often not economically feasible. Seasonality also has to be taken into account in cheap forage production for many warm-climate areas are characterized by distinct dry periods of 5 or more months. Consequently, western and northwestern South America and most of India and Southeast Asia are considered to be of less potential than semidry or wet areas (Davies and Skidmore 1966; Sanchez and Tergas 1979).

Livestock production specialists in the warm areas of the world generally look to improved varieties of perennial forages as a means of increasing beef production both in total and on a per animal basis. A partial list of the more common improved grasses found in warm climates is presented in Table 4.4. Most of the improved grasses originated in Africa but, as the common names indicate, have been adapted to Latin America.

Introducing legumes into improved or natural pastures and managing natural grasslands to allow their propagation are inexpensive methods of improving forage quality and quantity in warm climates (Hutton 1975). Legumes are especially useful in that they provide associated grasses with nitrogen since they are nitrogen fixers and thus contribute to the total nutritive value of the forage. Ingestion of legumes by animals does not decline with maturity. It is interesting that while legumes are

Table 4.4. A partial list of the major warm-climate grasses used in cultivated pastures

Scientific name	Common name(s)	Native to	Major use	Comments
Andropogan gayanus	Gambagrass	Africa	Grazing	Perennial 1–2m
Axonopus scoparius	Imperialgrass	Tropical South America	Soilage, silage	Perennial 0.6–2m
Brachiaria decumbens	Signalgrass, Surinamgrass (Jamaica)	Eastern Africa	Grazing	Perennial up to 1m
Brachiaria mutica	Parágrass	Africa	Grazing, greenchop	Perennial 1–2m
Cenchrus ciliarus	Buffelgrass	Africa, India, Indonesia	Grazing, hay	Perennial up to 1.4m
Chloris gayana	Rhodesgrass	Africa	Grazing, hay	Perennial 0.5–2m
Cynodon dactylon	Bermudagrass	Turkey, Pakistan, some varieties from Africa	Grazing, greenchop, hay, haylage	Perennial up to 0.6m
Cynodon nlemfuensis	Stargrass	Africa	Grazing, hay	Perennial 0.3–0.7m
Dichanthium aristatum	Angletongrass	Southeastern Africa	Grazing	Perennial up to 1m
Digitaria decumbens	Digitgrass, pangola	Africa	Grazing, greenchop, hay, silage	Perennial up to 0.5m
Echinochloa polystachya	Pasto Aleman	America	Grazing	Perennial 1–2.5m
Eragrostis curvula	Weeping lovegrass	Southern Africa	Grazing	Perennial 0.3–1.2m
Hyparrhenia rufa	Jaragua	Africa	Grazing, hay	Perennial 0.3–2.5m
Melinis minutiflora	Molassesgrass	Africa	Grazing, hay	Perennial up to 1.5m
Panicum coloratum	Kleingrass	Africa	Grazing	Perennial 0.4–1.5m
Panicum maximum	Guineagrass	Africa	Grazing, greenchop	Perennial 1–2.5m
Paspalum notatum	Bahiagrass	South America	Grazing	Perennial 0.15–0.70m
Pennisetum americanum	Pearl Millet	Western Africa, India	Grazing, hay, silage	Annual 1–3m
Pennisetum clandestinum	Kikuyugrass	East Africa	Grazing	Perennial 0.2–0.3m
Pennisetum purpureum	Napiergrass, Merker, Elephantgrass	Africa	Greenchop, silage, limited grazing	Perennial 2–4m
Setaria anceps	Setaria	Africa	Grazing, hay	Perennial 1–2m
Sorghum almum	Columbusgrass, Pasto Colon	Argentina	Grazing, hay, silage	Perennial 1–3m
Sorghum sudanense	Sudangrass	Sudan, Egypt	Grazing, greenchop	Annual up to 3m

Source: Developed in consultancy with Gerald Mott, University of Florida, and based on Bogdan 1977.

Zebu breeds are the base of the rapidly growing herds in Brazil. Zebu and Zebu crosses are adapted to much of the tropical and subtropical areas of the world.

widely used in temperate climates, they have only recently been accepted in the warm areas. Some common legumes are listed in Table 4.5.

Despite their numerous advantages, legumes must be used with care (Skerman 1977). One reason is that nitrogen fixation of legumes may be of little value in high rainfall areas because much of the nitrogen is removed before it can be utilized by grasses. In addition, legumes may not withstand grazing satisfactorily, may have poor drought resistance, and may not be able to establish stands. Although our knowledge of management is poor and more research is needed on legumes, the potential for legumes seems high, especally in view of anticipated high fertilizer prices owing to continual escalation of energy prices.

Almost any type of soil can be found in the warm-climate region, but nearly all soils in those areas have less natural fertility than do soils in temperate climates (Eyre 1963). Thus, while areas with heavy vegetative growth such as the Amazon region may appear to have great potential for intensive livestock production, the soils there are low in organic matter and are relatively fragile (Sanchez 1976). Many areas in northern Brazil, for instance, have a thin topsoil with sand below. If, in clearing, the protective cover is destroyed, rainfall quickly percolates through and subsequent forage yields will be low. Furthermore, these soils are often poor and require heavy applications of fertilizer. In times of high world fertilizer prices, such as the mid-1970s, intensive beef production is often not profitable even if beef prices are relatively high.

In many tropical or subtropical areas a hardpan lies below the surface. During seasons of heavy rainfall or spring runoff from mountains,

Table 4.5. A partial list of the major warm-climate legumes used in cultivated pastures

Scientific name	Common name(s)	Native to	Major use	Comments
Aeschynomene americana	Joint Vetch	Americas	Grazing, hay	Annual, possibly perennial
Arachis glabrata	Perennial Peanut	South America	Grazing	Perennial
Cajanus cajan	Pigeon Pea	Africa or India	Green manure, green-chop, grazing, grain	Perennial often grown as an annual
Centrosema pubescens	Centro	South America	Grazing	Perennial
Clitoria ternatea	Cordofan Pea	Africa, Americas, India, Arabia, Asia	Grazing, hay	Perennial
Desmodium heterophyllum	Hetero; Spanish Clover	Mauritius, S.E. Asia, Indo-china, Malaysia, Indonesia	Grazing	Perennial
Desmodium intortum	Greenleaf Desmodium	Central and South America	Grazing	Perennial
Glycine wightii	Perennial Soybean	Africa, Arabia, India	Grazing, green manure	Perennial
Lablab purpureus	Hyacinth Bean	Africa	Grazing and hay	Annual or short-lived perennial
Leucaena leucocephala	Leucaena	Mexico	Grazing, leaf meal	Perennial, shrub or tree
Macroptilium atropurpureum	Siratro	Central and South America	Grazing	Perennial
Macroptilium lathyroides	Phasey Bean	South America	Grazing, hay	Annual or short-lived perennial
Mucuna pruriens	Velvet Bean	South Asia	Grazing, hay	annual or short-lived perennial
Puereria phaseoloides	Tropical Kudzu	S.E. Asia, Malaysia, Indonesia	Grazing, hay	Perennial
Stylosanthes guianensis	Stylo	Central and South America	Grazing, hay	Perennial
Stylosanthes humilis	Townsville Stylo	Subtropical America, Mexico, Ivory Coast (Africa)	Grazing	Reseeding annual

Source: Developed in consultancy with Gerald Mott, University of Florida, based on Bogdan 1977.

these areas may be flooded for months at a time. In some areas, such as the Argentine or Paraguayan Chaco, expensive dikes have to be constructed to create a dry place on which cattle can lie down, in an effort to control disease and improve productive and reproductive efficiency.

Numerous mineral deficiencies, mineral imbalances, and toxicities severely hamper efforts to improve the cattle industry in most developing countries (McDowell 1976). Specific mineral requirements are difficult to pinpoint, however, since exact needs depend on various chemical and mineral interrelationships. Even though most tropical forages contain less minerals during the dry season, deficiencies are more prevalent during the wet season because cattle gain weight more rapidly during wet periods and thus require more minerals. Furthermore, cattle introduced into an area may show deficiency signs while indigenous breeds that are slow growing and late maturing may not exhibit these signs (Payne 1966). The mineral most likely to be lacking is phosphorus, followed by cobalt and copper. Deficiencies of these minerals, as well as selenium during the rainy (growing) season and potassium on dry forage in the cool or dry season, may lower production or may cause death of cattle. Mineral toxicities, such as selenium or fluorine in certain restricted areas, are very difficult to control.

From a technical point of view, however, mineral deficiencies are easily corrected, and mineral supplements have been shown to have a high economic return in both the intermediate and long run. Minerals can be fed directly to cattle through water, mineral licks, mixture, heavy pellets, or injections. They can be given indirectly in fertilizers or in certain grasses or legumes grown in soil whose pH has been altered. In general, mineral deficiencies can be corrected at little cost but cases of mineral deficiencies nonetheless abound because climatic conditions, beef prices, and management vary from year to year and confound the results from use of minerals. Many ranchers, for example, purchase mineral supplements with scarce cash, or borrow money at high interest rates, only to see steers actually lose weight, perhaps because of extremely dry conditions that year or failure to treat for parasites. But if minerals had not been added, the cattle might have lost even more weight. The benefits are even more difficult to measure in cows than in steers being fattened, as these benefits often are reflected in healthier and faster growing progeny (Preston and Willis 1970).

BEEF CATTLE BREEDS IN LATIN AMERICA

There were no cattle in the Americas until the Europeans arrived. Columbus, on his second voyage, brought the first cattle to the Western Hemisphere and unloaded them at the island of Santo Domingo. Twenty-eight years later (1521) Gregorio de Villalobos transferred cattle from Santo Domingo to Vera Cruz, Mexico. These were the first cattle to arrive on the North American continent. Nearly a century later, in 1607, cattle unloaded at Jamestown on the Atlantic coast formed a nucleus of native cattle in the eastern part of North America. By 1690 Spanish cattle had arrived in what is now the state of Texas. Around the mid-1800s improved breeds introduced primarily from the United Kingdom led to the extinction of native cattle in the Midwest and East. By 1900 Texas Longhorns derived from Spanish introductions were nearly gone, and only the Florida Scrub, also of Spanish origin, were left; there were substantial numbers of Florida Scrub until the 1960s.

The story of Latin American cattle is different from that of U.S. breeds since native cattle are still flourishing in all the Latin American countries despite considerable efforts since World War II to introduce new breeds and to upgrade through crossbreeding. The "criollo" cattle evolved during the past four and a half centuries by natural selection. Although these Iberian (origin) cattle are of the *Bos taurus* type, criollo cattle bear little resemblance to the European breeds. In fact, they are closer in appearance to N'Dama cattle of West Africa.

Of the vast number of cattle breeds and types in the world, only a few are found in appreciable numbers in Latin America. All cattle can be classified as beef, milk, draft, or a combination (dual- or triple-purpose) breed and type. Cattle are also classified into physiological types including: (1) *Bos taurus,* which is synonymous with European breeds; (2) *Bos indicus,* which is humped (the term generally refers to Zebu cattle originating from Southeast Asia and India); and (3) breeds recently developed by crossing *Bos indicus* and *Bos taurus* cattle. Payne (1970) classifies the native or criollo cattle of Africa and the Western Hemisphere as a separate type, but in reality they belong to the above categories. In this book we classify the breeds as (1) European, (2) native American, or (3) new breeds.

The predominant breeds of the world are listed in the Appendix, Tables A.10–A.14. The major uses of the breed—for beef, milk, or draft—are given in order of importance in the column titled "purpose." Thus, for example, a listing of "beef-milk" means that the cattle have been developed as a dual-purpose breed, the principal use being beef, and milk being a secondary objective. Utility of the cattle varies greatly among regions, of course, and the classification is meant only as a guide. The same caveat holds true for size of cattle. In addition, the list is not meant to be exhaustive; rather, our purpose is to present, in summary form, a panorama of the important breeds of the world.

Almost every country in the world has, or has had, some type of native cattle. A considerable proportion of the cattle inventory of many developing countries is still made up of native stock. As we have noted, in Latin America they are called "criollo" cattle, but in Africa they are merely called "natives," although a few types are named.

Animal husbandry researchers recognize that the special conditions existing in warm climate areas require more research work for a proper evaluation of the efficiency of native breeds, introduced breeds, and systems of crossbreeding. Researchers differ, however, in their ordering of priorities between emphasis on breed improvement and development of management skills. There is nonetheless wide agreement that it is important for a breed to be adaptable to the high temperatures of the 30°S to 30°N region, since distinct correlations have been observed between genotype and reaction to the environment. Zebu cattle, for example, can tolerate hot climates better than English breeds. All breeds known today, as well as their crosses, have certain strengths and weaknesses, and their adaptation to environmental and physical conditions varies substantially.

Some researchers have observed that a major advantage of indigenous cattle in hot climates is their small size relative to that of most breeds originating in cooler climates. Other specialists believe that promotion of large breeds in warm climates is justified if there are ample feed supplies, good to excellent measures of disease control, and some protection against the extremes of climate (Plasse 1976).

Anatomical traits of cattle have also been studied widely to measure the adaptability of certain breeds to warm climates. Correlations between individual characteristics such as growth and milk yield, or growth and reproductive efficiency, have been very low, however, ranging from 0.00 to less than 0.30. Researchers seem to agree that external body characteristics such as the large dewlap, long ears, and hump of Zebu cattle are not the principal factors responsible for their adaptability to warm climates, even though they can be helpful to the animal (McDowell 1972). It has been found that the hump is not an area of high fat concentration, as commonly believed, because the fat that does impede heat loss generally occurs immediately adjacent to the skin. Furthermore, color of hair does not appear to play a significant role in heat loss, although the consensus is that a short, sleek, light-colored glossy coat of hair over soft and pliable pigmented skin is most appropriate for warm climates since this type skin provides more surface area for heat dissipation.

Research also indicates that most livestock first respond to thermal stress by decreasing food intake, reducing milk yields in lactating cows, and reducing energy for reproductive processes. High temperatures may also influence the growth rate of cattle, but the extent is not known. There do seem to be negative correlations between reproductive efficiency and temperature, and between size of animal and temperature. For example, reproductive efficiency of both males and females may be retarded by increased temperatures, for it has been found that sperm formation in the male (spermatogenesis) can be interrupted under high temperature conditions, and that a high number of abnormal or dead sperm may be produced. Females may experience a cessation or shortening of heat (estrus) and ovulation.

AUSTRALIAN BEEF SYSTEMS

Australian beef production increased substantially in the 1960s and 1970s largely as a result of access to export markets in North America. During this time it replaced Argentina as the leading beef exporting country. The beef enterprise has generally been a high risk enterprise, and when an industry depends heavily on the international market, the level of risk is even higher because the international market is often more volatile than the market in individual countries.

In Australia, as in most countries, beef is generally part of a mixed-enterprise farming or ranching system. This spreads risks, allows for better utilization of land, labor, and capital, and results in a more stable cash flow. The more extensive, specialized type of cattle operations are mainly found on marginal lands not suitable for crops in both humid and arid areas. The nature of beef enterprises in Australia will be described in detail.

Australia's beef cattle industry is based on cattle that graze on pasture throughout the year, from breeding to fattening for slaughter (Australia BAE 1975). Although extensive and intensive fattening operations are found, dry-lot fattening of calves with little or no previous grazing after weaning accounted for only 5 percent of slaughterings in 1973–1974. Australia has four main production and fattening systems: (1) breeding and fattening calves (vealers); (2) breeding and fattening older cattle; (3) breeding and selling "stores" cattle, that is, cattle that require further fattening, and possibly growing, before slaughter; and (4) purchasing and fattening of stores cattle (Table 4.6). About 60 percent of Australian producers fall in the categories of "breed vealers" and "breed fats." The four types of systems are found in all states.

Because their cattle are fed-out on grass, Australian producers are locked into an extended planning horizon (from 1½ to 4 years) that restricts their management alternatives, especially in the production of cattle destined for the export market.[2] Export operations are an integral part of nearly all beef enterprises in all twenty-six beef producing regions of Australia (Figure 4.1), most of which depend on export to a high

2. Much of this section is from Australia BAE 1976.

Table 4.6. Characteristics of the Australian beef economy by type of system

Type of system	Total producers	Total beef cattle	Average herd size per property	Slaughter as a proportion of total turnoff (Avg. 1968–1971)	Composition of total returns by type of enterprise, 1968–1969 to 1970–1971					Rate of return to capital and management 1968–1969 to 1970–1971
					Beef	Dairy	Sheep	Other	Total	
	(%)		(head)				(%)			
Breed vealers	29	13	118	91	31	4	36	29	100	3.5
Breed fats	29	41	365	84	41	2	31	26	100	4.0
Breed stores	9	18	542	22	51	2	30	17	100	4.3
Fatten stores	5	3	147	98	32	12	17	39	100	4.4
Other	28	25	235	65	33	9	26	32	100	3.4
Total	100	100	261	71	38	6	31	25	100	3.9

Source: Adapted from Australia BAE, 1976.
Note: 1971 except as noted.

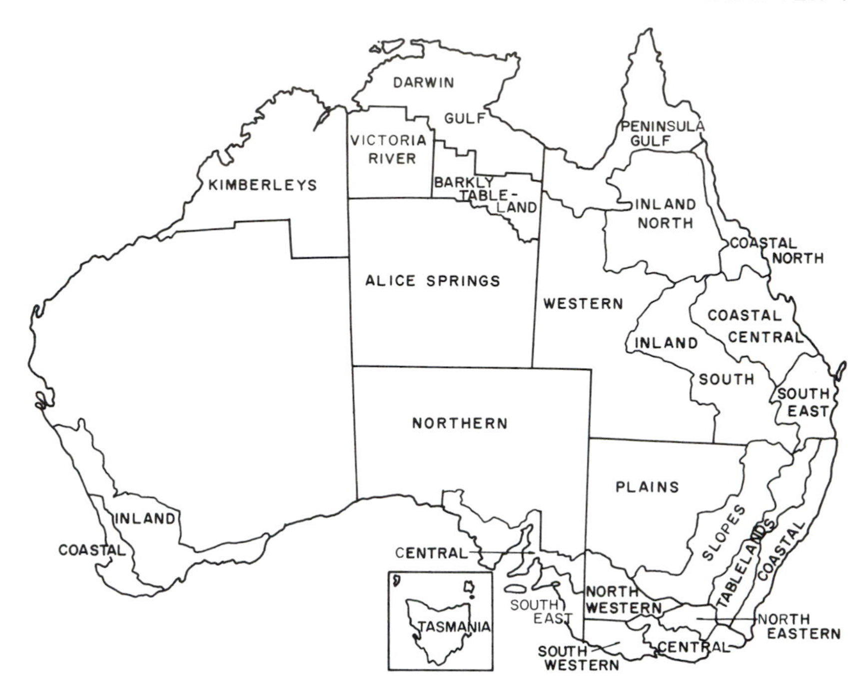

Fig. 4.1. Australian beef cattle regions.

degree (more than 66 percent of production of "turnoff" for slaughter). More than half of Australia's export beef, almost all of which is manufacturing grade, goes to the United States. Other substantial export markets are the United Kingdom, Japan, and Canada. Changes in export markets thus have a direct impact on nearly all beef cattle producers.

The breeding and fattening of vealers is, in reality, a type of beef system, as the fat young cattle of either sex are slaughtered at 6 to 12 months of age. True vealers, in contrast, are typically slaughtered at less than 3 months of age and have subsisted partially or entirely on feeds other than milk for a considerable period of time. Consequently, they have developed heavier middles and physical characteristics associated with maturity beyond the vealer stage. Australian "vealers" normally weigh between 170 and 240 kg at slaughter and have a carcass yield of about 53 percent. Most of the meat is consumed in the domestic market as table beef.

The second common system, breeding and fattening of older cattle—or the breed fats system as it is known in Australia—is oriented to male cattle older than 12 months. About 84 percent go to slaughter. Three main types of cattle 1-3 years of age are slaughtered: (1) yearlings aged 12–16 months with a live weight between 300 and 400 kg and a cutout or yield of about 55 percent; (2) steers from 16 months to 3 years of age with a live weight of 400 to 540 kg and a cutout of about 56 percent; and (3) bullocks, which are older than steers but have similar weights. In general, the older the animal, the more likely that its meat will be exported. Breed fats constitute about 40 percent of the national herd.

The third system, breeding and selling stores cattle, involves either production (turnoff to the Australians) and fattening, or the sale of lighter animals for further fattening. This system is unlike that in the United States, where most cattle from production operations are sold at weaning time. Systems are largely influenced by price of cattle—the higher cattle are priced, the more incentive there is to fatten them by some timesaving methods, perhaps using crop residue or feedlots. Australian cattle prices are low relative to those in the United States, and except for production on very marginal land, there is generally more profit in the grass fattening of steers on good quality land than in their production. The breed stores system accounts for about 18 percent of all cattle. Only 22 percent of sales are to slaughter. The final system, purchasing and fattening of stores cattle, accounts for only 3 percent of the national herd. Most of these cattle are sold within a 12-month period. Almost every animal sold goes to slaughter.

These four systems within the beef cattle industry are highly interdependent. All systems turn off some nonslaughter cattle that are purchased by operators of other systems. There are also considerable sales between operations within each system. Approximately 20 percent of nonslaughter turnoff moves among Australia's twenty-six regions.

Beef cattle production in Australia is generally part of a mixed farming enterprise. Only about 20 percent of operations receive 85 percent or more of their total returns strictly from beef cattle. A larger share, 26 percent, have both beef and sheep as well as cereal grains. About 4 percent are beef and dairy combinations. The remaining 50 percent combine beef with other types of operations. About 10 percent of Australia's total pasture area is in "improved" species. Most of it is used by the breed fats system.

During the 3-year period 1968/1969 to 1970/1971 about 3 percent of all total income of operations dealing with breed vealers was derived from sales of vealers or beef cattle. In contrast, the breed stores system averaged 57 percent of total return from beef sales. A significant proportion of income in all systems, except "fatten stores," is derived from sales of sheep.

The rate of return to capital and management is consistent among operations, ranging only from 3.4 percent (other) to 4.4 percent (fatten

On extensive ranches with lots of brush, a cowboy can use a helicopter to good advantage in the roundup.

stores). There is a positive correlation between export dependence and net farm income, but considerable variability in income because of changes in international prices. The average rates of return are comparable with those in the United States (see Chapter 3).

Dry-lot feeding of cattle has developed since the mid-1960s, primarily because of technology imported from the United States. The industries in the two countries differ significantly, however. Almost all dry-lot fed beef is consumed domestically in the United States whereas in Australia almost all of it is shipped to Japan (a few additional markets such as Iraq have opened up since the mid-1970s).

Another difference between the feedlot industries is length of feeding. Although both countries use concentrates, in the United States cattle are fed for 5 to 6 months, but in Australia they are generally held for only 3 months, so that a substantial proportion of the final weight is from grazing prior to reaching the feedlot. This indicates a less favorable beef/grain price ratio in Australia. Furthermore, while the United States has a relatively assured supply of suitable feeder cattle in large uniform lots at the appropriate time, the necessary links are not well established in Australia. The same is true of feed supplies and market outlets (Johnson and White 1976).

Feedlot size varies considerably in Australia, from a few hundred head to lots having more than 5,000 head one-time capacity. Most of the smaller lots are operated as part of a farming or ranching business on a regular seasonal basis or as opportunistic operations. Many large lots appear to be operated as distinct entities, despite intercorporate ownership or other less formal arrangements. Because many of the large lots are oriented to the Japanese market, important additional relationships in some cases are deals made with the Japanese meat trade in the form of joint ventures.

Prior to the arrival of the Europeans, there were no cattle in Australia or New Zealand. As the English colonized Oceania, they brought their breeds of cattle with them. Overall, the British breeds never became acclimatized in tropical and subtropical Australia and ranchers soon realized that more suitable breeds were needed; some of the more progressive operators imported small numbers of Zebu types prior to the 1850s, but significant imports were not made until after the turn of the century. By the 1930s considerable effort was being put into crossbreeding and experimentation with Zebu cattle.

Progress in Australian breeding is apparent from the establishment of various breed associations. The Australian Zebu Breeders Association was founded in 1946, and the Australian Brahman Breeders Association was established in 1954. Two years later the Tropical Beef Breeders Association was organized, followed in the early 1960s by associations of Brangus and Braford breeders. Breeders of the Droughtmaster, a composite breed that is an original Australian contribution, formed the Droughtmaster Stud Breeders Society in 1962. Even though substantial numbers of English breed cattle, especially Shorthorns, remain in the Australian tropics, they will probably be eventually replaced by straightbred or crossbred Zebu types.

The Australian beef cattle industry has improved and expanded greatly in the past twenty years and shows considerable promise of further expansion in the 1980s (Johns 1978; Longmire et al. 1979). The country has a comparative advantage in beef production, and with further improvements in production and marketing systems it can continue to be the world's leading exporter (Longmire 1979). New policies will be required, however, and are already being formulated (Miller 1979).

SUMMARY AND CONCLUSIONS

The factors influencing cattle systems such as land characteristics, production complementarity, tradition and social conditions, location, level of effective demand, price, and government policies, are discussed. They are then related to production systems in

Latin America. Efficiency measures and a discussion of the economics and production aspects of intensive and extensive cattle raising methods are also added. This includes a discussion of forages and cattle breeds found in Latin America.

The Australian beef system—breeding and fattening calves, breeding and fattening older cattle, breeding and selling stores cattle, and purchasing and fattening of stores cattle—are described. The parallels between Latin American and Australia, where cattle are generally fattened on forages rather than on concentrates in confinement feedlots, are shown. We may conclude that considerable productive potential exists in both areas. The Latin Americans will exploit this potential largely because of their rapid population growth. Australia—and to a large extent New Zealand, which displays similar characteristics to Australia—will expand beef production in response to world conditions.

We now turn to the beef production systems and breed types found in Africa and Europe. Whereas the Latin American and Australian systems have certain characteristics in common, the systems described in Chapter 5 are quite different from the ones we have just examined, as well as from each other.

REFERENCES

Australia, Bureau of Agricultural Economics. 1975. *The Australian Beef Cattle Industry*. Industry Econ. Monogr. no. 13. Canberra.

______. 1976. *Production Systems in the Australian Beef Cattle Industry*. Beef Res. Rep. no. 18. Canberra: Australian Government Publishing Service.

Bogdan, A. V. 1977. *Tropical Pasture and Fodder Plants*. New York: Longman.

Davies, W., and Skidmore, C. L. 1966. *Tropical Pastures*. London: Faber and Faber.

Eyre, S. R. 1963. *Vegetation and Soils: A World Picture*. Chicago: Aldine Publishing Co.

Hutton, E. M. 1975. Tropical Pastures and Beef Production. *World Anim. Rev.* 12:1–7.

Johns, M. 1978. *Production and Export Projections for the Australian Beef and Veal Industry 1978–82*. Occas. Pap. no. 50. Canberra: Bureau of Agricultural Economics.

Johnson, C. B., and White, G. M. 1976. Current Adjustment Problems in the Australian Beef Cattle Industry. *Quart. Rev. Agric. Econ.* 29:47–60.

Longmire, J. 1979. *Livestock and Meat Marketing in Australia: An Economic Evaluation*. Canberra: Bureau of Agricultural Economics.

Longmire, J., et al. 1979. *Agricultural Supply Projections: Australia 1982–83*. Tech. Suppl. Grazing Livestock. Canberra: Bureau of Agricultural Economics.

Maynard, L. A., and Loosli, J. K. 1969. *Animal Nutrition*. 6th ed. New York: McGraw-Hill.

McDowell, R. E. 1972. *Improvement of Livestock Production in Warm Climates*. San Francisco: W. H. Freeman and Company.

______. 1976. Mineral Deficiencies and Toxicities and Their Effect on Beef Production in Developing Countries. In *Beef Cattle Production in Developing Countries,* ed. A. J. Smith, pp. 216–41. Edinburgh: Centre for Tropical Veterinary Medicine, University of Edinburgh.

Miller, G. 1979. *Agricultural Marketing Policies for the 1980s: Future Agricultural Policy in Australia*. Canberra: Bureau of Agricultural Economics.

Morgan, Q. M. 1973. *The Beef Cattle Industries of Central America and Panama*. USDA, FAS-M-208.

Payne, W. J. A. 1966. Nutrition of Ruminants in the Tropics. *Nutr. Abstr. Rev.* 36:653–670.

______. 1970. *Cattle Production in the Tropics. Vol. 1: Breeds and Breeding*. Bristol: Longman Printing Services Ltd.

Plasse, D. 1976. The Possibility of Genetic Improvement of Beef Cattle in Developing Countries with Particular Reference to Latin America. In *Beef Cattle Production in Developing Countries,* ed. A. J. Smith, pp. 216–41. Edinburgh: Centre for Tropical Veterinary Medicine, University of Edinburgh.

Preston, T. R., and Willis, M. B. 1970. *Intensive Beef Production*. 2d ed. New York: Pergamon Press.

Preston, T. R., and Long, R. A. 1979. Sugarcane as Cattle Feed. *World Anim. Rev.* 28:44–48.

Ranjhan, S. K. 1979. Use of Agro-Industrial By-Products in Feeding Ruminants in India. *World Anim. Rev.* 28:31–37.

Rivas, L., and Nores, G. A. 1978. *Evolucion de la Ganaderia Bovina en America Latina*. Cali, Colombia: CIAT.

Sanchez, P. A. 1976. *Properties and Management of Soils in the Tropics*. New York: John Wiley and Sons.

Sanchez, P. A., and Tergas, L. E., eds. 1979. *Pasture Production in Acid Soils of the Tropics*. Cali, Colombia: CIAT.

Skerman, P. J. 1977. *Tropical Forage Legumes*. Rome: United Nations Food and Agriculture Organization.

United Nations Food and Agriculture Organization. 1964 and 1978. *Production Yearbook*. Rome.

Valdez, A., and Nores, G. 1979. *Growth Potential of the Beef Sector in Latin America—Survey of Issues and Policies*. Washington: International Food Policy Institute.

5

BEEF PRODUCTION IN EUROPE, AFRICA, AND ASIA

The production systems and related aspects of Europe, Africa, and Asia are the subject of this chapter. (See Appendix Table A.1.) These systems, which vary considerably from those found in the Western Hemisphere, are grouped in one chapter merely for convenience since Europe's beef production is intensive while Africa's is extensive. Africa has 0.37 head of cattle per person, for example, whereas Western Europe has 0.27 head per person. Production per head of cattle has hardly increased in Africa, but in Europe it is continually improving.

Western Europe has 12 percent of the population in the agricultural sector versus 66 percent in Africa. The population growth rates also differ: 0.5 percent and 2.8 percent, respectively. Western Europe has about 100 million head of cattle while Africa, with six times the land base and ten times as much area in permanent pastures, has only 170 million head of cattle. Asia has about 59 percent of the population in agriculture but only has 0.15 cattle per person. The population densities for Africa, Asia, and Europe are 15, 91, and 99 people per km^2, respectively.

EUROPEAN BEEF SYSTEMS

France, with 24 million head of cattle, has the largest inventory of the thirty-two European countries (see Appendix Table A.1). The ratio of cattle to people in Europe varies widely, from a low of 0.04 in Malta to 2.19 head of cattle per person in Ireland. Albania has the highest percentage of people in agriculture (61 percent), while the United Kingdom has the lowest (2.1 percent).

Europe's cattle inventory increased about 19 percent annually from 1960 to 1979, with the nine countries of the European Economic Community (EEC) registering 18 percent change (Table 5.1). Total production increased 50 percent in the EEC versus 63 percent in all Europe (see

Most of the section on production structure is based on personal interviews with international and European specialists. Marketing aspects are largely adapted from OECD 1978.

Table 5.1. Cattle inventory in Europe and USSR, 1960 and 1979

Region and country	Year 1960	1979	Total change
	(1,000 head)		(%)
Western, EEC			
Belgium/Luxembourg	2,846	3,085	8
Denmark	3,397	3,034	−11
France	18,735	23,510	25
Germany, Federal Republic	12,480	15,007	20
Ireland	4,741	7,178	59
Italy	9,399	8,556	−9
Netherlands	3,507	5,149	47
United Kingdom	11,711	13,534	15
Total	66,876	79,053	18
Western, other			
Austria	2,308	2,594	12
Finland	1,922	1,736	−10
Greece	1,046	973	−7
Norway	1,129	971	−14
Portugal	...	1,050	...
Spain	3,640	4,650	28
Sweden	2,501	1,911	−24
Switzerland	1,746	2,038	17
Total	...	15,923	...
Eastern			
Bulgaria	1,284	1,763	37
Czechoslovakia	4,303	4,887	14
Germany, Democratic Republic	4,465	5,572	25
Hungary	1,971	1,966	0
Poland	8,695	13,036	50
Yugoslavia	5,297	5,491	4
Total	26,015	32,715	26
Others[a]	6,217	6,844	10
Total, Europe	113,400	134,535	19
USSR	74,155	114,086	54
Total, Europe and USSR	187,555	248,621	33

Source: FAO *Production Yearbook,* 1962 and 1979.

[a]Includes Albania, Andorra, Faeroe Islands, Gibraltar, Liechtenstein, Malta, Monaco, Romania, San Marino, and missing countries.

Appendix Table A.2). France is Europe's largest beef producer, with 1.8 million metric tons in 1979, followed by West Germany with 1.5 million metric tons. The United Kingdom and Italy produce just over 1 million metric tons.

The European cattle industry is based almost exclusively on dairy production (Poitevin, Mallard, and Picon 1977). In fact, apart from England, France, Ireland, and Italy, beef breeds make up only about 2–5 percent of all cattle. In these four countries they account for about 20–25 percent of their combined cattle inventory, but this percentage is declining continuously. The southern countries of the East European Bloc have numerous relatively large economic units and, consequently, have tended to specialize more in either dairy or beef breeds. The northern East European Bloc countries, on the other hand, have more dual-purpose breeds.

The farm unit in Europe, apart from the cooperative farms of East Europe, tends to be a small family farm with 20–40 ha of land. Because of their size and the harsh winters, especially in the northern regions where animals have to be confined during the winter months, the operations are based almost exclusively on dairy herds as beef cattle would not provide a sufficiently high total farm income. In these operations beef is almost a by-product of dairying, although in some respects it must be considered a joint product.

Size and management of production systems vary greatly both between and within European countries, although there are no large breeding units of the type found in the United States or Australia. In fact, there are few cow/calf operations in which calves are sold at weaning time, except for a few large "ranches" in southern Spain and southern Italy which raise cattle on an extensive scheme rather than the intensive system that is typical in Europe. In general, producers are numerous, or very numerous, and dispersed. As a rule the same people do not breed, rear, and finish cattle. Also, concentration in cattle feedlots is considerable, just as it is in the midwestern United States. Although almost all cattle are fed-out on family-owned farms, one exception is northern Italy where many nonresident landowners have feedlots. A large proportion of the cattle and much of the feedstuffs are either imported or transported long distances.

Europe's cattle production structure underwent considerable change during the 1960s and 1970s when small-scale confinement cattle feeding became a major industry. With it, beef production has increased dramatically, from 6.1 million metric tons in 1960 to 10.0 million metric tons in 1979 (see Appendix Table A.2). Among several factors that accounted for this development were the generalized introduction and improvement of corn and grass varieties, which are now heavily used for the production of silage, the principal ingredient in most rations.

Two types of cattle were sold for slaughter until the early 1960s: cull cows and bulls, and veal calves. By the late 1970s veal constituted only 5–10 percent of beef and veal production, but it is still important in France, the Netherlands, and Switzerland. Almost all European countries tend to feed calves out to the limit at which they can still be sold for veal (about 150 kg), whereas previously they were slaughtered at a few weeks of age.

Another breakthrough apart from feedstuffs that led to confined cattle feeding was an improvement in methods for assembly of calves, largely through cooperatives. These associations of cattle producers are much more important in Europe than in other parts of the world, principally because most cattle in Europe are used for dairy production. Cooperatives are most prominent in Sweden, followed by Norway, Denmark, France, Ireland, and West Germany. Little use is made of cooperatives in Belgium, Italy, and the United Kingdom. The growth of cooperative producer organizations—which include group purchase of inputs, cooperative rearing of animals, and association marketing of animals—can be traced to the existence of many small production units and governmental incentive policies. Credit institutions specializing in financing cooperatives have also influenced the producer groups.

Europeans utilize cooperative production agreements for beef to a greater extent than do other countries of the world. In 1976, for example, about 14–15 percent of all calves in West Germany were produced under a previously signed contract. In Belgium about 90 percent fell into this category, in France 25–30 percent, in Italy 5 percent, and in the Netherlands 70 percent. This type of contract is hardly used in other European countries or the rest of the world.

Feed manufacturers were the first promoters of production type contracts; they were quickly followed by livestock merchants, wholesalers, vertically integrated cooperatives, and supermarket chain stores. The most common contract in Europe involves an agreement by the farmer to purchase all milk powder for calf feeding from the milk replacer manufacturer, as well as to comply with certain technical requirements. In this resource-providing contract, the manufacturer invariably supplies the feed, sometimes the calves, and occasionally the marketing function. The main advantages to the producer are financial and technical assistance. Another less common contractual link is fattening of cattle to slaughter weights.

During the 1950s and 1960s considerable emphasis was placed on improving the genetic potential of Europe's dairy breeds, and North American Holsteins and semen were imported in large quantities. As fattening became a major business, beef bulls, especially Charolais, were used on dairy cows with the intention of producing a beef-type calf, either heifer or bull, for sale to feedlots. Because of persistent calving difficulties, however, farmers began turning to both crossbred dairy cows, especially those crossed with Brown Swiss, and dual-purpose cattle to such an extent that concern now exists that pure lines will be lost. Paradoxically, there is also considerable concern about the rapid disappearance of the diverse original European cattle breeds.

Since milk is the most important farm commodity of the EEC, the dairy herd makes a major contribution to beef supplies. There are excess supplies of milk so that considerable sums of money have to be spent on supporting farm prices, while the surpluses are disposed of at a loss. The high prices stimulate high levels of production, but one cannot infer that the industry is necessarily inefficient, for milk yield per cow is high—about 3,321 kg per cow in lactation for Europe as a whole compared with 5,098 kg in the United States. In the Netherlands it is 5,094 kg while it is 4,760 kg in the United Kingdom. A program is clearly needed to solve the surplus milk problem. Given the social implications, one solution would be to maintain or even possibly reduce cow numbers and to put increased emphasis on crossbreeding—and consequently pay more attention to improving pure breeds—to provide more beef-type calves for fattening. The world demand for surplus European dairy products is also growing, but because of exports being subsidized. The milk pricing problem in Europe is a classic case of the conflicts which arise in meeting social, economic, and political goals.

The great variety of breeds found around the world is well-represented in Europe, which has more than seventy breeds (see Appendix Table A.12). This rich diversity of breeds in Europe is no accident; rather, it is the outcome of a long history of selection and breed improvement as well as the relative isolation of different agricultural areas. Most of the original breeds are dual-purpose types, with the exception of several breeds developed in the United Kingdom, where selection for single-purpose beef cattle began over 150 years ago. This development corresponded with a demand by breeders in the burgeoning cattle industries of Argentina, Uruguay, Canada, the United States, Australia, New Zealand, and South Africa. Breeders in Great Britain were so successful in meeting the increasing demand for beef-type animals that cattle of English and Scottish origin are now found throughout the world. Among other benefits, the purebred business has earned considerable amounts of foreign exchange for the United Kingdom.

The European "Green Revolution" in forages is continuing with hybrid varieties of maize being introduced as far north as Finland, a country that now commonly uses a ration of urea mixed with grass forage. Now that more and better forage varieties are being developed, farmers pay close attention to balancing least cost rations and average daily gain. The fattening of heifer calves has become increasingly important with the result that in countries such as Sweden, there is little price discrimination between sexes.

Most grain used in Europe is imported, however, as little progress has been made in increasing feed grain yields at a reasonable cost. For this reason most rations are based on forages. Another factor is that European policy has allowed soybeans, a good source of protein, to be imported at very low or no duties, whereas imported feedgrains fall under a complicated series of tariff and nontariff barriers. Since another policy has aimed at encouraging imports of nontraditional feedstuffs such as citrus pulp and cassava (yuca or manioc), feed manufacturers have a wide range of ingredients to choose from; the ones chosen depend on price relationships and availability.

The livestock/feed policies, combined with a highly effective extension service and research organization, have helped Europe move from being a large deficit market to being self-sufficient in beef and veal. Pasture fattening has practically disappeared and thus this resource is free for grazing by more dairy cows and other types of animals. Although these improvements have greatly increased the efficiency of European cattle and milk production, the industry continues to be a high-cost operation that is heavily subsidized. Nevertheless, the potential for continued improvements in production efficiency and output seems promising. Considering that hogs and poultry are almost exclusively fed on grain, which is becoming increasingly costly relative to forages, the cattle industry (with beef as one of its main products) will no doubt continue to expand rapidly in the mix of meat supply. Production of beef and veal per head of cattle inventory has already shown a strong increase, from 55.7 kg in 1960 to 74.1 kg per head of inventory in 1979 for Europe as a whole. The EEC's output per capita increased from 66.1 kg to 84.0 kg over this same time period (Table 5.2).

AFRICAN CATTLE SYSTEMS

Domestic cattle were first introduced into Africa from western Asia about 5000 to 2300 B.C. Three types were imported by nomadic people, or migrated there: the Hamitic "Longhorn" of Egypt, the "Shorthorn" of Egypt, and the Zebu. These types are not related to the Longhorn of the United States or the Shorthorn breed developed in England. The first admixtures of Zebu with Longhorn and/or Shorthorn probably took place as early as 2000 to 1500 B.C. at various centers in North and East Africa. Cattle that resulted from the indigenous admixtures between humped and humpless cattle can be classified as Sangas (Payne 1970).

The original country or area in which the different breeds were found is given in Appendix Table A.13 along with other information on breeds and types of cattle in Africa. In some cases the origin encompasses various countries, but an effort has been made to define a center of concentration in the development and diffusion process. Description of the principal distribution follows the same approach. East Africa, for example, is probably the center of origin for all Sanga types. In addition, there is evidence that Sanga cattle were widely distributed in Africa until modern times (Payne 1970). Estimations of cattle weights, also given in

Table 5.2. Production of beef and veal per head of cattle inventory, 1960 and 1979

Region and country	Year 1960	1979	Change 1960–1979
	(kg)		*(%)*
Western, EEC			
Belgium / Luxembourg	75.9	94.2	24
Denmark	70.1	80.8	15
France	79.3	76.2	−4
Germany, Federal Republic	71.8	101.3	41
Ireland	53.2	52.7	−1
Italy	48.5	124.5	157
Netherlands	65.9	66.4	1
United Kingdom	66.3	74.6	13
Average	66.1	84.0	27
Western, other			
Austria	62.5	76.5	22
Finland	37.5	60.5	61
Greece	...	100.7	...
Norway	44.3	72.1	63
Portugal	...	85.7	...
Spain	40.7	83.2	104
Sweden	51.6	78.9	53
Switzerland	56.7	78.6	39
Eastern			
Bulgaria	28.8	60.7	111
Czechoslovakia	...	71.6	...
Germany, Democratic Republic	...	71.3	...
Hungary	68.5	68.8	0
Poland	38.3	56.2	47
Yugoslavia	29.5	62.8	113
Average, Europe[a]	55.7	74.1	33
USSR	...	57.1	...

Source: Calculated from data in Appendix Tables A.1 and A.2.

[a]Includes Albania, Andorra, Faeroe Islands, Gibraltar, Liechtenstein, Malta, Monaco, Romania, San Marino, and missing countries.

the Appendix Table A.13, show that cattle size varies substantially—cows of some native breeds weigh as little as 220 kg, while cows of other breeds, such as the Nguni, average as much as 485 kg.

Two distinctive beef production systems are found in Africa: (1) a modern system, in which the cattle and land are usually owned by a person of European ancestry; and (2) a traditional system, which is based on pastoral grazing. Naturally, there are also numerous subsystems and areas in transition, but these two are the dominant systems.

In the modern system in Africa cattle are fed intensively on pasture in combination with a cow-calf operation. In many cases, cattle form part of a mixed agriculture enterprise and are fattened on homegrown grains or forage residue. These systems are prevalent in the temperate climates, and on fertile land. Most of the improved or imported cattle breeds are owned by Europeans.

Communal grazing, a livestock production system in which cattle belong to individuals or a group of people and are grazed on a common property, forms the basis for the traditional African cattle production system. Communal grazing of cattle is widespread, being the dominant system in most of Africa, the Near East, much of Asia, and the Eastern Bloc countries. It is found to some degree in every country of the world, but is most common where livestock raising is based on a tribal or indigenous social structure that is still operating at or near the subsistence level, and where there is no well-developed private property system. Cattle on several Indian reservations in the United States, for example, are still herded in a communal manner bearing a striking resemblance to methods used in the heart of Africa (Simpson 1970). An abundant literature adequately demonstrates that minimal problems occur under low usage of the commons, for there are sufficient resources to satisfy most of the user's desires. Difficulties arise when the resource becomes scarce or overutilized, as has happened in many communal grazing areas.

The numerous technical problems of cattle production and marketing are compounded under a communal grazing system in that a variety of additional social and cultural factors have to be considered. Attempts to modernize cattle production compound inherent difficulties, for improved output implies a shift from a survival orientation to one that encompasses aspirations for improved standards of living. Furthermore, increased productivity means greater use of purchased inputs, which requires a shift to a business orientation. Bringing about and maintaining this change can create a critical management problem; there is little incentive for anyone to use improved management techniques, inputs, or better stock to any degree when cattle are herded communally since the benefits are usually enjoyed by other members of the group rather than just being captured by the investor. If one person buys a good bull and turns it into the herd, for example, the other common's users share the benefits unless the investor can separate out his or her cows. In a communal system intense social pressures, as well as practical ones, preclude such actions.

Perhaps the most serious aspect of communal cattle grazing is the extreme difficulty of controlling stocking rates on common property rangelands. Since few inputs and management techniques are feasible, the system is an extractive one in which the number of cattle rather than production efficiency becomes the decision-making criterion. In this atmosphere, if outside agencies or the livestock people themselves make investments such as water wells, the end result will be greater overgrazing, further ecological deterioration, and benefits that do not cover costs (Simpson 1970).

This does not mean that communal grazing systems are all bad and should be replaced immediately by other land tenure systems. On the contrary, in many cases, communal systems are the best alternative at a particular time. The point is that while communal grazing of livestock can be successful under certain conditions, there are serious drawbacks if the goal is modernization and increased productivity. Where stocking rates and other management and investment decisions can be made and enforced by the group or an agency, good management is possible and the system will work (Spicer 1952).

Communal grazing systems in part account for the fact that pastoralists hold the highest numbers of cattle possible rather than operate with fewer numbers and a higher offtake. In addition, the simple ownership of cattle becomes an end in itself in many pastoral societies, especially in Africa, as cattle are a symbol of prestige as well as an economic asset. Cattle are also considered a type of savings or bank account. Recognizing the difficulty of accumulating an inventory owing to a communal system, low calf crops, and the ease of spending money, a substantial proportion of African herders are wise to remain within their cultural barriers and to market cattle only for special social or ceremonial occasions, or emergency needs (de Wilde 1967).

African pastoralists are often accused of not being price responsive, but Doran, Low, and Kemp (1979) have shown that for the Swazis an inverse relationship exists between sales and both cattle price and rainfall. Results of their regression analysis indicate that about 65 percent of the variation in offtake is accounted for by price and rainfall. Another characteristic response is that, if further price increases are expected a rational stockman will hold back females and bulls from sales to make more cattle available for later sales.

A number of generalizations can be made about African pastoral grazing systems, but since this discussion concentrates on nomadic systems over a wide variety of climate and social conditions, care must be taken in applying the generalizations to particular situations. One common feature that does stand out, however, is that most cattle in Africa are owned by individuals although they are grazed communally (Baker 1976). The animals are a source of prestige and are important in tribal customs and ceremonies (Dalton 1967; Bohannan 1968). Cattle are an integral part of daily life in Africa since they provide sustenance in the form of milk and—to a much reduced extent today—blood, and are also a central object within social customs (Dalton 1967; Kilma 1970). The pastoral people eat little beef and slaughter an animal only when it is near death.

The heads of many pastoral families own 40–50 cattle, although 10 or 15 are more typical. Herds of 100–125 head are common, and leaders in countries such as Senegal own more than 1,000 head (Schneider 1968). Bulls run with cows all year. Few of the males are castrated, and selective breeding is almost nonexistent. Cattle are usually corralled at night to permit milking and to protect them from wild animals and thieves (Schneider 1957, 1968). The daytime grazing is overseen mainly by stock boys, although some adults are also used. Because the animals are enclosed at night, severe overgrazing occurs near the corrals. Calves are rarely weaned until the next calf is born with the result that only small amounts of milk are available for human consumption. In more settled areas, oxen are ridden and used as draft animals.

African systems are also greatly affected by climate. Seasonal factors, heightened by periodic droughts, force some pastoral people to move regularly. Availability of drinking water is obviously a limiting factor. Since no one takes responsibility for more than nominal improvement of an area, resultant overgrazing leads to poor health, low calving rates (45–55 percent), and attendant low productivity. Improvements in health care come mainly from government sanitary programs, but such programs lead herders to become dependent on government as they become more involved with a market economy.

In general there is no specialized effort in Africa to fatten animals for eventual sales to traders, government, or packers (Sullivan et al. 1977). Cattle are usually trekked long distances to assembly points or to the market itself (Forlag 1974). There are few direct sales in which owners drive cattle to a central market or packing plants, although in some areas such as Zimbabwe government-sponsored auctions are held in tribal areas. In some regions, especially where the herdsmen are seminomadic, cattle are driven to cultivated areas during the dry season.

KENYA, TANZANIA, AND UGANDA. The region known as East Africa comprises six countries, of which three are the tropical countries of Kenya, Tanzania, and Uganda. These countries have dry areas that are of little use except for grazing, as well as vast stretches of land with high productive potential. The majority of the 31 million head of cattle in the three countries are kept by pastoralists and small holders still operating in their traditional way (Dyson-Hudson 1969; Dunmore 1978). In all three countries about 80 percent of the population is involved in agriculture; each country has a population growth rate of more than 3 percent annually. The number of cattle per person for Kenya, Tanzania, and Uganda is 0.66, 0.88, and 0.42, respectively, while the population density is 27, 18, and 54 people per km^2 (Appendix Table A.1). About 25 percent of the cattle in Africa are found in these three countries.

No systematic grazing management is practiced and heavy losses occur because of disease and parasites (FAO 1967). Government control of meat and live animal exports, combined with control of domestic prices at low levels, have discouraged production increases (Farris and Sullivan 1976). In an effort to raise production, all three countries have set up "grazing schemes" and parastatal ranches since the early 1950s. Since the grazing schemes have not worked out well, many have been completely abandoned (Saunderson 1977). They cannot be regarded as an optimum ranching organization as they lead to constant financial and personal obligations by the government. Furthermore, the livestock owners fail to accept any degree of responsibility in caring for the facilities since they consider themselves to be merely tenants (Meyn 1970). All three governments are trying to stratify the beef industry by channeling feeder cattle to fattening ranches but, as with the parastatal cow-calf units, they have been less than successful (Lele 1975). On commercial operations, primarily those belonging to people of European descent, exotic single- and dual-purpose breeds have been introduced and some crossbreeding is being practiced.

Traditional tribesman control most of the cattle in Kenya; the Masai are the foremost cattle raisers. There are only a few European ranchers. The Kenyan government is using most of its cattle funds to develop tribal areas for pastoral people. Although land tenure problems are being addressed by directed companies, communal cooperative societies, and other means, and although Kenya has a good potential for expanded livestock production, it appears that communal ownership and grazing practices will prevent it from reaching that potential (Saunderson 1977). It has been suggested that despite social constraints one way to overcome the low productivity is to develop cattle feeding systems in specialized areas (FAO 1977). However, such finishing enterprises need to be linked to growing systems.

In East Africa, and Africa as a whole, rapid expansion of beef production can be achieved only through the implementation, on a large scale, of intensive growing and finishing operations (Auriol 1974). Feedlots would not only increase meat production but would also make it much more efficient. In addition, a well-planned feedlot industry can be an important catalyst to the overall development of the beef industry.

Despite various attempts throughout Africa to feed cattle in confinement facilities and to encourage short-term feeding on specialized pasture operations, cattle raising is changing very little on this continent. A sound feedlot beef industry cannot be created unless certain prerequisites are met. One is the assurance of favorable cattle and input price relationships. In general, feed prices have been relatively high in Africa, while strict controls on meat prices have led to uneconomic relationships. Perhaps the major difficulty has been the encouragement of government

This herd, belonging to a Masai tribesman, is hunting grass on the drought stricken area west of Mt. Kilimanjaro. This is a multipurpose herd with the first purpose to provide milk for the family and extended family of the herdsman.

or parastatal ownership of feedlots at the expense of private initiative. The result has been poor management, attendant losses, and the closing of lots. There has been a general failure to recognize that management is all-important and that therefore the lots have to be either small (25 to 50 head) so they can be run by a farmer owner-operator, or very large so that a competent experienced manager can be employed. Numerous decisions have to be made daily on feed purchases, ration adjustments, cattle purchase and sale, herd health, and personnel management; if a manager is not free to make these decisions when needed, efficiency suffers.

Past experience in Africa and the highly successful feeding history of Europe, parts of Latin America, and the United States suggest that even though confinement feeding of cattle is needed in Africa and could become a good business, little progress will be made until cattle feed/price ratios improve to make feeding profitable. In addition, the need is urgent to adopt policies which will provide incentives for increased slaughter early in the dry season and/or make provisions for placing cattle in confinement lots at the beginning of the dry season.

Sullivan et al. (1980) state that livestock systems based on uncontrolled communal grazing result in inefficient utilization of native forages and low livestock production. Using a forage/livestock herd simulation model adapted to East African conditions, a fenced hay enterprise for lactating cows (not possible with uncontrolled communal grazing) was found to increase nutritional and monetary welfare of a typical village.

Medium to high levels of foot-and-mouth disease, anthrax, blackleg, brucellosis, anaplasmosis, liver flukes, and intestinal worms are reported in all three countries. Rinderpest is reported in Kenya. One of the most serious health problems is trypanosomiasis, a disease caused by a protozoon of the genus *Trypanosoma*. The pathogenic trypanosomes, which are transmitted by tsetse flies (*Glossina* spp.), also transmit sleeping sickness in humans. Eight species of the fly are found in East Africa alone.

ETHIOPIA, SOMALIA, AND SUDAN. Ethiopia, Somalia, and Sudan are the other three countries of East Africa. They are similar to Kenya, Tanzania, and Uganda in that about 80 percent of the population are in the agricultural sector and many of these are pastoral cattle raisers, but there are also some major differences. Population density and population growth rates are lower, the number of cattle per person is higher, and they have slightly more cattle. Nearly 30 percent of Africa's cattle are in Ethiopia, Somalia, and Sudan.

Ethiopia has the dubious distinction of being the only country in the world to report the incidence of all twenty diseases listed in Appendix Tables A.15 and A.16. Somalia and Sudan have fewer diseases and lower incidences of them than Ethiopia. All three countries have hoof-and-mouth disease, but since the tsetse fly is prevalent in lower Africa, they have only a low incidence of trypanosomiasis. That advantage is offset somewhat by the occurrence of rinderpest, which is endemic to all three countries. Major health problems in Ethiopia are distomatosis—or the common liver fluke—and scabies.

Pastoralists own about one-fourth of Ethiopia's cattle and their principal occupation is maintaining cattle. Livestock are a large, underutilized or misused resource. This underutilization is especially discouraging since the cattle show a high productive capability when modern management is applied. In addition to pastoralists there are small crop and plantation farmers who own cattle for draft purposes. In the opinion of an FAO study team (1977), of all the countries of the Middle East and East Africa, Ethiopia probably possesses the greatest natural resources for livestock development, but it has serious problems to solve in its organizational, institutional, and service structures.

The major economic activity in Somalia is subsistence-oriented livestock production. Various livestock projects such as parastatal ownership of large ranches and feedlots have been initiated but little change has taken place. Sudan has one of the highest livestock potentials in the Near East and East Africa but its vast national grazing resources are underutilized. At present, no more than 3 percent of the 161 million ha considered agriculturally usable are under cultivation. Exploitation of even a small percentage of the area for intensive livestock production and feeding would mean that meat production could easily be increased five- or tenfold. In view of the massive amounts of capital and manpower required for expansion of the livestock industry, it will probably expand only in specialized areas, and the pastoral system will no doubt remain for many years.

BOTSWANA, SOUTH AFRICA, ZAMBIA, AND ZIMBABWE. Botswana, South Africa, Zambia, and Zimbabwe are located in the southern part of Africa. The population density of Botswana is extremely low, one person per km^2, while Zimbabwe has 18 persons per km^2. South Africa and Zambia have 23 and 7 persons, respectively. With the exception of Botswana, the percentage of people in agriculture is lower than in most other countries of Africa; these percentages are 81, 29, 67, and 59, respectively. South Africa has the greatest number of cattle, about 13 million; the total for all four countries is about 23 million head, 14 percent of the total cattle in Africa. Botswana has 4.14 head of cattle per person, which places this little African country first in cattle/people density in the world (Uruguay is second at 3.44 head). South Africa has 0.46 head, Zambia 0.33, and Zimbabwe 0.70 head per person. (See Appendix Table A.1.)

Foot-and-mouth disease is prevalent in all four countries. Brucellosis, anaplasmosis, and liver flukes are particularly troublesome. Heartwater, a disease found in most African countries, is a widespread problem in Botswana, South Africa, and Zambia. Another especially troublesome disease in these four countries of southern Africa is intestinal salmonella, or enteric salmonellosis, which is caused by bacteria of the genus *Salmonella*. Intestinal salmonella infection is worldwide in distribution. The only country reporting a high incidence is South Africa.

About a third of the cattle in Zimbabwe belong to people of European descent, a higher percentage than in any other African country. These people have established breed societies, and many of the herds based on exotic breeds are comparable to those found in England or the

United States. The Europeans fatten cattle in feedlots, market boxed beef, and in general have done an outstanding job of producing and marketing beef. The ranching methods of European descendants are in stark contrast to the traditional methods employed by the indigenous Africans.

In South Africa, as in Zimbabwe, there is a complete dichotomy between advanced and traditional pastoral systems. About two-thirds of the cattle belong to Europeans, who average about 200 brood cows on their 4,000–7,000 ha family-type, mixed farm operations. Considerable emphasis has been given to livestock research and there are numerous experimental stations. About 80 percent of the cattle in Zambia are owned by native African pastoralists. The Europeans in the country have well-developed ranches. The government also owns several large operations. There is considerable potential for expansion in all four countries, but the tsetse fly is a major drawback. Cultural mores and political strife are probably the chief impediments to development in Zimbabwe and South Africa.

MADAGASCAR AND MOZAMBIQUE. Mozambique, which lies just east of Zimbabwe, is an interesting contrast to Madagascar, a large island off the eastern coast of Africa in the Indian Ocean. Beef production in both countries is based on a pastoral system, but the people of Madagascar take more of a business approach toward cattle than do people in Mozambique. The population densities are about the same, 14 people per km^2 for Madagascar and 13 people per km^2 for Mozambique. There are 84 and 65 percent of the population in agriculture and 1.03 and 0.14 cattle per person, respectively.

The island of Madagascar has 8.7 million head of cattle, which constitutes about 5 percent of Africa's inventory. Whereas Mozambique has nearly all of the twenty diseases listed in the Appendix tables with many at high levels of incidence, Madagascar has only eight of the twenty diseases reported, with liver flukes being the main problem, followed by blackleg and tuberculosis. Madagascar is free of hoof-and-mouth disease.

Despite its disease problems, Mozambique probably has the better potential of the two countries for developing a cattle industry, especially if an economical means of eliminating the tsetse fly can be found. Increased production of cattle in both countries is seriously inhibited by traditional pastoral practices.

CAMEROON, CHAD, AND NIGERIA. Cameroon is on the western coast of Africa where the continent begins to jut out into the Atlantic Ocean. Nigeria is northwest of it, also on the coast, while Chad is inland, just to the east of Nigeria. The population densities for Cameroon, Chad, and Nigeria are 17, 3, and 81 persons per km^2, respectively, while the percentages in agriculture are 81, 84, and 54. Although Chad and Nigeria have about the same land area, Nigeria has three times as many cattle. All three countries have a substantial percentage of cattle maintained in pastoral systems, but Chad certainly maintains the greatest degree of traditionalism. There are few cattle producers of European descent in any of the three countries.

All three countries have foot-and-mouth disease, pleuropneumonia, anthrax, blackleg, tuberculosis, brucellosis, mange, and stomach worms as enzootic diseases. There is a high incidence of trypanosomiasis and liver flukes. All three countries are free of rinderpest.

ANGOLA, SENEGAL, AND OTHER COUNTRIES. An appropriate way to conclude a brief overview of African cattle production systems is to compare two countries that are geographically remote from each other. Angola and Senegal are about 4,000 km apart, both are on the west coast and both have had considerable opportunity for contact with the outside world. Despite the fact that Angola is six times as large as Senegal, the cattle populations are about the same, 3.1 and 2.8 million head, respectively. Population densities are 6 and 28 people per km^2, and 58 and 75 percent of the population are in agriculture. Both have about the same number of cattle per person, 0.45 and 0.51, respectively.

About the same diseases are present in both countries. Hoof-and-mouth disease is enzootic, the tsetse fly is a major problem, and many other diseases are found here and there. Perhaps most interesting is the fact that most cattle in the two countries are owned by nomadic and seminomadic pastoral tribesmen. Probably the greatest difference is that in Senegal, where there are more pronounced seasons, cattle are trekked to cultivated areas during dry periods and grazed on crop residues such as peanut vines.

These two countries further illustrate how entrenched the pastoral system really is in Africa. Although some countries have not been discussed here, it is fair to generalize that over almost the entire continent the majority of cattle are maintained under a communal grazing system. The extent to which the cattle are kept in a nomadic, seminomadic, or settled manner varies as much within countries as it does among them. The higher productivity of South Africa and Zimbabwe as well as the experience of systems in other parts of Africa where noncommunal grazing methods have been used demonstrates the sharp improvement where grazing is controlled and incentives for increased productivity are present.

ASIAN CATTLE SYSTEMS

There are forty-four Asian countries that are heterogeneous in terms of climate, ethnic groups, and geographic factors. For the purposes of this discussion, the countries have been divided into five groups: (1) large temperate, (2) large warm, (3) other temperate, (4) other warm, and (5) small countries. The five large temperate countries (Afghanistan, China, Iran, Mongolia, and Turkey) cover 51 percent of the total area, while 34 percent of the land area is in the large warm Asian countries (Table 5.3). About 60 percent of the cattle and buffaloes and 45 percent of the camels are found in the large warm countries. Approximately 86 percent of the total cattle, buffalo, and camel inventory in Asia is found in the eleven large countries. These countries account for about 78 percent of all beef, veal, and buffalo meat production. The Asian countries account for about 97 percent of the world's buffaloes, but only 28 percent of the camels. About 37 percent of the world's cattle, buffalo, and camel inventory is located in Asia, compared with second-place South America, which has 16 percent of the inventory (Table 5.4). The breeds found in Asia are listed in Appendix Table A.14.

The large temperate climates have a productivity of about 23 kg per head of inventory, while the small and medium-size temperate countries have a much higher average, 58 kg per head. The extremely low productivity in some areas such as the large warm countries where there is a high proportion of camels and buffaloes to cattle, mainly because of the need for transport and draft power, reduces the average for all Asia to 10 kg per head of inventory (Table 5.3). In much of Asia cattle are kept for draft or milk primarily, with meat as a by-product.

Simpson's (1980) detailed study on beef, veal, buffalo meat, and camel meat production and trade in Asia indicates that despite rapid population growth, inventories need not increase significantly if meat productivity per head can be increased to any extent. The extreme importance of buffaloes as work rather than meat animals is clearly recognized, but it is also understood that new and innovative ways can, and have, been found for fattening excess animals and for reducing death loss, especially among calves. In addition, significant increases in productivity per head of inventory can be made by reducing the incidence of internal and external parasites, by the innovative use of feedstuffs, by improved methods of disease control, and by improved production and marketing methods (see Appendix Table A.14).

SUMMARY AND CONCLUSIONS

The various beef production systems described here and in Chapters 3 and 4 have evolved as the result of a series of

Table 5.3. Land area, permanent pastures, inventory of cattle, buffaloes, and camels, and indigenous production of cattle and buffaloes by region, Asia, 1979

Region and country	Land area	Permanent pastures	Permanent pastures as % of land area	Inventory: Cattle	Inventory: Buffaloes	Inventory: Cattle and buffaloes	Inventory: Camels	Inventory: Total	Buffaloes as % of all inventory	Indigenous production of cattle and buffaloes	Production of beef, veal, and buffalo meat per head of cattle and buffalo inventory
	(1,000 ha)		(%)	(1,000 head)					(%)	(1,000 metric tons)	(kg)
Asia, large countries											
Temperate[a]	1,423,804	466,163	33	92,981	31,314	124,295	1,990	126,285	25	2,870	23.1
Warm[b]	933,612	115,165	12	216,144	81,519	297,663	2,097	299,760	27	1,020	3.4
Asia, other temperate[c]	134,034	13,994	10	10,540	226	10,766	266	11,032	2	621	57.7
Asia, other warm[d]	264,234	27,615	10	46,850	14,021	60,871	277	61,148	23	500	8.2
Asia, small[e]	1,758	104	6	64	18	82	1	83	22	4	48.8
Total Asia	2,757,442	623,041	23	366,579	127,098	493,677	4,631	498,308	26	5,015	10.2
Asia, large countries											
Temperate	51	75		25	25	25	43	26		57	...
Warm	34	19		59	64	61	45	60		21	...
Asia, other temperate	5	2		3	...	2	6	2		12	...
Asia, other warm	10	4		13	11	12	6	12		10	...
Asia, small	...	...		...	...	...	...	...		...	...
Total Asia	100	100		100	100	100	100	100		100	

Source: Compiled from FAO *Production Yearbook*, 1979.
[a]Includes: Afghanastan, China, Iran, Mongolia, and Turkey.
[b]Includes: Burma, India, Indonesia, Pakistan, Saudi Arabia, and Thailand.
[c]Includes: Iraq, Israel, Japan, Jordan, Korea-DPR, Korea-Rep., Lebanon, and Syria.
[d]Includes: Bangladesh, Bhutan, East Timor, Kampuchea DM, Kuwait, Laos, Malaysia, Nepal, Oman, Philippines, Qatar, Sri Lanka, United Arab Emirates, Viet Nam, Yemen AR, and Yemen Dem.
[e]Includes: Bahrain, Brunei, Cyprus, Hong Kong, Macau, Maldives, and Singapore.
[f]Missing data for many countries prevent estimation of production and production per head.

Table 5.4. Comparison of world cattle, buffalo, and camel inventory, 1979

Region	Cattle	Buffaloes	Camels	Total
		(1,000 head)		
Africa	170,110	2,321	11,983	184,414
Asia	366,579	127,098	4,630	498,307
Europe	134,535	458	...	134,993
Oceania	36,203	...	...	36,203
North and Central America	174,385	8	...	174,393
South America	216,119	310	...	216,429
USSR	114,086	360	220	114,666
Total	1,212,017	130,555	16,833	1,359,405
		(%)		
Africa	14	2	71	14
Asia	30	97	28	37
Europe	11	1	...	10
Oceania	3	...	...	2
North and Central America	14	...	...	13
South America	18	...	...	16
USSR	10	...	1	8
Total	100	100	100	100

Source: United Nations, Food and Agriculture Organization, 1979.

specific circumstances. Continental European production consists of dairy operations largely because it is based on small holdings in cold climates. The United Kingdom has developed a system of single-purpose breeds in addition to dual-purpose cattle such as the Shorthorn as a result of high availability of pastures and a strong export demand for specific breeds. In contrast, northern Mexico has a well-defined system of producing calves for fattening in feedlots owing to an abundance of land suitable mainly for cow-calf production systems. Countries such as Argentina, Australia, Paraguay, and Uruguay, on the other hand, have developed an extensive type of beef production system characterized by relatively low production rates per hectare and per unit of livestock inventory, largely because their slaughter cattle are fattened on forages rather than in feedlots. We can conclude, then, that systems evolve in relation to various factors such as climate, topography, trade, and point in history. For example, the formation of the European Economic Community (EEC), with its protective trade policies, and conditions relating to hoof-and-mouth disease have caused Argentina's production and trade to remain static, while production in the Central American countries has doubled in the past 20 years, and their exports have increased fivefold, in large part because of their access to the U.S. market.

Throughout the Near East and in much of Africa pastures are overgrazed, and in many countries there is little prospect of bringing additional land into pastures without substantial investment. Livestock husbandry in Africa and much of Asia is rarely geared to efficient meat production methods, and animal nutrition and health are very poor. Marketing arrangements are so rudimentary that in all probability almost no African or Asian countries will be able to develop or expand their beef export trade. Some will probably even become beef importers as domestic prices will no doubt be set at levels that will discourage industry expansion, and social biases about communal grazing and cattle as a store of wealth seem impossible to overcome.

Worldwide beef production could be doubled with little difficulty, from a technical point of view, but the cruel conclusion is that only a gradual increase in beef production can be expected before the end of the twentieth century. In the meantime, beef production in economically developed countries will continue to expand at a rapid rate through adoption of technological innovations and attractive public policies. We now turn our attention to a more detailed discussion of animal health in an effort to better understand technical factors that influence economic considerations in beef production and marketing.

REFERENCES

Auriol, P. 1974. Intensive Feeding Systems for Beef Production in Developing Countries. *World Anim. Rev.* 9:18–23.

Baker, P. R. 1976. The Social Importance of Cattle in Africa and the Influence of Social Attitudes on Beef Production. In *Beef Cattle Production in Developing Countries,* ed. A. J. Smith, pp. 360–74. Edinburgh: Centre for Tropical Veterinary Medicine, University of Edinburgh.

Bohannan, Paul. 1968. Some Principles of Exchange and Investment among the Tiv. In *Economic Anthropology,* ed. E. E. LeClair and H. K. Schneider, pp. 300–11. New York: Holt, Rinehart, and Winston.

Dalton, G. 1967. The Development of Subsistence and Peasant Economies in Africa. In *Tribal and Peasant Economies,* ed. G. Dalton, pp. 155–68. Garden City, N.Y.: The Natural History Press.

de Wilde, J. S. 1967. *Experiences with Agricultural Development in Tropical Africa.* Vols. 1, 2. Baltimore, Md.: Johns Hopkins University Press.

Doran, M. H.; Low, A. R. C.; and Kemp, R. L. 1979. Cattle as a Store of Wealth in Swaziland: Implications for Livestock Development and Overgrazing in Eastern and Southern Africa. *Am. J. Agric. Econ.* 61:41–47.

Dunmore, C. 1978. Kenyan Pastoralists Enter 20th Century. *For. Agric.* 16:10–11.

Dyson-Hudson, R., and Dyson-Hudson, N. 1969. Subsistence Herding in Uganda. *Sci. Am.* 22:76–89.

Farris, D. E., and Sullivan, G. 1976. *Tanzanian Livestock-Meat Subsector.* 4 vols. Washington, D.C.: Texas A & M University and USAID.

Forlag, A. 1974. *Change and Development in East African Cattle Husbandry.* Copenhagen: Akademisk Forlag.

Kilma, J. 1970. *The Barabaig: East African Cattle Herders.* New York: Holt, Rinehart, and Winston.

Jollans, J. L. 1979. Milk Production in the European Community—Its Implications for Breeding Policy. *World Anim. Rev.* 30:11–17.

Lele, U. 1975. *The Design of Rural Development: Lessons from Africa.* Baltimore, Md.: Johns Hopkins University Press.

Meyn, K. 1970. *Beef Production in East Africa.* Munich: Institute for Wirtschaftsforschung.

Organization for Economic Cooperation and Development. 1978. *Structure, Performance and Prospects of the Beef Chain.* Paris.

Payne, W. J. A. 1970. *Cattle Production in the Tropics, Vol. 1: Breeds and Breeding.* Bristol: Longman Printing Services Ltd.

Poitevin, J.; Mallard, M.; and Picon, E. 1977. *Beef Production in Southern Europe.* Paris: OECD.

Saunderson, M. H. 1977. Some Observations on East Africa Livestock Ranching Development. *The Florida Cattleman,* pp. 48–50, 52, 107.

Schneider, H. K. 1957. The Subsistence Role of Cattle among the Pakot and in East Africa. *Am. Anthropol.* 58:278–300.

———. 1968. Economics in East African Aboriginal Societies. In *Economic Anthropology,* ed. E. E. LeClair, and H. K. Schneider, pp. 426–44. New York: Holt, Rinehart, and Winston.

Simpson, James R. 1970. Uses of Cultural Anthropology in Economic Analysis: A Papago Indian Case. *Hum. Organ.* 29:162–68.

———. 1980. Developments in Asian and Australasian Beef and Buffalo Production, Consumption and Trade: Implications for Asian and Oceanic Animal Scientists. *Proceedings: First Asian-Australasian Animal Science Congress.* Kuala Lumpur, September 1–6.

Spicer, E. H. 1952. Sheepmen and Technicians. In *Human Problems in Technological Change,* ed. E. H. Spicer, pp. 185–207. New York: Russell Sage Foundation.

Sullivan, G. M.; Farris, D. E.; Yetley, M. J.; and Njukia, W. J. 1977. A Socio-Economic Analysis of Technology Adoption in an African Livestock Industry. Texas Agric. Exp. Sta. Tech. Artic. no. TA 13513.

Sullivan, G. M.; Stokes, K. W.; Farris, D. E.; Nelson, T. C.; and Cartwright, T. C. 1980. Transforming a Traditional Forage/Livestock System to Improve Human Nutrition in Tropical Africa. *J. Range Mgt.* 33:174–78.

United Nations, Food and Agriculture Organization. 1967. *East African Livestock Survey-Regional—Kenya, Tanzania, Uganda,* 3 vols. Rome.

United Nations, Food and Agriculture Organization and the World Bank. 1977. *The Outlook for Meat Production and Trade in the Near East and East Africa,* 2 vols. Washington, D.C., and Rome.

United Nations, Food and Agriculture Organization. 1962 and 1979. *Production Yearbook.* Rome.

6
ANIMAL HEALTH

Animal health is of interest to a wide range of individuals and interest groups for a variety of reasons. Governmental bodies protect both producers and consumers through quarantine, inspection, and treatment programs that are designed to limit the spread of diseases and parasites. Consumers depend heavily on suppliers and government for assurance that safe meat is sold. Producers likewise rely on government to carry out measures to minimize the spread of disease, and to prevent infection. For example, to protect domestic livestock from infection, countries free of foot-and-mouth disease have an embargo against importation of live animals (except in certain instances) from countries where the disease exists. Some countries, such as the United States and Canada, do not even allow uncooked meat to be imported from infected areas. Most European countries do permit fresh, chilled, or frozen meat from nations known to have foot-and-mouth disease, but it must be deboned.

The United Kingdom provides an interesting case in which tighter regulations had to be imposed. Prior to 1968 it permitted carcass beef and lamb to be imported from several South American countries in which foot-and-mouth disease was endemic. An outbreak in England forced producers to slaughter a large number of valuable breeding cattle, and subsequently import regulations were changed to allow only deboned meat from infected areas in an effort to reduce the risk of reinfection. If the United Kingdom did not depend so heavily on beef imports, it might have placed an import embargo on all uncooked meat from countries reporting incidences of the disease.

To assure protection of consumers and domestic producers, most importing countries permit beef to be supplied only from certain countries and, in many cases, only from approved slaughter plants within those countries. Because no slaughter plants in Africa have been approved by the U.S. Department of Agriculture, the United States does not permit (as of 1979) beef or meat of any kind to be imported from that continent. The EEC rules are less strict in that certain approved plants in Africa can export canned meat to EEC countries, but these plants must regularly meet certain standards to maintain their export permits. Health

Disease prevention is basic to economical beef production. There are better ranch facilities than these for working cattle, but the all-important consideration is to get to the cattle and handle them at the proper time.

and sanitation regulations are among the primary obstacles to expanded international trade. Likewide, the prevalence of diseases and their control must be considered in any strategy aimed at reducing beef prices for consumers while simultaneously attempting to improve producer incomes. Well-designed and conscientiously operated disease control and prevention programs are obviously essential to a reasonably efficient beef business, but the high prevalence of numerous diseases around the world reduces the chances for easy solutions and administration (Griffiths 1976). The following section examines data on the prevalence of twenty major diseases affecting cattle. This topic is discussed in an effort to develop an appreciation for the magnitude of the problem. Control of diseases in livestock and limiting the geographic spread of animal diseases are key factors in expanding trade and improving living standards.

PREVALENCE OF DISEASES

The prevalence of twenty principal bovine diseases in forty-seven temperate countries is given in Appendix Table A.15, and the prevalence in 113 warm-climate countries is presented in Table A.16. A single "x" signifies low prevalence, two "x"s mean moderate prevalence, and three "x"s indicate a high level of the disease. Blank spaces indicate the disease has either not been recorded or has not been confirmed. There are, of course, many more diseases than those described here, but these were selected because they provide a representative cross section of health problems among the cattle of the world.

Twenty major bovine diseases in warm and temperate countries found throughout the world in 1978 are summarized in Table 6.1. In that year there were, for example, 62 warm-climate countries reporting hoof-and-mouth disease compared with 17 temperate-climate countries. This is 78 and 22 percent of the world incidence. The greatest percentage of incidence of diseases is in the warm-climate countries, but this is because there are about twice as many warm-climate countries as there are temperate ones.

The problem of differences in country numbers has been partially overcome in the lower part of Table 6.1 where the incidences are reported as a percentage of the possible incidences in each region. Foot-and-mouth disease (called hoof-and-mouth disease in some countries) is reported, for example, in 55 percent of the warm-climate countries and 36 percent of the temperate countries. About half of the countries in the world report the disease.

There are some interesting contrasts brought out in Table 6.1. Bovine tuberculosis, for example, is reported in about 80 percent of both the warm- and temperate-climate countries. Heartwater, on the other hand, is only found in the warm-climate countries. Liver flukes, a parasitic disease often thought of as a warm-climate disease, is reported in 91 percent of the temperate countries compared with 79 percent of the warm-climate areas.

DESCRIPTION OF DISEASES

FOOT-AND-MOUTH DISEASE (APHTHOUS FEVER, AFTOSA, EPIZOOTIC APHTHAE). Foot-and-mouth disease (FMD) is an acute, highly communicable viral disease that is confined to cloven-footed animals (Siegmund 1977). Cattle and swine seem to be most susceptible, and sheep and goats are only slightly less sensitive. FMD is endemic (that is, constantly present in a locality) in certain parts of Europe, Asia, Africa, and South America. Spread is by contact with infected animals or with contaminated fomites. The virus is found in,

Table 6.1. Number and percentage of countries with low, moderate, and high levels of 20 bovine diseases, 1978

Region and number of countries	Foot-and-mouth disease	Rinderpest	Bovine rhinotracheitis	Contagious bovine pleuropneumonia	Rabies	Heartwater	Leptospirosis	Anthrax	Blackleg	Intestinal *Salmonella* infection	Bovine tuberculosis	Johne's disease	Actinomycosis (lumpy jaw)	Brucellosis (Bang's disease)	Anaplasmosis	Trypanosomiasis (insect borne)	Mange and scabies	Warble infestation	Distomatosis (liver fluke)	Echinococcosis-Hydatidosis
	1	2	3	4	5	6	7	8	9	10	11	12	13	14	15	16	17	18	19	20
	(Number of countries reporting the disease)																			
Warm climate (113)																				
Low	33	13	16	20	47	20	37	52	37	50	64	32	61	53	43	26	64	26	26	47
Moderate	21	3	1	6	9	6	6	16	27	14	24	2	3	26	22	13	20	4	29	17
High	8	1	1	2	2	4	4	5	12	1	2	0	0	12	14	19	4	1	34	6
Total	62	17	18	28	58	30	47	73	76	65	90	34	64	91	79	58	88	31	89	70
Temperate (47)																				
Low	12	1	19	1	25	0	24	37	24	30	32	28	38	25	12	1	24	28	22	25
Moderate	4	0	6	1	1	0	3	0	8	5	5	0	1	6	2	0	5	7	15	8
High	1	0	0	0	0	0	0	0	0	0	1	0	0	2	0	0	1	1	6	3
Total	17	1	25	2	26	0	27	37	32	35	38	28	39	33	14	1	30	36	43	36
World (160)																				
Low	45	14	35	21	72	20	61	89	61	80	96	60	99	78	55	27	88	54	48	72
Moderate	25	3	7	7	10	6	9	16	35	19	29	2	4	32	24	13	25	11	44	25
High	9	1	1	2	2	4	4	5	12	1	3	0	0	14	14	19	5	2	40	9
Total	79	18	43	30	84	30	74	110	108	100	128	62	103	124	93	59	118	67	132	106

Source: Compiled from FAO *Animal Health Yearbook*, 1978 (see Appendix Tables A.15 and A.16 for individual countries).

Table 6.1. (continued)

Region and number of countries	Foot-and-mouth disease 1	Rinderpest 2	Bovine rhinotracheitis 3	Contagious bovine pleuropneumonia 4	Rabies 5	Heartwater 6	Leptospirosis 7	Anthrax 8	Blackleg 9	Intestinal *Salmonella* infection 10	Bovine tuberculosis 11	Johne's disease 12	Actinomycosis (lumpy jaw) 13	Brucellosis (Bang's disease) 14	Anaplasmosis 15	Trypanosomiasis (insect borne) 16	Mange and scabies 17	Warble infestation 18	Distomatosis (liver fluke) 19	Echinococcosis-Hydatidosis 20
	(% of total incidences reported in the world)[a]																			
Warm climate	78	94	42	93	69	100	64	66	70	65	70	55	62	73	85	98	75	46	67	66
Temperate	22	6	58	7	31	0	36	34	30	35	30	45	38	27	15	2	25	54	33	34
World	100	100	100	100	100	100	100	100	100	100	100	100	100	100	100	100	100	100	100	100
	(% of possible incidences in each region and the world)[b]																			
Warm climate	55	15	16	24	51	27	42	65	67	58	80	30	57	81	70	51	78	27	79	62
Temperate	36	2	53	4	55	0	57	79	68	74	81	60	83	70	30	2	64	77	91	77
World	49	11	27	19	53	19	46	69	68	63	80	39	64	78	58	37	74	42	83	66

[a]For example, there were 62 warm-climate countries out of 79 countries in the world reporting the disease. The warm-climate countries thus account for 78 percent of the reported worldwide incidences.

[b]For example, 62 warm-climate countries report foot-and-mouth disease out of a possible 113 countries; this is 55 percent. Similarly, 79 out of a possible 160 countries in the world report the disease; this is 49 percent.

and on, blisters on the feet and in the mouth. On breaking, the blisters form erosions. Inanimate objects such as feed, harness, vehicles, clothing, and other articles contaminated with the virus are instrumental in spreading the disease.

There are seven distinct types of FMD: A, O, C, South African Type (S.A.T.) No. 1, S.A.T. No. 2, S.A.T. No. 3, and Asian Type 1. The mortality rate from FMD is low (5 pecent), but great economic loss results because animals deteriorate severely. Young animals in excellent health and high nutritional state appear to be most susceptible. Although control depends on the conditions in each country, a practical approach is through slaughter if outbreaks are relatively small, and through vaccination if the disease is endemic. Control will be discussed later in this chapter.

North America, Mexico, and Central America are free, while nearly all of the South American countries are infected. Many European countries have at least a low prevalence. South America and Asia are regions of hyperendemicity (i.e., high levels of incidence), while Africa is an area of hypoendemicity (low levels of prevalence).

RINDERPEST (CATTLE PLAGUE). Rinderpest is an acute, febrile (of or relating to fever), highly contagious viral disease found primarily in cattle and buffaloes. Characteristics include inflammation, hemorrhage, erosion and necrosis of the digestive tract and mucous membranes, with diarrhea. The morbidity rate (i.e., number displaying clinical levels of the disease) approaches 100 percent and the mortality rate may exceed 90 percent in the most susceptible animals. Transmission is by means of direct contact. Affected animals experience sharp rises in temperature followed by oral lesions and diarrhea. Dehydration then sets in and is followed by death within 8 to 12 days. Small areas of infection are eliminated by strict quarantine and slaughter. Larger geographic locales are controlled through vaccination.

Rinderpest is not present in the Western Hemisphere or Europe, but is confined mainly to parts of Africa and Asia. Only one temperate-climate country, Lebanon, reported the disease in 1978, but other temperate countries have reported the disease in the past. As of 1978, only eighteen countries reported having the disease. Only four countries reported moderate or high levels.

INFECTIOUS BOVINE RHINOTRACHEITIS (IBR, RED NOSE). IBR is an acute febrile disease characterized by inflammation of the upper respiratory tract that leads to coughing and a profuse nasal discharge. Pneumonic symptoms develop as the viral disease advances. In the United States it is most often observed in feedlot cattle. Morbidity ranges from 15 to 100 percent. An initial temperature rise is followed by coughing and accelerated respiration. Pregnant animals often abort. A modified live virus vaccine is used for control.

IBR is reported in forty-three countries, of which 58 percent are in temperate areas. There are no high levels reported, but moderate levels are found in Canada, France, Japan, the Netherlands, United Kingdom, United States, and Mexico.

CONTAGIOUS BOVINE PLEUROPNEUMONIA. This highly contagious disease, caused by a microorganism called *Mycoplasma mycoides,* results from inhalation of small droplets of moisture breathed out by infected cattle. An infection rate of 100 percent may occur in susceptible herds, and mortality may be correspondingly high. The symptoms are typical of pneumonia and pleurisy. The disease was eradicated from the United States by slaughter of all infected herds but immunization with live vaccines is practiced in areas where the disease is endemic (Lindley 1979).

Contagious bovine pleuropneumonia is reported in 28 warm-climate and two temperate-climate countries (Cyprus and Mongolia). No countries of the Western Hemisphere report the disease; most of those infected are in Africa.

RABIES. Rabies, an acute encephalomyelitis caused by a virus, is almost always fatal. The virus travels up nerves to the spinal cord and then to the brain. The resulting damage to the brain cells causes animals to attack other animals and humans. Economic effects of this disease are most serious in South America, where it is transmitted to cattle through bites of vampire bats.

The disease is widespread: out of 160 countries researched by FAO, 84 report rabies. It is found on all continents and in all regions except Oceania. The United States is the only temperate-climate country with a moderate level; the twenty-five other temperate countries reporting the disease have low prevalences.

HEARTWATER. This disease is caused by a rickettsial microorganism called *Cowdria ruminantium,* which is transmitted by the "bont" tick. Infected animals suddenly develop a fever, collapse, and die in convulsions, usually discharging froth from the mouth as the virulent microorganism enters the bloodstream. Unless treatment is instituted early, death generally supervenes whether the case is acute or chronic. Control is by vaccine.

Heartwater is confined to warm-climate countries, all reported cases being in Africa. Mozambique and South Africa report high levels.

LEPTOSPIROSIS (REDWATER OF CALVES, ASYMPTOMATIC ABORTION, LEPTOSPIRAL MASTITIS). Leptospirosis is caused by a variety of leptospiral serotypes (subdivisions within a species distinguishable on the basis of antigenic character). The usual route of infection is via intake of urine-contaminated food or water. Primarily calves are affected, with morbidity approaching 100 percent; mortality rates in affected herds range from 5 to 15 percent. Wet weather and waterlogged pastures are conducive to rapid spread of infection. The urine of cattle affected by leptospirosis is clear red in color, thus the name "redwater." Calves develop fever, prostration, and anemia. The disease may be overlooked in older cattle, although milk production will drop off and abortion may be common. Control is by elimination of carriers and vaccination.

Leptospirosis is reported in 42 percent of the warm-climate countries and 57 percent of the temperate ones. The vast majority of countries report low prevalence. It occurs primarily in the Latin American countries, but high prevalence is reported only in Bolivia and the Dominican Republic.

ANTHRAX (SPLENIC FEVER, CHARBAN, MILZBRAND). Anthrax is an acute, infectious, febrile disease caused by the bacterium *Bacillus anthracis.* It is essentially a septicemia having a rapidly fatal course. Infection is usually the result of grazing on infected pastureland, of ingesting other contaminated feedstuffs, or of drinking from befouled pools. In the acute form, which is most common in cattle, the first symptoms are a rise in body temperature and a period of excitement followed by depression, stupor, spasm, respiratory or cardiac distress, staggering, convulsions, and death in a day or two. In the subacute form death takes up to 5 days or longer, although some animals recover. Anthrax can be controlled largely by annual vaccination of all animals in endemic areas.

Distribution is worldwide—110 of the 160 countries listed report the disease. About 79 percent of the temperate-climate countries and 65 percent of the warm-climate countries have the disease. The prevalence is low in all temperate-climate countries.

BLACKLEG (BLACK QUARTER). Blackleg is an acute, febrile disease of cattle and sheep caused by the soil-borne bacterium *Clostridium chauvoei.* If a pasture becomes heavily contaminated, the disease may appear regularly in susceptible animals year after year. It is believed that the organisms enter through the digestive tract, multiply in it, gain access to the bloodstream, and are deposited in the various tissues throughout the body. Affected animals, oddly enough, are generally in good or high condition. Cattle usually affected are between the ages of 6 and 12 months. The first symptom is acute lameness followed by depression and fever. Characteristic swelling develops in the muscles of the hip, shoulder, chest, back, neck, tongue, and pharynx. The disease is usually fatal 12 to 48 hours after onset. Routine vaccination of calves is the best control measure.

Blackleg, like anthrax, is worldwide and has a similar distribution between the warm and temperate countries, except that there is greater moderate and high prevalence in warm countries. In fact, it is a serious disease in the twelve warm climate countries, which report high levels of infection.

ENTERIC SALMONELLOSIS. This disease is caused by bacteria of the genus *Salmonella*. There are various types, all of which can cause weakness, recumbency, increased temperature, and diarrhea. Pregnant animals may abort. Control is best achieved by raising animals under sanitary conditions and in surroundings where they are not subject to environmental stress.

Intestinal *Salmonella* infection is worldwide in distribution. The only country reporting a high prevalence is South Africa.

BOVINE TUBERCULOSIS. Bovine tuberculosis is a chronic disease caused by infection with acid-fast organisms belonging to the genus *Mycobacterium*. Primary lesions usually take place in the lungs, but regardless of where the organisms localize, their activity stimulates the formation of tumorlike masses that are called tubercles. Invasion of the bloodstream by tubercle bacilli is usually fatal. The disease is more prevalent in cattle kept for milking purposes because they are raised in greater confinement. Tuberculosis is spread by the expiration of infected droplets from the lungs of infected animals during coughing. The droplets contaminate feed and water troughs. Pastures are infected via the feces.

Tuberculosis is controlled by: (1) test and slaughter, (2) test and segregation, and (3) immunization. The first method is used in the United States and Canada, and the second in the Scandinavian countries. Immunizations have been little used.

About four-fifths of the countries in the world report tuberculosis, but high levels are reported only in Chile, Guatemala, and Uganda. About 81 percent of the temperate countries and 80 percent of the warm countries report low prevalences.

JOHNE'S DISEASE (PARATUBERCULOSIS). This is a chronic infectious disease of cattle, and occasionally sheep and goats, characterized by thickening of the intestinal wall and a recurrent fetid diarrhea that may persist for months and cause a gradual loss of condition. With few exceptions, animals showing clinical symptoms eventually die. Most cases occur in females from 2 to 4 years of age. Johne's disease is almost always brought into a clean herd by the introduction of an infected animal, but manure is the primary source of infection after the herd is contaminated. There is no satisfactory treatment. Control is by test and slaughter.

High or moderate levels of Johne's disease are reported in only two countries, Uganda and India, both of which are in the warm areas. Almost as many temperate countries (28) as warm countries (34) report low levels.

ACTINOMYCOSIS (LUMPY JAW). Lumpy jaw is a chronic disease of the bony tissue of the head. The lesions last for many months. Various treatments are possible.

High or moderate levels are reported in only three warm countries. Nearly 80 percent of the temperate-climate countries report low levels, while only about 50 percent of the warm-climate countries report low levels of prevalence. It is distributed throughout the world.

BRUCELLOSIS (BANG'S DISEASE). Abortion is the most obvious manifestation of the disease although milk production of infected cows also declines. In uncomplicated abortions the cow's general health is seldom impaired. Infected cows do not look sick and abortion may be the first sign. Brucellosis is one of the most serious diseases of cattle in terms of economic loss and is also a major public health hazard as humans can develop undulant fever by ingestion of organisms in contaminated carriers such as milk.

Brucellosis is spread mainly by unrestricted contact between infected and susceptible animals. The appearance of the infection in a herd that has been free of the disease is characterized by rapid spread and many abortions. The rate of abortion is low in herds where the disease is endemic. Natural transmission is mainly through ingestion of the organisms, which are present in large numbers in the aborted fetus, membranes, and uterine discharge. The organisms can live in a cool damp environment for more than two months, but exposure to direct sunlight kills them in a matter of a few hours.

Eventual eradication of brucellosis depends on testing and elimination of reactors. Clean herds must be protected from reinfection and the greatest danger is via replacement animals. Both cows and calves can be vaccinated. A combination of vaccination and elimination of reactors by slaughter is the most widely used method of control.

Brucellosis is distributed throughout the world and occurs in 78 percent of the 159 countries surveyed. Only two temperate countries, Chile and Mongolia, report high prevalences, but high levels are found in twelve warm-climate countries. About 70 percent of the temperate-climate countries report it, versus 81 percent of the warm-climate countries.

ANAPLASMOSIS (GALLSICKNESS). Anaplasmosis is an acute or chronic infectious disease of cattle characterized by the presence in the red blood cells of bacterial organisms known as *Anaplasma marginale.* The most common mode of transmission is by vectors such as ticks and biting flies. The disease is most severe in adult cattle. Infected cattle display marked anemia with a variable mortality rate. The disease can vary from acute to chronic. Control has been difficult as animals that recover can become carriers of infection (Bram and Gray 1979). As a result, the usual recommendation is that they be sold for slaughter as soon as possible. Vaccines are available and animals respond to treatment.

The disease is widespread and endemic in warmer parts of the world. There are no high levels in temperate areas. Only two temperate countries, Israel and Syria, report moderate levels. Other than Bulgaria and Greece, Europe is free of it.

TRYPANOSOMIASIS. A protozoon of the genus *Trypanosoma* is responsible for an entire group of diseases (Murray et al. 1979). The pathogenic trypanosomes are transmitted by tsetse flies (*Glossina* spp.) and, wherever these flies occur, there is also trypanosomiasis. The tsetse fly transmits sleeping sickness in humans and various forms of trypanosomiasis in cattle. Trypanosomiasis, or *nagana,* is the most serious disease in Africa because it prevents the keeping of domestic animals in much of Africa between 15° and 20° S. Tsetse fly areas follow geographic bounds, although the bounds are not solid. Eight species of tsetse fly are found in East Africa alone. They seriously limit cattle production because their habitat is open woodland and bushy grassland.

There are two major approaches to eradication of trypanosomiasis: (1) eradication of the tsetse fly, and (2) the development of curative and prophylactic drugs that will enable cattle to survive and produce in infected areas (Murray 1979). A third approach, but one showing little promise, is to breed resistance. Only a few breeds of cattle, notably the N'Dama and Muturu of West Africa, have managed to acquire naturally a degree of immunity to the disease. Considering that the tsetse fly also transmits sleeping sickness, the ultimate answer will be complete eradication of the fly.

Various methods have been used for tsetse fly control. Large areas of Uganda have been cleared by the elimination of game, which are carriers in the wild. Wholesale clearing in countries such as Uganda has met with strong criticism from wildlife enthusiasts and is probably not a viable alternative because of negative repercussions on the general economy from loss of tourism. A number of prophylactic and curative drugs have been developed, but their effectiveness is limited by the need for periodic revaccination of cattle, which is prohibitive except in intensive management situations. Thus, despite the impressive figures that may be compiled about the benefits of controlling the disease, it will be many years before substantial control is brought about (Hendry 1979).

Trypanosomiasis is reported in numerous countries outside Africa, but the diseases elsewhere are different from those found in Africa.

MANGE OR SCABIES. Scabies is an itch or mange with exudative crusts. It is a contagious skin disease caused by sarcoptic or demodectic mites. Natural infection with mange mites may take place either by direct contact with diseased animals or by means of various objects that have been in contact with diseased animals. About 4 to 6 weeks elapse from contact to appearance of the first visible skin lesions.

There are three types of mange, or scabies, in cattle. Sarcoptic mange first appears on the head and neck and then spreads to other parts of the body. The lesions cause the skin to thicken and form large folds. The second type, demodectic mange, is rare. The third type, chorioptic mange, is the most common. In this type, lesions develop chiefly in the tail region, spreading to other parts of the body. Control is effected by careful examination of cattle and attendant shipping precautions. Treatment is by dips and sprays.

Scabies is found in 64 percent of the temperate countries and 78 percent of the warm ones. Argentina, Ghana, Mali, Zaire, and Indonesia are the only countries reporting a high prevalence of the disease.

WARBLE INFESTATION (CATTLE GRUBS). Two species of botflies attack cattle. The smaller species (*Hypoderma lineatum*) develops into the common or early cattle grub, while the other (*H. bovis*) emerges as the late or northern cattle grub. In both cases the adult flies fasten their whitish eggs to hairs of the legs and lower portions of the animal's body. The eggs hatch in 2 to 6 days, at which time the young maggots crawl down the hair to the skin and burrow directly into the tissues. The larvae travel for about 5 months in the animal until they finally form a breathing hole in the back. Cysts are formed around the holes and after about 40 to 60 days, the larvae squeeze out, drop to the ground, and pupate. By this time they are about 25 mm long and 8 mm wide. The number of warbles or cattle grubs in an infested animal ranges from 1 to 300 or more. Calves are the most heavily infested. Numerous external or internal treatments are used.

Warble infestation is most common in the Northern Hemisphere, but is distributed all over the world. Incidences are reported in 77 percent of the temperate countries versus 27 percent of the warm countries. Only two nations, Mongolia and Ecuador, report high levels.

DISTOMATOSIS (LIVER FLUKE). The common liver fluke, which is about 25 mm long, infects domestic and wild animals, including cattle, sheep, goats, rabbits, deer, and, rarely, pigs, horses, and humans. Eggs eliminated in animals' feces hatch in water, multiply, and develop in snails, then emerge and encyst promptly on aquatic vegetation. Animals become infected by eating that vegetation. The larvae stage of the fluke is freed from the cyst in the digestive tract, then penetrates the intestine, enters the liver, and proceeds to the bile duct to complete its development. Damage depends on the duration and degree of infection. Death may occur in calves with acute infection. Animals with chronic infections may display no symptoms apart from anemia, general weakness, and emaciation.

Control of liver flukes is difficult, as it may not be economically possible to destroy the snail habitats. In some cases animals can be treated with medication to destroy the adult flukes so that fewer eggs will be passed onto the pastures. There are numerous treatments.

Liver flukes are found worldwide; 83 percent of the 160 countries surveyed report at least low prevalence. About 91 percent of the temperate countries report it, versus 79 percent of the warm countries. Over 40 percent of the warm-climate countries report high levels of flukes.

ECHINOCOCCOSIS-HYDATIDOSIS. Hydatid disease is caused by a small tapeworm, *Echinococcus granulosus,* whose definitive host is the dog or other canine species such as wolves or dingoes. The worm attaches itself to the host's gut wall and the most distal segment—which is filled with eggs—is shed in the dog's excreta; the distal segments are then renewed within the host. A gravid segment may contain as many as 800 to 1,000 eggs, which in time dry up and are scattered about the pasture.

The eggs, which remain viable for at least 3 months after passing from the dog, may be swallowed by animals or humans. The shell is destroyed by digestive juices and the freed embryo burrows through the intestinal wall, then enters the bloodstream, and finally the liver or perhaps other organs such as the lungs. Embryos that survive form a fluid-filled cyst that develops tapeworm heads. The cycle is completed when dogs eat cysts contained in the organs of an infected animal.

Control is most effective through treatment of dogs to rid them of tapeworms. Reinfection of dogs is controlled by preventing them from eating infected organs of stock animals. Unlike many other diseases, this one could be eliminated at low cost if dog and stock owners understood the problem and cooperated.

Echinococcosis-hydatidosis is reported at high levels in sheep rearing countries such as Argentina, Chile, and Uruguay. Even though FAO does not report this disease in the cattle section of the *Animal Health Yearbook* (1978), it is endemic in sheep raising areas of the United States such as Arizona, California, New Mexico, and Utah. It is found in some Central American and Caribbean countries and is common in Asia, Europe, and Africa.

HEALTH CONTROL IN DIFFERENT TYPES OF SYSTEMS

Health patterns vary greatly among herds, regions, countries, and continents according to intensity of the production system. In nomadic herds, food and water are the chief limiting factors as their scarcity reduces the resistance of animals to disease. Long treks and intermingling provide excellent opportunities for spread of infectious diseases such as rinderpest or contagious bovine pleuropneumonia. Excessive concentration of stock is an ideal condition for worm infestation and foot-and-mouth disease.

Nomadic herd structure and traditional management practices require health programs to be administered largely by the government, since individuals are powerless to improve the health of their own animals without the consensus of the whole group, which is usually difficult to obtain. As cattle owners become settled, disease patterns change. In Central Africa the incidence of trypanosomiasis would probably decrease, for example, if production became more intensive. On the other hand, diseases associated with more intensive systems—for example, brucellosis—would probably increase. In any case, as communal grazing is replaced by individuals locating their herds on enclosed pastures, exposure to disease will be reduced considerably and more effective control will become possible. Preventive medicine can be instituted and the term "management" can take on real significance. Under settled conditions, it also becomes possible to isolate herds during epidemics of infectious disease.

To increase beef production cattle systems have to become more intensive; that is, efforts have to be made to increase carrying capacities and to replace or upgrade native breeds with improved stock (Ellis and Hugh-Jones 1976). But some introduced breeds suffer greatly from many health disturbances (Callow 1978). New diseases such as salmonellosis or leptospirosis may develop, and the likelihood of worm parasitism may increase with more intensive use of grazing facilities. Preventive measures thus become all-important in meeting health expectations. Owners of cattle must assume greater responsibility as beef production intensifies. The outlook shifts from one that regards cattle as chattel or a symbol of prestige to one that views cattle production as a business. The government's role in disease control can shift from one of physical involvement in health control to one of regulation and provision of information. Most important, government can begin to plan for disease control rather than reacting to crisis situations.

ECONOMICS OF DISEASE CONTROL

Veterinarians have carried out some economic analyses of animal diseases, but their efforts up through the 1960s only touched on minimal parts of the problem (Rosenberg 1976). In part, they concentrated on schemes for elimination of the disease and showed little or no concern for costs or the possibilities of control at a subelimination level (Johnston and Mason 1973). In addition, they used economic data to support predetermined conclusions rather than to determine optimum policies (Power and Harris 1973).

The 1970s have witnessed adoption of benefit-cost techniques by veterinarians as well as economists, and a few major studies have been carried out (Hugh-Jones et al. 1975). Despite an understanding of the techniques, until the late 1970s analytical studies were restricted to measuring gross benefits of control or eradication, comparing ex post facto two alternatives, or estimating net benefits of one alternative under restrictive assumptions (Morris 1975). In the remainder of this section the elements for economic analysis will be set forth in some detail as the same tools have wide applicability in livestock economics. The concepts will be especially useful in the last chapter, which concentrates on factors involved in improving the world's beef business.

Benefit-cost analysis is a technique that was originally devised to evaluate large projects such as dams or irrigation canals. The projected stream of benefits, along with the stream of costs for the project's life, are placed on a present benefit and cost basis by using an appropriate rate of discount, a percentage that is essentially the opposite of interest rates. The benefits and costs must be discounted to the present in order to place them on a comparable basis as they are almost always noncoincident. An exception to the discounting procedure is given as a case study in this section. A more detailed explanation of discounting appears in Chapter 7.

The net benefits are calculated by subtracting the benefits without the project from the benefits with the project. The same procedure is done on the cost side. In other words, the objective is to determine additional benefits and costs of a project. The sum of discounted benefits is divided by the sum of discounted costs to derive a ratio. If it is larger than one, the project is considered viable. If more than one project is being considered, the project having the highest ratio is the most acceptable one from an economic standpoint. Other criteria of choice, such as comparisons of net present costs of expenditures and the internal rate of return, are also used.

Benefit-cost analyses are carried out from the points of view of society (often called the economic analysis) and those affected by the project (the private or financial analysis). Both analyses are necessary to assure that society as a whole benefits, that the project will not place undue financial hardship on the people directly affected by it, and that the affected group will cooperate. These aspects are especially important to consider in projects having a national perspective such as programs for the control or treatment of brucellosis or foot-and-mouth disease.

Analysts have only recently recognized the need to compare various alternative measures for combating a disease or sanitation problem. The growing emphasis on cost effectiveness has been acccompanied by awareness of the need to rank potential projects. Should control or eradication of brucellosis have priority over swine fever, for example, or is there some other disease or health problem that deserves greater priority? One way of setting and evaluating these priorities is to rank the projects.

Foot-and-mouth disease serves as a good example of a health problem involving considerable worldwide costs. In fact, FMD control programs commanded more resources in the 1970s than any other animal disease (Rosenberg 1976). We now describe the disease in more detail and present the results of an economic study on controlling it as an example of how optimization can be employed in actual practice.

The incidence of FMD varies around the world. At present, North and Central America (from Panama north), Australia, New Zealand, Iceland, Japan, and the Irish Republic are free of the disease. FMD is endemic to central and southern Asia, central and southern Europe, Africa, and South America, where occasional widespread epidemics take place. Animals that have been exposed to the disease apparently become immune for a certain period of time. Flare-ups occur if there is an increase in the number of young, susceptible animals or if a different type or subtype of the virus is introduced.

FMD programs have a number of options: (1) virtually no control, a practice followed in some Asian and African countries; (2) limited vaccination for protection of more valuable or accessible animals along a country's border; (3) large-scale vaccination schemes and quarantine measures; and (4) prevention of the disease or eradication of the disease if it does occur. Rubinstein evaluated the various possibilities through the use of a computer simulation model in her pathbreaking work, *The Economics of Foot-and-Mouth Disease Control and Its Associated Externalities*. Her results for Colombia constitute our example.

The Darien region of southern Panama, which includes the border area of Colombia and Panama, is the last remaining barrier between the FMD-endemic countries of South America and the disease-free areas of Central and North America. Political pressure to complete the 156 km road through the Darien Gap has made it imperative that some method be devised to effectively control FMD in the departamento (state) of Cordoba in Colombia just south of the gap to prevent the disease from spreading north. Rubinstein's study encompasses that region.

The approach used by Rubinstein was first to calculate net benefits from a program limited to vaccination. By simulating herd development and cash flows (with and without FMD) for growing and/or fattening operations, it was discovered that a producer would maximize net benefits by a vaccination level of about 70 percent of the region's cattle. It was determined by observation that for regional vaccination levels up to 60 percent, the profit-maximizing level of vaccination for each farm is 100 percent. On the other hand, if regional vaccination levels are 80 percent or more, it would be most rational for livestock owners to eliminate their vaccination programs as they would be protected by their neighbors. An important point here is that, whereas earlier studies concentrated on deriving benefit-cost ratios for different programs, such as disease elimination or all vaccination, Rubinstein determined the optimum level of vaccination for a particular program.

The social analysis was carried out by considering the region under study as a "consolidated" farm that internalizes (takes advantage of) the external benefits arising from FMD control. These benefits consist of reduction in the outbreak probability on the other cattle farms and ranches in the region when one farm vaccinates its herd against the disease. Calculations of the net annual benefits indicate that the highest net annual social benefit is achieved at 90 percent vaccination. This result confirms a general hypothesis that stockmen tend, in general, to underinvest in FMD control. The results also show that a vaccination program should not necessarily aim at a 100 percent vaccination coverage.

The next part of the analysis carried out by Rubinstein was the social evaluation of eradication. The departure point for implementation of slaughter is arrived at when the steady-state situation is achieved with the social optimum (90 percent) vaccination strategy. The alternative is to continue the vaccination program indefinitely. One major consideration is the potential for access to disease-free markets as the price of beef has traditionally been higher in those countries than in ones accepting meat from FMD countries. Another consideration is the social discount rate, or in other words, the opportunity cost of capital at the national level. This rate roughly corresponds to the average return from public and private investments.

Results of the analysis indicate that at a discount rate of 10 percent, net benefits from a continual vaccination program are less than half those expected from the eradication program if there is access to world markets importing from FMD-free countries. The two-step approach is also the best solution up to a social discount rate of 20 percent. Thus, following achievement of a steady situation involving a 90 percent vaccination level, eradication is the optimum control strategy for the region if there is access to the FMD-free market. If the region cannot gain such access, the eradication option is less preferable than 90 percent vaccination for discount rates of 10 percent or more.

A difficulty surfaces that is typical of most disease control situations. The results clearly show that even though a 90 percent level of participation by stockmen is necessary for a social optimum, in all likelihood only 70 percent of the producers would voluntarily participate. Thus, some appropriate government policy is required. Two possibilities are: subsidies to farmers, with an additional subsidy for slaughter after the steady state is reached; and a compulsory vaccination program. Rubinstein points out that compulsory vaccination conducted by government-hired professionals (a form of subsidy) is now being used to combat FMD in South America. The degree of success that has been attained is directly related to the availability of funds.

Rubinstein's study is a major contribution for it provides a method for carrying out ex ante evaluations of disease control. The analysis also demonstrates that alternative ways of combating the disease should be assessed from an economic as well as a technical point of view. We may conclude that the benefits depend on numerous assumptions, including access to a high-priced meat market. Since there is only limited potential for such an activity, it appears that the benefits of disease control should be considered within the framework of the entire world rather than in just one country when one or more of the benefits can be affected by the actions of other countries.

SUMMARY AND CONCLUSIONS

Health and sanitation regulations are among the primary obstacles to expanded international trade. Likewise, the prevalence of diseases and their control is of fundamental importance in planning strategies aimed at reducing beef prices to consumers and simultaneously attempting to improve producer incomes. A list of twenty principal diseases affecting the world's cattle industry is presented in this chapter along with a discussion about them. The data indicate that for the twenty diseases examined, contrary to popular belief, the prevalence on a country basis is as high in temperate countries as in warm ones.

A review of health control in different production systems indicates that as production sophistication improves, government involvement can shift from one of physical involvement in health control to one of regulation and provision of information. The outline of a method, benefit-cost analysis, for evaluating disease control and eradication measures is given. A case study on foot-and-mouth disease in Colombia is provided as it serves as a good example of the means for planning a program for control of a major disease.

Up to this point we have discussed animal health and approaches to analysing various production systems, and systems found around the world. Let us now consider economic and financial aspects of cattle production.

REFERENCES

Bram, R. A., and Gray, J. H. 1979. Eradication—An Alternative to Tick and Tick-Borne Disease Control. *World Anim. Rev.* 30:30–35.

Callow, L. L. 1978. Ticks and Tick-Borne Diseases as a Barrier to the Introduction of Exotic Cattle to the Tropics. *World Anim. Rev.* 28:20–25.

Ellis, P. R., and Hugh-Jones, M. E. 1976. Disease as a Limiting Factor to Beef Production in Developing Countries. In *Beef Cattle Production in Developing Countries,* ed. A. J. Smith, pp. 105–17. Edinburgh: Centre for Tropical Veterinary Medicine, University of Edinburgh.

Griffiths, R. B. 1976. The International Control of Animal Diseases. In *Beef Cattle Production in Developing Countries,* ed. A. J. Smith, pp. 43–57. Edinburgh: Centre for Tropical Veterinary Medicine, University of Edinburgh.

Hendry, P. 1979. After the Tsetse. *CERES.* Pp. 13–18.

Herlich, H. 1978. The Importance of Helminth Infections in Ruminants. *World Anim. Rev.* 26:22–26.

Hugh-Jones, M. E.; Ellis, P. R.; and Felton, M. R. 1975. *An Assessment of the Eradication of Bovine Brucellosis in England and Wales.* Dep. of Agric. Study no. 19. Reading, England: University of Reading.

Johnston, J. H., and Mason, G. 1973. *A Cost-Benefit Study of Alternative Poli-*

cies in the Control or Eradication of the Cattle Tick in New South Wales. N.S.W. Dep. of Agric. Div. Mark. Econ. Mimeo.

Lindley, E. P. 1979. Control of Contagious Bovine Pleuropneumonia with Special Reference to the Central African Empire. *World Anim. Rev.* 30:18–22.

Morris, R. 1975. *The Integration of Economic and Epidemiological Methods in the Study of Animal Disease.* Melbourne, Australia: University of Melbourne, School of Veterinary Science.

Murray, M. 1979. A Review of the Prospects for Vaccination in African Trypanosomiasis: Part I. *World Anim. Rev.* 32:9–13.

Murray, M.; Morrison, W. I.; Murray, P. K.; Clifford, D. J.; and Trail, J. C. M. 1979. Trypanotolerance—A Review. *World Anim. Rev.* 31:2–12.

Power, A. P., and Harris, S. A. 1973. A Cost-Benefit Evaluation of Alternative Control Policies for Foot-and-Mouth Disease in Great Britain. *J. Agric. Econ.* 34:573–97.

Rosenberg, F. 1976. *El Conocimiento de la Epidemiologia de la Fiebre Aftosa con Particular Referencia a Sudamerica.* Rio de Janeiro, Brazil: Centro Pan Americano de Fiebre Aftosa.

Rubinstein, Eugenia M. de. 1977. The Economics of Foot-and-Mouth Disease Control and Its Associated Externalities. M.S. thesis, University of Wisconsin.

Siegmund, O. H. 1977. *The Merck Veterinary Manual.* 8th ed. Rahway, N.J.: Merck and Co.

United Nations, Food and Agriculture Organization. 1978. *Animal Health Yearbook.* Rome.

7 ECONOMIC AND FINANCIAL ASPECTS OF CATTLE PRODUCTION

Beef cattle economics can be approached from two viewpoints: that of the individual rancher or that of the industry as a whole. The direct interests of ranchers or potential investors (a micro-viewpoint) are seldom the same as those of individuals—for example, legislators or breed associations—working on the national (macro) level. The individual rancher is concerned mainly with daily operating problems such as determining the type of bull that will yield the highest return or how application of fertilizer affects profit rates. Organizations or individuals working at the macrolevel, on the other hand, have a much broader outlook. Lobbyists of the National Cattlemen's Association in the United States need information on overall rates of return from ranching or conduct studies on the type of research that will potentially provide the highest benefit-cost ratio to cattle producers. Legislators and breed associations are more interested in determining how their constituents can be more competitive with cattlemen in other nations, or in making 5-year projections of supply and demand than, say, in the economics of fence maintenance.

Recognizing that interest in economic questions varies, this chapter and the next one contain information about both the micro- and macrolevels of the cattle industry. We begin by explaining how ranch size can be measured using animal units. We then consider the methodology for developing ranch budgets, followed by cost and return information for three quite different ranch situations: Texas in the United States, Paraguay, and Tanzania. Budgets evaluating long-term profitability of ranching are then set forth, and a Central American beef cattle investment is evaluated. The chapter concludes with a brief explanation of investment evaluation by capital budgeting techniques.

MEASUREMENTS WITH ANIMAL UNITS

Before proceeding further, we should discuss the method for placing cattle on an animal unit basis, as this is especially useful in defining size of "large" or "small" ranches or livestock farms. Given the diversity of land types, livestock classes, climatic conditions, land values, input costs, and cattle prices, a size classification may be difficult to establish. Consider climate, for example. It is intuitively obvious that a ranch in the desert area of Peru would have to be much larger than a farm in the midwestern United States to yield the same total net income. The same line of reasoning holds true for using cattle numbers as a measurement of size. The difficulty is, of course, that herd composition changes according to type of production method. Livestock owners in many areas of the world fatten stockers to slaughter weight rather than selling them as calves, so their mix of cattle will differ from that of a cow-calf operation selling calves at weaning.

Recognizing the drawbacks of comparing enterprises having different types of livestock, researchers have developed a system of classifying operations according to the number of animal units. The total number of animal units is calculated by multiplying the numbers of each type of animal by the appropriate coefficients (Gray 1968, p. 122). Although the coefficients are determined by the amount of forage an animal eats, which in turn is related to the body surface, the following coefficients generally hold for Brahman or English breed cattle.

Type of animal	Animal unit (A.U.) coefficient
Mature cow	1.00
Long yearling	0.80
Weaned calf	0.60
Unweaned calf	0.40
Pregnant heifer	1.00
Cow with calf	1.40
Mature bull	1.25
Two-year-old steers	1.00
Horse	1.25
Sheep	0.20

Researchers in the United States have attempted to standardize ranch classifications by referring to operations with less than 150 animal units as "small," those with 150 to 400 A.U.s as "medium," and those with more than 400 A.U.s as "large." In some cases, ranches with above 800 A.U.s are called "extra large."

The system of animal units is also helpful in measuring carrying capacity of land on a comparable basis because time as well as type of livestock can be transcended. The total carrying capacities can be calculated by multiplying the annual number of days animals are on a pasture by the appropriate coefficients. If, for example, a livestockman had 400 long yearlings and 200 mature cows, and if these were carried all year, that person would have 400 × 0.8 = 320 A.U.s plus 200 × 1.0 = 200 A.U.s, or a total of 520 A.U.s. Carrying capacities can be determined by dividing the pasture area by the total animal units.

ECONOMIC BUDGETS FOR THREE CATTLE SYSTEMS

An excellent way for an individual to evaluate his or her particular ranch, or to make comparisons between enterprises, is to construct cost and returns budgets (Westberry 1979, 1979a). The data required are basically those kept by accountants plus some other information about the operation. A major difference between economic budgets and accounting records is that economic budgets are used for analytical purposes and decision making, while accounting budgets are kept primarily for control and for tax purposes, to balance income and expenses, and to derive various business statements such as profit and loss or net worth.

The format for economic budgets depends on the intended purpose. One method, set forth in Table 7.1, is to first present data about numbers of mature cows, amount of pastureland, calf crop, death loss, and so forth to provide an overall picture of the operation. Because units vary considerably in size, it is desirable to place the budgets and related information on a 100-cow basis if the purpose is to compare operations. This procedure has been carried out for nearly all the information in Table 7.2. In addition, the inventory information has been set forth on an animal-unit basis. The next step involves financial information such as investment, cash and noncash expenses, and income. From this data base net income and a variety of information for analytical purposes can be calculated. Budgets for three entirely different types of cattle operations are now presented.

RIO GRANDE PLAINS. The Rio Grande Plains area of southern Texas is a semiarid region covered with brush and scrub trees and some open savannas. Ranches there tend to be family operations that hire outside labor mainly for special projects. Some of the larger ranches have one or two permanent employees. The typical medium operation has about 600 A.U.s, small operations carry about 200 A.U.s, and large ranches stock approximately 2,400 units. Nearly all the cattle are some type of Zebu-English breed, for example, Brangus or

Table 7.1. Inventory and production measures for typical cattle operations in Texas, Paraguay, and Tanzania, 1971

Item	Units	Rio Grande Plains area of Texas (USA)[a]	South-eastern Paraguay[b]	Dodoma seminomadic tribesmen, Tanzania[c]
Actual conditions at time of study				
Total animal units	A.U.	618	631	22
Range or pastureland	ha	4,253	1,112	...
Cropland	ha	...	...	...
Mature cows	no.	393	256	11
Total animals (not including calves)	no.	527	672	23
Production measures				
Calf crop (weaned)	(%)	82.0	50.0	45
Production per A.U.	kg	140	...	...
Death loss (weaned calves and older)	(%)	2.2	5.0	10.0
Replacement rate	(%)	11.5	...	18.0
Land per A.U.	ha	6.9	1.8	...
Inventory, 100-cow basis				
Livestock				
Animal units	A.U.	157	250	204
Mature cows	no.	100	100	100
2-yr. heifers	no.	13	23	18
Yearling heifers	no.	13	24	27
4-yr. steers or bulls	no.	...	21	...
3-yr steers or bulls	no.	...	22	18
2-yr steers or bulls	no.	...	23	9
Yearling steers or bulls	no.	...	24	18
Bulls	no.	6	6	18
Horses	no.	3	8	...
Total animals	no.	135	251	208
Number of animals marketed (100-cow basis)				
Cows	no.	9	...	9
Calves	no.	82	...	...
Steers and heifers	no.	...	...	18
Bulls	no.	1	...	...
Consumption, home	no.	...	...	18
Total animals marketed	no.	92	...	45

[a]Adapted from Boykin, Forest, and Adams, 1972.
[b]Adapted from Mitchell and Casati, 1971.
[c]Adapted from Farris and Sullivan, 1976.

Beefmaster. Ranch owners usually live on the ranch, so their school-age children either live in town or travel long distances on a school bus. Ranching in this area is both a business and a dearly held way of life.

The medium-size ranch analyzed in Table 7.1 has 618 A.U.s, or a total of 527 animals not including calves. The 393 cows graze on 4,253 ha divided into pastures of various sizes to allow for herd rotation. Bulls are purchased rather than being raised on the ranch. Cows are culled rather heavily to maintain the 82 percent weaned calf crop. Nearly all calves are

Table 7.2. Costs, returns, and analytical information for typical cattle operations in Texas, Paraguay, and Tanzania, 1971

Item, 100-cow basis	Rio Grande Plains area of Texas (USA)[a]	South-eastern Paraguay[b]	Dodoma seminomadic tribesmen, Tanzania[c]
		(dollars)	
Investment			
Owned land and permits	189,034	13,850	...
Buildings and improvements	10,366	2,522	...
Machinery and equipment	1,298	299	...
Livestock	29,018	19,445	7,831
Total investment	229,716	36,116	7,831
Cash expenses			
Grazing fees and rent	...	...	...
Labor, hired	1,739	269	...
Feed purchased	659	...	...
Repairs and maintenance	...	...	...
Buildings and improvements	563	...	...
Machinery and equipment	48	...	...
Veterinary services and supplies	158	...	...
Taxes	791	...	...
Seed and fertilizer	...	...	...
Machinery, operating	254	...	...
Machinery, hired	...	...	...
Transportation	241	...	...
Insurance	52	...	...
Utilities	108	...	...
Miscellaneous	56	654	100
Subtotal	4,669	923	100
Noncash costs			
Depreciation			
Buildings and improvements	374	126	...
Machinery and equipment	239	27	...
Bulls	76	...	...
Bull death loss	63	...	...
Interest on cash expenses	163	76	...
Subtotal	915	229	...
Total expenses	5,584	1,152	100
Gross income			
Cattle sales and home consumption	13,405	2,204	2,090
Crop sales	...	...	...
Total income	13,405	2,204	2,090
Net income	7,821	1,052	1,990
Minus operator labor value	464	376	...
Return to capital and management	7,357	676	1,990
Percent return on investment	3.20	1.87	25.4
Average sale price of cattle (kg)	0.60	0.17	0.21

[a]Adapted from Boykin, Forest, and Adams, 1972.
[b]Adapted from Mitchell and Casati 1971.
[c]Adapted from Farris and Sullivan 1976.

Table 7.2. (continued)

Item, 100-cow basis	Rio Grande Plains area of Texas (USA)[a]	South-eastern Paraguay[b]	Dodoma seminomadic tribesmen, Tanzania[c]
		(dollars)	
Per breeding cow economic measures			
Total investment	2,297	361	78
Total ranch income	134	22	21
Total ranch expense	56	12	1
Net ranch income	78	10	20
Return to capital and management	74	7	20
Per A.U. economic measures			
Investment in land	1,204	55	...
Total investment	1,463	144	38
Total ranch income	85	9	10
Cash expenses	30	4	0.5
Noncash expenses	6	1	...
Total ranch expenses	36	5	0.5
Net ranch income	58	4	10
Return to capital and management	47	3	10
Actual situation			
Income (actual and imputed)	52,816	5,563	230
Expenses	22,001	2,908	11
Net income	30,815	2,655	219

sold at weaning time. There is a 2.2 percent death loss of animals above weaning age. A few pastures have improved grasses especially selected because they continue to produce, albeit at reduced rates, during dry periods when native range species have been grazed off. Some ranchers in the area irrigate part of their improved pasture, but this is not true of our Rio Grande case study.

The discussion now shifts to analyzing the operation on a 100-cow basis to facilitate comparison with the other two case studies. On this basis there are 135 total animals including 3 horses, 6 bulls, 13 two-year-old heifers and 13 yearling heifers. At the time of the study (1971) a 100-cow unit required an investment of about a quarter of a million dollars, 80 percent of that being in land (Table 7.2). The major cash expense is hired labor, followed by feed, repairs, and taxes. Total annual cash expenses are $4,669, which increase to $5,584 when noncash costs such as depreciation are included. Gross income for the Rio Grande ranch on a 100-cow basis is $13,405 while return to capital and management is about $7,800. The live-weight price of $0.60 per kg prevailing in 1971 when the study was carried out is used in the income computations. The rancher has a 3.2 percent return on investment.

SOUTHEASTERN PARAGUAY. Our second case study is set in Paraguay, a subtropical developing country located approximately in the middle of South America. About 50 percent of Paraguay's 2.8 million people live in the agricultural sector. Paraguay has traditionally been a large cattle producing country, and currently has 1.75 head per person. This figure may be compared with 0.50 head per person in the United States or with 0.88 head per person in Tanzania. A substantial percentage of the ranches are owned by absentee landlords who depend on a foreman, or "mayordomo" to run the operation. It is becoming common in this area, as well as in most of Latin America, to hire a college graduate or experienced individual to visit the ranch 2 or 3 days a month on a consulting basis.

The southeastern part of Paraguay, in which our case study is set, has gently rolling hills and occasional heavy forest. In general, it is an area of open savanna. The ranches all have perimeter fences, but effective range management is hampered by the large size of pastures. Much of the area is burned after the dry season to stimulate new forage production. Large pastures, lack of machinery, and a generally low level of efficiency make the operation labor intensive. In the Texas case study involving 618 A.U.s, hired labor was used only on a periodic basis, whereas the Paraguay operation (which has 631 A.U.s) has two or three permanent laborers in addition to the foreman.

The case study operation has a total of 672 animals, of which 256 are cows. The land is more productive than the Texas case, so that only 25 percent as much land is required (1,112 ha versus 4,253 ha). The carrying capacity is 1.8 ha per animal unit versus 6.9 in Texas. The low level of hoof-and-mouth disease that prevails is controlled by vaccination. A high level of brucellosis is found. Other diseases at moderate levels of incidence are rabies, anthrax, tuberculosis, and scabies. Internal and external parasites are a major problem because of extensive prolonged flooding during the wet season.

On a 100-cow basis, there are twenty-three two-year-old heifers and twenty-four yearlings. There are also about twenty to twenty-four head each of yearlings, 1-, 2-, 3-, and four-year-old steers or bulls, as all cattle are fattened on the ranch. Whereas the calves in Texas are weaned at about 6–8 months of age and sold at 180–200 kg, the steers in Paraguay will be sold at 2½ to 4 years of age weighing about 450 kg. There is a 50 percent calf crop and about a 5–8 percent death loss. Our southeastern Paraguayan rancher in 1971 was estimated to have an investment of about $32 per ha versus $54 for the Rio Grande ranch. This difference may not seem great, but on an animal-unit basis, the Paraguayan operation has only about 10 percent as much investment as the Texas rancher.

The low investment per animal unit, plus low expenses (20 percent of those in Texas), lead to the hypothesis that ranching in Paraguay should be a profitable business. This hypothesis is disproven, however, by extremely low cattle prices. Whereas the average price in 1971 of all cattle sold was $0.60 per kg in Texas, it was only $0.17 per kg in Paraguay. On a 100-cow basis, total returns were $2,200 versus $13,400 in Texas. The final result was that the rancher in the Parguayan case study had a 1.87 percent return on investment in 1971 versus 3.20 for the Rio Grande operation in the same year.

TANZANIA. The Dodoma region in central Tanzania is charactrized by dry plains and low rainfall, similar to much of the brush area of south Texas included in the first case study. The Masai tribe, one of the dominant groups in this area, consists mainly of seminomadic herdsmen. Cattle are individually owned but grazed in a communal fashion. The traditional livestock herd evolved as a multipurpose enterprise for subsistence and survival of the family. Cattle are first used for milk production, wealth, and prestige, and second as a source of cash. The cattle owners upon which this case study is based fit the characteristics for native African cattle owners described in Chapter 5.

The average cattle owner in Dodoma has twenty-seven animals, of which there are eleven cows, four calves, and five yearling and 2-year-old heifers. The remainder are bulls or steers.[1] Since calves are not counted as part of the inventory in Table 7.1, only twenty-three animals are listed. The calf crop is about 45 percent with an average death loss of 10 percent for weaned calves and older stock. The replacement rate is nearly 20 percent. The typical individual sells one cow and two steers or heifers annually with two more animals left for home consumption.

A major difficulty is poor nutrition and very low levels of management. For example, calf mortality is about 28 percent and calves are 11 months old at weaning. Water facilities are poorly developed, and because cattle are generally corralled at night, herding cattle back and forth takes so much time that cattle often have only a minimal period of grazing. The lack of adequate forage is compounded in the dry season.

External parasites present such a problem that cattle are supposed to be dipped every week. Despite a government-sponsored program of free dipping facilities, only 19 percent of the respondents in a recent survey had dipped their cattle in the last week prior to the interview, and less than half of the respondents had vaccinated their cattle within the previous year (Farris and Sullivan 1976). Over 75 percent of the respondents reported that health care centers were within a half day's

[1]The majority of this section is taken from Farris and Sullivan 1976.

trek, but only 5 percent of all respondents had drenched their cattle for internal parasites in the past 12 months.

The lack of orientation toward commercial beef raising is reflected by the fact that only 15 percent of the respondents had sold cattle in the previous 12 months, even though nearly all respondents indicated that the time required to trek cattle to market was a half day or less. Nearly all sales occurred at marketplaces, even though water and forage there were estimated to be almost nonexistent, especially during drought or dry periods.

Dodoma area livestockmen have almost no actual investment, as they graze on nationally owned land, and own virtually no facilities. Consequently, we assumed in Table 7.2 that investment cost in these items was zero. However, an imputed value of $0.21 per kg is assigned to cattle. The livestock are quite small and consequently have a relatively low inventory value. Cows and 3-year-old steers, for example, weigh only about 200 kg, while older bulls reach 250 kg. A 2-year-old steer weighs about 150 kg. Consequently, total investment on a 100-cow basis is only $7,830, compared with $230,000 and $36,000 for the Texas and Paraguayan cases.

Expenses are minimal for the seminomadic herds, primarily for some veterinary products. Most health care is provided free of charge by the government. Gross income on a 100-cow basis is $2,090, compared with $13,405 and $2,204 for Texas and Paraguay, respectively. But the gross income from the actual head of twenty-two A.U.s is estimated at only $230 per year, of which $138 is from cash sales, while $92 is imputed as a value for home consumption. There are about $11 of expenses so that net cash income from cattle sales is about $127 total. Each family also raises goats so that total net family income is slightly higher. The above calculations indicate a 25 percent return on investment.

COMPARISON OF THE THREE SYSTEMS.

The Tanzanian seminomadic herder has the highest percentage of return on investment because capital outlays are restricted to livestock and operating costs are minimal. Furthermore, income was calculated using a price of $0.21 per kg, a large part of which includes a subsidy from the government. Both the international and domestic meat market were below this price in 1971. A price of $0.17 per kg was used for Paraguay.

Comparison of the economic data in the latter part of Table 7.2 is facilitated by presenting data on a per breeding cow basis. The differences between systems are astounding. Total investment in Texas is $2,297 per breeding cow versus $78 in Tanzania. Total income for the Texas ranch is $134 per cow versus about $22 each for Paraguay and $21 for Tanzania, but return to capital and management is $47 for Texas versus $3 for Paraguay. In other words, a Paraguayan would have to have about 10 times as many breeding cows as a rancher in Texas to make a comparable return on capital and management.

The differences among the three systems are equally striking when they are compared on an animal-unit basis. In Paraguay there are fewer cows as a percentage of all breeding animals because cattle are fattened out on the ranch. Thus, net income on an animal-unit basis is lower in that country. In fact, the Paraguayan would require about 14 times as many total cattle as the Texan to receive the same net ranch income. It is little wonder that South American ranches tend to be large operations!

The last part of the table has income and expenses carried back to the original ranch sizes (where the Texan had an inventory of 527 animals, the Paraguayan 672, and the Tanzanian 23). Net income for the Texan is $30,815, for the Paraguayan $2,655, and for the Tanzanian $219. But, the actual cash income for the Tanzanian is only $138 since two of the five animals potentially available for sale are consumed rather than sold.

A word about economies of size is in order. Extensive studies on ranching indicate that total average costs per animal unit from ranches in the same area of the United States decrease only about 10–15 percent by moving from small- to large-scale operations. In addition, differences between regions within a country where capital is free to move are relatively small. Naturally, very small operations would have higher unit costs and reduced net incomes. Ranch size is, for the most part, limited by the available parcels of land. Nevertheless, the method just described of budgeting on a costs-and-returns basis is a useful tool for determining whether a ranch is of sufficient size to provide an adequate income.

The land resources available are extremely limited if we think in terms of livestock production in relation to an adequate standard of living. For this reason, potential entrants to the ranching industry are well advised to consider the low return and high amounts of credit and cash reserves necessary for a successful operation. Unfortunately, economic planners in developing countries frequently fail to take into consideration the available resources and future improvements in the standard of living. Thus, much extensive agrarian reform in areas suited only to livestock has caused ranches to be subdivided to such an extent that they become suboptimal units. In developing countries that practice extensive economic planning, planners must take into account the economies of size associated with ranching both as a means for holding down domestic prices and for maintaining low costs so that their country can compete on the international market.

LONG-TERM COSTS AND RETURNS FROM CATTLE RAISING

Most studies indicate low returns on investments in both ranching and farming compared with the risks involved (Boykin 1968, Gray 1969, Goodsell 1974). Theoretically, capital should abandon agriculture because of these lower returns, but it does not, and for several good reasons. One is that beef cattle have typically provided a low return to land mainly because they are confined to areas that are marginal for crop production (Gray et al. 1965). Another reason is that ranching has been considered a "good" way of life or a status symbol in most cattle producing regions. Furthermore, in many areas such as Latin America, ranching has traditionally provided a good income with little managerial effort because the units have been large. A third reason is that land has tended to appreciate rapidly in most countries since World War II. In the United States and elsewhere there have also been substantial tax advantages to owning ranches or cattle. In times of high inflation, such as the late 1970s, agricultural land proves to be a good hedge against inflation.

Annual incomes from cattle raising can fluctuate considerably with changing weather conditions and price movements. In many countries, such as those of the EEC or the USSR, prices are either partially controlled or fixed and thus reduce the impact of this factor. Also, because livestock raising is generally part of a farming operation in those areas, weather conditions are less important than on ranches found in dryer areas of the world. The extreme drought in northern Africa during the middle 1970s is a stark example of the way weather affects the pastoral people of that area.

Low profitability also holds for the extensive cattle raising operations typical of South America, Australia, and the United States. Because of the cattle cycle, even a well-managed ranch can experience prolonged periods without profit. Naturally, total long-term profit or loss depends on the entry point of the business. Starting an operation when cattle and land prices are depressed, and selling at the high point of prices, would yield a result entirely different from the opposite pattern of entry and exit.

Several examples are now developed for a typical ranch in the Rio Grande Plains of south central Texas to demonstrate the effect of entry at different periods. The area is somewhat better ranching country than that in the earlier Rio Grande case study, as our current example requires only 5.4 ha per animal unit whereas the other operation required 6.9 ha. Other than that, the operations are similar.

A number of factors have made ranching appear to be a good business venture in the United States. The demand for beef has grown rapidly, and there has been an abundance of land that is best suited for beef production. Also, beef cattle require little labor, and investment in machinery and equipment is low. But these attractive features are offset by seasonal and cyclical fluctuations in cattle prices, which have resulted in highly variable annual returns to capital and management. The long-term effect on profitability is now measured by following through a case study.

STARTING IN 1950 WITH A PRODUCING HERD. Assume that in 1950 a rancher purchased a 1,620-ha ranch in the Rio Grande Plains area of south central Texas. Assume further that the ranch was operating, and that the total cost for cattle, equipment, and land was $237,000 (Farris and Mallet 1973). Even though 1950 was a period of favorable beef prices, net farm income to the operator's labor, management, and capital would have been negative for 15 of the 27 years from 1950–1976 (Table 7.3). Furthermore, an average annual loss of $106 would have been incurred over the period 1950–1976 (Bentley 1978).

Let us examine individual periods within the time frame more closely by adding the assumption that the rancher borrowed half of the land purchase capital and two-thirds of that required for cattle purchase. The business would have covered cash costs and provided the operator income for the first 3 years, but for the next 5 years would not have yielded sufficient return to pay all interest charges on the land purchase.

Low cattle prices during the mid-1950s resulted in 6 consecutive years (1952–1956) of no return to the operator's equity, labor, or management. Prices improved in the late 1950s and there was a small return in 1958 and 1959 ($1,146 and $1,141). During the next 8 years the ranch failed to show a profit. Then, beginning in 1968, the business began to provide the operator with income above costs—1972 was a good year and 1973 even better. The conclusion is that this operator could have stayed in business only by deriving additional income from hunting leases, crop farming, outside income, or by initially owning the ranch free of debt. In this latter case there would have been an average annual return of $7,500 on equity, labor, and management.

Net income is highly correlated with the income side as ranch operating expenses have little annual variation (Oppenheimer 1970 and 1972). The high correlation with feeder prices is depicted in Figure 7.1. Another major factor influencing profitability is weather, as severe droughts can force ranchers to liquidate part of their herds. If this liquidation occurs during periods of low prices, returns can be severely affected (Williams and Stout 1964).

Table 7.3. Estimated returns for a budgeted cattle ranch with 300 cows on 1,619 ha, 1950–1976

Year	Gross income from cattle	Income above cash operating cost[a]	Income above cash operating cost and interest on cattle[b]	Income above total cash cost[c]	Income to operator, capital, and management[d]
			($)		
1950	21,202	12,991	9,972	4,704	1,032
1951	26,115	17,049	14,030	8,762	5,090
1952	20,439	11,373	8,357	3,086	−586
1953	13,717	5,221	2,202	−3,066	−6,738
1954	14,577	6,081	3,062	−2,206	−5,878
1955	14,097	5,793	2,774	−2,494	−6,166
1956	13,239	4,935	1,916	−3,352	−7,024
1957	15,786	7,194	4,175	−1,093	−4,765
1958	20,236	11,455	8,436	3,168	1,146
1959	20,327	11,450	8,431	3,163	1,141
1960	17,868	9,087	6,068	800	−1,222
1961	18,036	9,159	6,140	872	−1,150
1962	19,365	10,382	7,363	2,095	73
1963	17,665	8,596	5,577	309	−1,713
1964	15,282	6,309	3,290	−1,978	−4,000
1965	17,264	8,099	5,080	−188	−2,210
1966	19,787	10,337	7,318	2,050	−29
1967	19,072	9,526	6,507	1,239	−783
1968	19,929	10,191	7,172	1,904	−118
1969	22,557	12,438	9,419	4,151	2,129
1970	23,287	12,787	9,768	4,500	2,478
1971	24,565	13,588	10,569	5,301	3,279
1972	29,598	17,953	14,934	9,666	7,644
1973	37,697	24,142	21,123	15,855	13,833
1974	28,205	12,837	9,818	4,550	2,528
1975	25,490	8,594	5,575	307	−1,715
1976	29,065	11,120	8,101	2,833	811
Annual average	20,905	10,692	7,673	2,405	−106

Source: Extended from Farris and Mallet, 1973.

[a]Does not include interest payments on ranch and cattle.

[b]Includes 8 percent interest on two-thirds of the original cost of the cattle.

[c]Includes all cash costs plus interest payments on cattle and 6 percent interest on one-half of the purchase price of the ranch.

[d]Income above total cash costs and nonoperating (depreciation) charges.

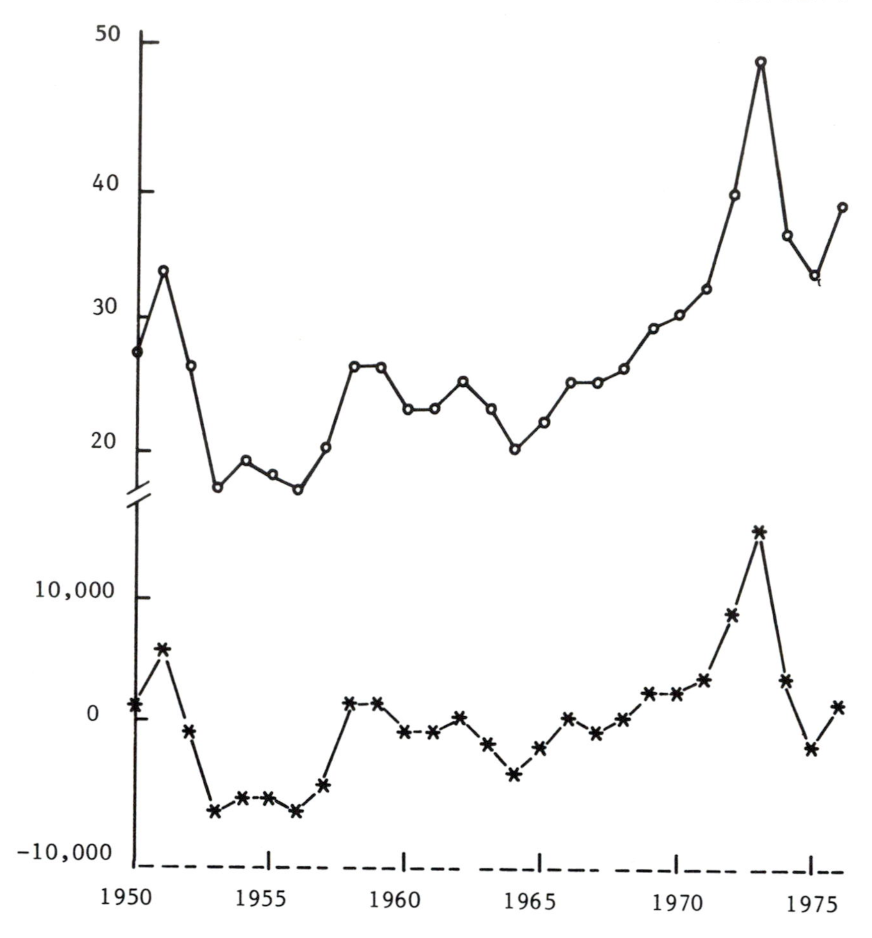

* = estimated income to operator's labor, capital, and management for a one-person ranch in Texas, 1950–1976 (US$). o = average price of feeder steers, all weights and grades. Kansas City, 1950–1976 ($/cwt).

Fig. 7.1. A comparison between returns to operator's labor, capital, and management and cattle prices, 1950–1976 (extended from Farris and Mallet 1973 by Bentley 1978).

Changes in age composition of herds is also a source of income variability. Productivity of cows increases during their first few calving intervals, reaches a peak, and declines over the rest of their lives. If culling is heavier than normal, actual productivity may increase for the remaining cows because their average age decreases. When herd expansion takes place, however, the higher proportion of heifers and old cows increases, and thus overall productivity and gross revenue are lowered.

If the same operations described previously are examined, but under the assumption that the rancher starts with unbred heifers rather than with producing cows, at least one year without a profit is added at the beginning. Such an analysis demonstrates clearly that because of the cattle cycle, a concept discussed at length in Chapter 9, even a well-managed ranch can be expected to undergo long periods with no profit or, if the land is owned free of debt, a very small return on investment.

DECISION MAKING AT THE RANCH LEVEL

We now turn our attention to ranch management, the micro viewpoint, by considering some decisions ranchers face in managing their firms. The following list of decisions adapted from Gray (1968) is included here as it indicates the number of decisions ranch operators must make. The broad classes are:

1. Level of production—What size calf crop should a rancher attempt to reach in terms of net return?
2. Method of production—Should ranges be grazed seasonally or during the entire year; should pastures be improved; should cows be palpated; and is crossbreeding a better approach than straight-breds?
3. Combination of production factors—What is the optimal combination of land, labor, capital, and management? How much should be spent keeping books as opposed to fixing a fence or making decisions about land improvements? How much credit should be used?
4. Combination of enterprises—How much time should be devoted to each enterprise in mixed farming-cattle raising? Should the operation be strictly a commercial cow-calf enterprise or should a purebred operation be added? Should stockers be raised in connection with a cow-calf operation? During periods of extremely high feed prices, should steers be grass fattened?
5. Size of business—How large should the ranch be? If the ranch is small, can the individual make an adequate total net income? If the ranch is quite large are their "diseconomies" of size (points at which the cost per unit increases as volume increases)? If land is appreciating in value should more land be purchased for speculation?

The above questions provide a framework for analysis of an ongoing operation. What happens, however, if we think about buying a ranch or associated business? It is to that problem that we now address our attention.

FEASIBILITY OF INVESTING IN A CENTRAL AMERICAN CATTLE RANCH

The low price of ranchland in Latin America has traditionally lured outside investors. In the early 1970s beef prices in the developed nations were increasing rapidly, as were production costs. It appeared that investments in the beef cattle industry of Latin America would yield profit rates in excess of those found in the industrialized countries. The situation in the early 1980s will probably parallel that of the early 1970s, with the result that entrepreneurs will once again look to Latin American ranching as a potential investment. This section is presented in the form of an abbreviated feasibility study to show how one situation was analyzed.

The ranch is first assumed to have 1,000 mother cows of mixed breeding, to have a 60 percent calf crop, and to be operating at an optimal carrying capacity. Heifers are bred as 2-year-olds and enter the herd at 3 years of age. Cows are sold at 12. The ranch is all open range with a stocking rate of 0.5 A.U. per ha on which steers average a daily gain of 0.28 kg and are sold at 42 months. There are 232.1 kg of steer beef produced annually per hectare. Calves are weaned at 6 months of age. There are 2,231 A.U.s (3,049 animals), which means that 4,462 ha are required since the stocking rate is 0.5 A.U. per ha. Miscellaneous land adds another 25 ha for a total of 4,487 ha, which at $100 per ha is about $445,100 (Table 7.4).

Calculating all individual capital investment items yields a total of $770 per cow. This amount can be compared to the estimate of $2,300 for the Texas rancher in the earlier case study and $361 for the Paraguayan case. The individual operating expenses are then calculated. The total of $31.81 is considerably more than the $9.23 for the Paraguayan case because the Central American countries are tied more closely to the United States. Wage rates and social security are also much higher. Adding noncash costs to the operating expenses yields a total of $47.32. Credits from the sale of cull cows, old bulls, and horses have been subtracted from the expenses as our intention is to determine the cost per kg of beef produced. The net adjusted expenses of $35.21 are divided by the kilograms of beef produced (164.43 kg) to arrive at the cost per kg of $0.21.

The ranch in the example also may have some improved pastureland. Results from that analysis are given in the second column of Table 7.4 under the assumption that the ranch has 300 ha of Pangola grass pasture with a stocking rate of 4 A.U.s per ha. The steers are now sold at 24 months of age, and 1,200 kg of beef are produced per ha. The average daily gain is 0.55 kg per day. Value of the improved pasture is $400 per ha. Total investment is much lower as fewer hectares are required for 1,000 mother cows.

Table 7.4. Estimated annual costs of producing 1 kg of beef on unimproved and improved pastures on a typical Central American ranch, 1973

Item	Cost per cow	
	Unimproved	Improved
	($)	
Capital investment		
Land	445.10	282.90
Capital improvements	80.00	80.00
Equipment and horses	20.00	25.00
Cows	200.00	200.00
Bulls	25.00	25.00
Total investment	770.10	612.90
Operating expenses—cash		
Wages, permanent labor	4.86	4.86
Wages, temporary labor	0.96	0.96
Social security, etc.	0.73	0.73
Administrative salaries	10.00	10.00
Meat consumed on ranch	1.30	1.30
Utilities, including telephone	1.75	1.75
Repairs and materials—capital improvements	1.25	1.25
Repairs and materials—machinery	0.88	1.12
Insurance	0.30	0.30
Fuel and lubricants	1.08	1.60
Office expenses	0.53	0.53
Fertilizer and lime	...	62.70
Miscellaneous	1.52	1.38
Concentrated feed	0.07	0.07
Veterinary and medicine	5.58	2.50
Taxes—property and vehicles	1.00	1.00
Ranch rent	...	...
Total cash expenses	31.81	92.05
Operating expenses—noncash		
Interest on operating capital	3.18	9.21
Depreciation on capital improvements	5.33	5.33
Depreciation on equipment and horses	2.00	2.00
Depreciation on bulls	5.00	5.00
Total noncash expenses	15.51	21.54
Total expenses	47.32	113.59
Credits	12.12	12.12
Net adjusted expenses	35.21	101.48
Cost per kilogram of beef	0.21	0.58
Cost per pound of beef	0.10	0.26

Source: Based on an unpublished feasibility study by Simpson and Sprott, 1973.

Note: No charge is made for opportunity cost on land. Including this item at 10 percent would increase cost on the unimproved model $44.50 so that net adjusted expenses would be $79.71. Cost per kilogram would soar.

Operating expenses are substantially higher owing to increased use of fertilizer and lime. In fact, fertilizer is the critical factor, accounting for about two-thirds of operating expenses. The result is that even though initial investment is lower, cost per kilogram of beef at $0.58 is nearly triple that of the operation with no improved pasture. Even if fertilizer cost were half that assumed from 1973 prices, cost would still be about 40 cents per kg, or nearly double that of the open range operation. The effect of different fertilizer prices on net income at various beef prices is given in Table 7.5. Comparison with the unimproved pasture model clearly shows the effect of this major input. Naturally, ranchers can and do use less than the optimal amount of fertilizer as costs of this input increase. Furthermore, if the price of fertilizer increases enough, as it did during the early 1970s, ranchers will simply stop using it and will accept lower offtake per hectare. That strategy will necessitate more frequent renovation of pasture to counteract the weed and brush invasion that could have been avoided by heavier fertilization. Unfortunately, little data are available on long-term pasture output under varying pasture maintenance plans.

Table 7.5. Estimated alternative net returns for unimproved and improved pasture ranches in Central America with two fertilizer costs, 1973

Selling price per kg of beef	Beef sold	Total return	Net expenses	Total net return	Net expenses	Total net return
($)	*(kg)*		*($)*			
Unimproved pasture model						
0.30	164,430	49,329	35,206	14,123		
0.40	164,430	65,772	35,206	30,566		
0.50	164,430	82,215	35,206	47,009		
0.60	164,430	98,658	35,206	63,452		
Improved pasture model			Fertilizer at $90.00 per ton[a]		Fertilizer at $45.00 per ton	
0.30	174,750	52,425	101,475	−49,050	74,475	−22,050
0.40	174,750	69,900	101,475	−31,575	74,475	−4,575
0.50	174,750	87,375	101,475	−14,100	74,475	12,900
0.60	174,750	104,850	101,475	3,375	74,475	30,375

Note: Opportunity cost on land not included in expenses.
[a]Cost per ton utilized in original budgets, Table 7.4.

The effect of changing the calf crop on cost per kilogram is illustrated in Table 7.6 with data from the previous model. Increasing the calf crop from the assumed basic 60 percent to 90 percent lowers the cost per kilogram from 21 cents to 14 cents on unimproved pastures, and from 58 cents to 39 cents on improved pastures, which is a considerable reduction considering that management is the principal input required. This table clearly demonstrates the importance of the calving rate.

Table 7.6. Effect of various calf crops on cost of producing beef on a hypothetical Central American ranch

	Calf crop			
	60%	70%	80%	90%
		($)		
Unimproved				
cost/kg	0.21	0.19	0.16	0.14
cost/lb	0.10	0.08	0.07	0.06
Improved				
cost/kg	0.58	0.51	0.46	0.39
cost/lb	0.26	0.23	0.21	0.19

Note: Assumes no land cost.

DISCOUNTING PROCEDURES FOR INVESTMENT EVALUATION

One-year budgets provide useful information for comparing different projects or businesses. For example, the information about the three different beef systems is a good tool. As the discussion on long-term returns from cattle ranching indicated, however, calculations of return on investment can be misleading if they are based on just 1 year. Thus, the potential for profit must take into account the potential net return during the entire period of the project or investment. The problem has been overcome by the development of capital budgeting techniques (Hopkin et al. 1973).

The heart of capital budgeting, and of benefit-cost analysis (described in Chapter 6), is recognition that a given sum of money is worth more today than at some point in the future. The reason is that the money can be invested so that, at a future time, the investor would have the original sum plus interest. This concept is known as the time value of money. As an example, assume that $1.00 is invested at 6 percent interest compounded annually for 10 years. The investor would receive $1.79 at the end of that period. Discounting is basically the reverse of compounding, since the objective is to bring a future stream of costs or benefits back to the present. The parallel question from the compounding example is: What would $1.79 received 10 years from now be worth today if your opportunity cost of capital, i.e., the amount you could receive on money invested, were 6 percent? The answer is $1.00. The stage is now set for using discounting as a tool in investment planning.

The rate of return on an investment is determined by discounting the projected stream of net benefits and costs back to the present in order to place them on a common denominator. Let us take a simple example. Suppose that a rancher invests $6,000 in a new well on an isolated part of the ranch where previously there was little or no water for cattle. It is determined that, as a result of the investment, additional annual benefits of $2,000 per year will be gained. Let us also assume that $500 annual maintenance is required, that the well has a useful life of 5 years, and that the equipment has a salvage value of $500 at the end of year 5. Should the well be installed, and if so, what is the potential rate of return? The computations presented in Table 7.7 show that the net benefits in year 1 are a negative $4,000 because of the investment that year. The net benefits are $1,500 in all other years except for the last one, when the salvage value of $500 is also included.

Table 7.7. Calculations of the pretax internal rate of return from a hypothetical ranch investment

				20%		25%	
Year	Investment and maintenance	Annual benefits	Net benefits	Present value factor	Present value	Present value factor	Present value
			($)				
1	6000	2000	−4000	0.833	−3332	0.800	−3200
2	500	2000	1500	0.694	1041	0.640	960
3	500	2000	1500	0.579	869	0.512	768
4	500	2000	1500	0.482	723	0.410	615
5	500	2000	1500	0.402	603	0.328	492
5 (salvage)			500	0.402	201	0.328	164
Net present value					+105		−201

Note: IRR = 20 + 5(105/105 + 201) = 21.7.

The first step in determining the internal rate of return is to choose a discount rate that is thought to about equalize discounted costs and benefits. In this case, 20 percent was chosen. For the first year the present value factor (given in financial tables) is 0.833. Multiplying that coefficient times a minus $4,000 yields a minus $3,332. This multiplication procedure is carried on for each year, the column summed, and a net present value calculated. It is a positive $105, which indicates that a higher discount rate must be tried (Table 7.7).

The second step involves using a higher discount rate. Using 25 percent yields a net present value of minus $201. Interpolating gives an internal rate of return of 21.7 percent, which can be interpreted to mean that the well project will yield an average net return of 21.7 percent each year during the project life for all the money tied up in the project, including maintenance costs. If the rancher has the opportunity to earn more than 21.7 percent in alternative projects, then they should be considered rather than the well.

The example is useful for it demonstrates the ease with which discounting can be used. The method is superior to others for it always gives correct ranking of projects. The payback, benefit-cost ratio, net present worth, or rate of return methods, for example, can give incorrect rankings. The internal rate of return method also accounts for the time-cost of money and clearly demonstrates the importance of including time and interest rates in investment decision making. Inflation is incorporated as part of the internal rate of return for it is implicit when the potential investor weighs his or her opportunity cost of capital against the internal rate of return. High inflation rates tend to drive up the interest rate, so that the opportunity for investing at a higher rate in alternative projects is also much higher (Gittinger 1972).

SUMMARY AND CONCLUSIONS

A number of factors relating to economics of the beef business have been discussed in this chapter. Comparison of ranch budgets between countries showed that returns on investment in the early 1970s were similar—and low. Profitability from ranching was further analyzed by looking at long-term costs and returns for a Texas ranch from 1950 to 1976, from which it was concluded that there are enough periods of low or negative net returns to make it difficult or impossible to meet interest payments if half the original investment capital is from borrowed sources. A further analysis carried out for a hypothetical Central American ranch indicated that returns there are also much lower than commonly believed. Other short analyses such as effect of calf crop on gross returns and the use of discounting for investment analysis were also presented to show the simplicity, yet effectiveness, of these analytical tools. Our conclusion is that, given the intense competition in the world's beef business resulting from rapid technological advances, the vagaries of the cattle cycle, and makeup of interregional competition for beef, economic and financial analysis deserves careful attention from private parties and government or lending agencies where investment in beef production is concerned.

REFERENCES

Bentley, Ernest. 1978. Adaptive Planning for Texas Cattle Producers Facing Uncertain Prices. Ph.D. diss., Texas A & M University.

Boykin, C. C., Jr. 1968. *Economic and Operational Characteristics of Cattle Ranches.* Texas Agr. Exp. Sta. Bull. MP-866.

Boykin, C. C., Jr.; Forest, N. K.; and Adams, J. 1972. Economic and Operational Characteristics of Livestock Ranches—Rio Grande Plains and Trans-Pecos of Texas. College Station: Texas Agricultural Experiment Station.

Brumby, P. L. 1973. International Lending for Livestock Development. *World Anim. Rev.* 5:6–10.

Farris, D. E., and Mallet, James I. 1973. Effect of Variability in Prices and Production Cycles on Profitability in Beef Cattle. Paper presented on behalf of the National Livestock Tax Committee for Tax Reform Hearings of the House Ways and Means Committee, March 12, Washington, D.C.

Farris, D. E., and Sullivan, G. 1976. Tanzanian Livestock-Meat Subsector, 4 vols. Washington, D.C. Texas A & M University and USAID.

Gittinger, J. P. 1972. *Economic Analysis of Agricultural Projects.* Baltimore, Md.: Johns Hopkins Press.

Goodsell, Wylie D. 1974. *Southwest Cattle Ranches, Organization, Costs and Returns, 1964–72.* USDA, AER no. 255.

Gray, James, R. 1968. *Ranch Economics.* Ames: Iowa State University Press.

______. 1969. *Production Practices, Costs and Returns of Cattle Ranches in Northeastern New Mexico.* New Mexico Agric. Exp. Sta. Res. Rep. 56. Las Cruces: New Mexico.

Gray, James R.; Stubblefield, Thomas M.; and Roberts, Keith N. 1965. *Economic Aspects of Range Improvements in the Southwest.* New Mexico Agric. Sta. Bull. 498. Las Cruces: New Mexico.

Hopkin, John; Berry, P.; and Baker, C. 1973. *Financial Management in Agriculture.* Danville, Ill.: Interstate Printers and Publishers.

Mitchell, G. H., and Casati, E. 1971. *Costs and Returns of Selected Beef Cattle Ranches in Paraguay, 1971.* Centro de Economia Rural y Estadistica. Asuncion, Paraguay: Universidad Nacional de Asuncion.

Oppenheimer, Harold L. 1970. *Cowboy Arithmetic.* Danville, Ill.: Interstate Publishers and Printers.

______. 1972. *Land Speculation.* Danville, Ill.: Interstate Publishers and Printers.

Simpson, J. R., and Sprott, M. J. 1973. Estimations of Costs for Beef Production in Central America. Unpublished feasibility study. College Station, Texas.

Westberry, George O. 1979. *Partial Budgeting.* Food Resour. Econ. Dep., Staff Pap. 130. Gainesville: University of Florida.

______. 1979a. *Enterprise Budgets: What, How and Why?* Food Resour. Econ. Dep., Staff Pap. 131. Gainesville: University of Florida.

Williams, Willard F., and Stout, Thomas T. 1964. *Economics of the Livestock and Meat Industry.* New York: Macmillan Co.

8

PRODUCTION ECONOMICS APPLIED TO CATTLE RAISING

Cattlemen and other decision makers will find the budgeting process described in Chapter 7 an effective tool for making useful comparisons because it includes information about ranch economics. Of course costs and returns budgets in themselves merely help describe a particular situation and consequently are somewhat limiting for designing and determining, from an economic point of view, optimum ranch management strategies. Optimal strategies are determined within the context of production economics, which is one part of the subdiscipline called farm management. This subdiscipline includes other areas such as finance and marketing strategies as they relate to the production unit.

Modern production economics really began with Heady's (1952) study of the use of resources in agriculture just after World War II. Since then there have been significant advances in clarifying theory and developing tools for evaluating a set of extremely diverse problems at the management level (Martin 1977). We begin with an introduction to these tools.

INTRODUCTION

Production economics is based on the assumption that all cattle raisers consciously or subconsciously must make three major production decisions, whether they operate under the most primitive nomadic conditions or under extremely sophisticated management: how to produce, how much to produce, and what to produce. The first decision, "how to produce," provides the necessary information for determining what and how much to produce. Also, after a firm has begun production, entrepreneurs should regularly reevaluate their management decisions in light of new information and changing input relationships.

Suppose that a decision is made to use land resources for a cow-calf unit or a steer fattening operation. Assume further that one of the best production methods for such an operation is thought to be improved pastures and supplemented feed. There are, of course, various supplements, but research at the local experiment station indicates that grain concentrate and molasses mixed with urea (called input or factor X_1) have given good results. This factor, along with roughage, which is called X_2, is the factor-factor relationship shown in Figure 8.1. The two factors may be substituted in different proportions and amounts to provide various levels of output which, represented by the curved lines called isoquants, are denoted as Y_{1a}, Y_{1b}, Y_{1c} in the figure. The lowest output is Y_{1a} and the highest is Y_{1c}. The straight lines between the two axes are called isocost lines or constant cost lines, as they represent various combinations of inputs that can be purchased by a given outlay.

As greater amounts of the two factors are used, higher levels of output—say kilograms of beef per steer or hectare—are obtained. There are, of course, an infinite number of isoquants (because there are an infinite number of output levels) just as there are an infinite number of output combinations (along any one isoquant). For simplicity, only two inputs are shown in Figure 8.1. The tangency point between an isoquant and isocost line is economically the most efficient (least-cost) combination of inputs for producing that level of output. The tangency points may be connected by a line indicating the least-cost combination of inputs (at the given input prices) for producing any particular level of output. Any combinations of inputs (at the given prices) other than those where

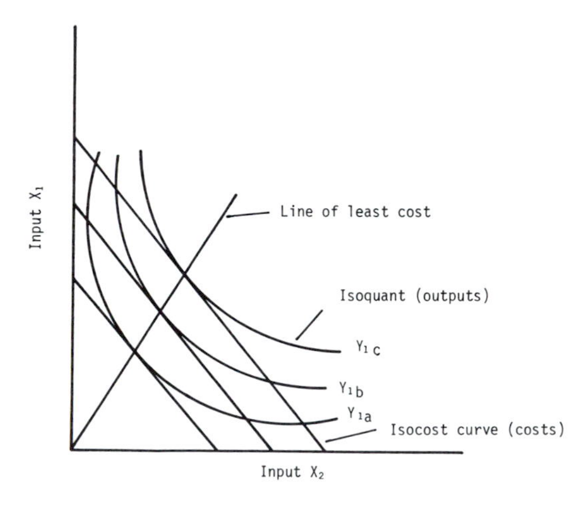

Fig. 8.1. Example of the production economic decision about how to produce.

$\partial x_1/\partial x_2 = -Px_2/Px_1$ (the tangency point) are not optimal solutions. The points may be connected by a line of least cost, which represents the most efficient input combinations for any given set of outputs. The optimal solution to the question is: Produce where marginal cost (*MC*) equals marginal revenue (*MR*), where the term "marginal" means the last additional increment.

Fortunately, cattlemen or other businessmen do not have to understand the refinements of economics to use it to good advantage in production decisions. Moreover, since the advent of home computers and programmable calculators, use of production economics can be reduced to a set of rather simple instructions. Furthermore, the perceptive operator often achieves the same results as that person who knows how to use the concepts of marginal revenue and marginal cost but has little practical knowledge. Good managers intuitively approximate the optimal condition (where $MR = MC$) when they reason, "Will it cost me as much extra (*MC*) to winter these calves as the added return (*MR*) from the extra weight and age in the spring? I might as well send them to market this fall and avoid the risk of a loss." If, for example, he applies 200 kilograms of fertilizer per hectare at an added cost of $20 (*MC*) he could expect an extra 100 kilograms of gain at an added return of $40 (*MR*). Here *MR* exceeds *MC* by $20. The manager would need to calculate the *MC* and *MR* from 300 and 400 kilograms to determine if chances are good for *MR* to equal or exceed *MC*. When all added costs including risk and returns for the activity are considered, the precise optimal condition is the input level where the extra cost (*MC*) equals extra revenue (*MR*). When certain inputs are limited, such as capital, and several enterprises or activities need the inputs, then the activity where *MR* exceeds *MC* by the largest amount should be allocated the scarce input first. This is exactly the way the computer proceeds to maximize profit in a complex decision model with many alternative activities.

The second economic question concerns how much to produce. The vertical line on the horizontal axis of Figure 8.2 between X_1 and X_2 means that all inputs other than the supplement (X_1) are being held constant. The production function *OA* shown in Figure 8.2 bends as output is now represented on the vertical axis while inputs are shown on the horizontal axis. Output, represented on the vertical axis, increases up to a certain point (*C*), after which it begins to decline because too much input (shown on the horizontal axis) is used.

The straight line *OB* represents a line in which the price of the input being varied (X_1) is divided by the price of the output (P_y). In effect, it is Px_1/P_y. Optimum output in this factor-product relationship is where the marginal physical product (*MPP*) is equal to the inverse of the price ratios, that is, price of the supplement divided by price of the output.

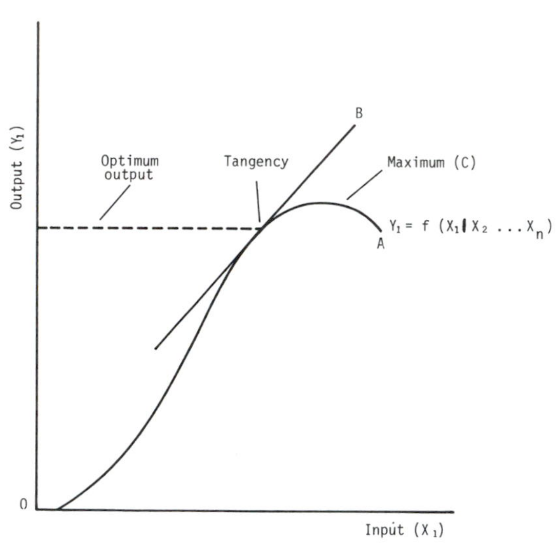

Fig. 8.2. Example of the production economic decision about how much to produce.

The equality is determined by the point of tangency. This relationship is also said to be producing where marginal revenue just equals marginal cost. In other words the "economically rational" cowman would continue increasing the quantity of supplement along with the molasses-urea mixture until the additional input cost just equals the additional income. This assumes that the other input(s) such as pasture (X_2) is held constant at some given level.

It is important to recognize that even though the economically optimal production level can be determined, in practice this entrepreneurial management objective is seldom achieved exactly as desired because interpersonal, culturally oriented constraints lead producers to use other input levels. For example, some people feel compelled to maximize production per unit of land and thus inevitably employ too high an input level. In many cases quantity of the input is too low because supplies are unavailable or because an individual failed to calculate the optimal production level. In some areas, especially Africa, social constraints caused by communal grazing systems and status lead to high cattle inventory levels and low offtake.

After calculating the optimal production level with this additional information, we can return to another important aspect of "how to produce" (a factor-factor problem). By holding constant all inputs except the ones being evaluated, it is possible to portray the law of diminishing returns (also called the law of variable proportions), which states that if an input is applied to a fixed factor, eventually, if enough input is applied, total output will increase at a decreasing rate (the inflection point) and may approach a maximum point when output could begin to decrease (point *C* in Figure 8.2). If the combined concentrate and molasses-urea mixture in our example is fed at too high levels (relative to the ability of the animals to digest it properly), especially in recently weaned calves, many of them would probably become sick or go off their feed, and some might even die. The result would be a decrease in total output.

The third decision, "what to produce," is made after sufficient knowledge is obtained about the relevant production functions. Some production functions of interest to cattlemen are gains among cattle on different forages relative to grade classes, weight, sex, and breed. Other useful information would be alternative forage yield responses with different input combinations such as fertilizer, mowing, or water, as well as average daily gain from supplemental feeding and various carrying capacities. Seasonality effects are also evaluated within the framework of this decision.

Assuming that some estimates of the production function have been obtained, the next step is to relate input costs to output prices. In the graphic presentation of this relationship in Figure 8.3, the curved lines represent two possible output relationships. For example, a rancher has the option of using land for a cow-calf operation and producing calves (Y_1), of fattening steers (Y_2) or of combining both enterprises in some way. The amount of potential production depends on the production functions, which in turn depend on the amount of input use, which finally is constrained by resource availability. The curved lines are thus called isoresource or production possibility curves.

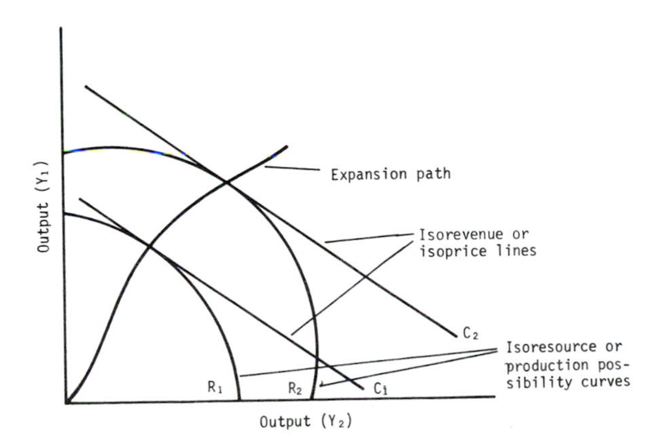

Fig. 8.3. Example of the production economic decision about what to produce.

The optimal amount of each possible output is determined by the tangency of the isorevenue with the production possibility curve. In other words, the criteria for determining the optimal amount of outputs depend on a combination of the physical, cost, and price relationships. If more inputs are used in the production processes, the isocost curve is shifted out, say from C_1 to C_2 so that more of each potential output—for example, calves or fed steers—could be produced. The location of C_1 and C_2 is determined by the input constraints, such as the amount of land available. In Figure 8.3 only one resource constraint can be shown. The tangency of the new isoprice lines with the new isocost curves may or may not lead to the same proportion of outputs. An expansion path can be drawn through the tangency points to show the optimal combinations at different output levels. Given the production possibilities curve associated with the given constraint, the most rational economic decision is to maximize revenue, that is, to produce at the tangency points, where the slope of the isoprice lines (C_1 and C_2) is the inverse of the relative output prices.

HOW MUCH TO PRODUCE: AN EXAMPLE

As indicated, the decision criteria determining the optimal output and use of inputs in the question how much to produce, are where marginal revenue equals marginal cost ($MR = MC$) or, alternatively, where the marginal value product (MVP) equals the input cost. The example presented in this section explains both the method of arriving at the optimal level and the concept of marginal revenue and marginal cost.

Consider again the cattleman who is interested in fattening steers on pasture. Our problem will be to determine whether supplementing them with the concentrate and urea-molasses mixture (the least-cost combination of X_1 and X_2) will yield a profit and how much feed should be used per head.

The analysis begins by budgeting a steer operation with no supplemental feeding (Table 8.1). It is assumed that 100 steers with an initial weight of 200 kg are purchased at $1.20 per kg. An anticipated average daily gain of 0.30 kg per head would result in 6,100 kg net gain. Cash expenses are $1,210. Adding $195 for noncash costs raises the total cost to $1,405. The cost is thus $0.23 per kg.

The next step is to determine the optimal level of supplemental feeding. The computations for this analysis are shown in Table 8.2. The first two columns indicate that when 3.80 kg of concentrate are fed, beef production increases by 0.2 kg over the no supplement level (0.50 − 0.30). A

Table 8.1. Computation of costs for 100 steer fattening operations

Items	No supplemental feeding	With supplemental feeding
	($)	
Investment		
Owned land	150,000	150,000
Buildings and improvements	2,000	2,000
Machinery and equipment	1,000	1,000
Livestock[a]	24,000	24,000
Total	177,000	177,000
Cash expenses (annual)		
Labor hired	...	...
Supplemental feed[b]	...	15,945
Salt and minerals	40	40
Repairs and maintenance		
Buildings and improvements	50	50
Machinery and equipment	50	65
Veterinary services and supplies	400	400
Taxes	300	300
Seed and fertilizer	...	...
Machinery, operating	255	350
Machinery, hired	...	...
Transportation	50	50
Insurance	...	...
Utilities	...	...
Miscellaneous	65	75
Subtotal without supplemental feed	1,210	1,330
Subtotal with supplemental feed	...	17,275
Noncash costs		
Depreciation		
Buildings and improvements	50	50
Machinery and equipment	75	75
Interest on cash expenses[c]	70	75
Subtotal	195	200
Subtotal without supplemental feed	1,405	1,530
Total with supplemental feed	...	17,475
	(kg)	
Kilograms of beef produced[d]	6,100	20,750
	($)	
Cost per kilogram without supplemental feed[e]	0.23	0.25
Cost per kilogram with supplemental feed	...	0.84
Total income[f]	7,320	24,900
Net income	5,915	7,425

[a]200 kg steers purchased at $1.20 per kg = $240 per head × 100 head = $24,000.

[b]5.90 kg of feed per head per day × 99 head = 584.1 kg per day × 210 days = 122,661 kg total × $0.13 = $15,945.

[c]No interest charge on cattle as they are assumed to come from a cow-calf enterprise of the same owner's operation. Interest on supplemental feed is included in its cost.

[d]Beef production with no supplemental feed is 0.30 kg of gain per day × 210 days = 63 kg per animal × 100 steers = 6,300 kg − 200 kg for 1 death = 6,100 kg net. With supplemental feeding production it is 1.00 kg per day × 210 days = 210 kg per animal × 100 steers = 21,000 kg − 250 kg for 1 death = 20,750 kg net.

[e]Cost without supplemental feeding divided by 6,300 kg.

[f]$1.20 per kg.

Table 8.2. Example of determining where $MC = MR$ in the decision about how much to produce

Example A. Decision Rule: Produce where input price $= MPV$				Example B. Decision Rule: Produce where $MC = MR$						
Daily per head ration of supplement Q_x	Beef production Q_y	Marginal physical product (MPP) $\Delta Q_y / \Delta Q_x$	Marginal value product (MVP) $P_y \cdot MPP$	Supplement variable cost (VC) $P_x \cdot Q_x$[a]	Fixed[b] cost (FC)	Total cost (TC) $VC + TC$	Total revenue (TR) $P_y \cdot Q_y$[c]	Profit per kg produced $TR - TC$	Marginal cost (MC) $\Delta TC / \Delta Q_y$	Marginal revenue (MR) $\Delta TR / \Delta Q_y$
0.00	0.30			0.00	0.23	0.23	0.36	0.13		
3.80	0.50	0.05	0.06	0.49	0.25	0.74	0.60	−0.14	2.45	1.20
4.00	0.60	0.50	0.60	0.52	0.25	0.77	0.72	−0.05	0.30	1.20
4.25	0.70	0.40	0.48	0.55	0.25	0.80	0.84	0.04	0.30	1.20
4.50	0.80	0.40	0.48	0.59	0.25	0.84	0.96	0.12	0.04	1.20
5.00	0.90	0.20	0.24	0.65	0.25	0.90	1.08	0.18	0.60	1.20
5.90	1.00	0.11	0.13	0.77	0.25	1.02	1.20	0.18	1.20	1.20
7.00	1.05			0.91	0.25	1.16	1.26	0.10		
8.00	1.05	0.04	0.05	1.04	0.25	1.29	1.26	−0.03	2.80	1.20

[a]Price of concentrate (P_x) = \$0.13 per kg.

[b]The fixed cost, i.e., the one that does not vary, is from Table 8.1. The cost is \$0.23 (first column) without supplement feeding. It increases slightly to \$0.25 (second column of Table 8.1) to account for the additional costs incurred in feeding the supplemental feed.

[c]Price of feed (P_y) = \$1.20 per kg, live weight.

daily ration of 5.0 kg per head allows production to increase up to 0.90 kg per day per head, and so forth.

Two closely related methods can be used in determining the optimal production level. The first one (A) is producing where the input price just equals the value of the last unit of product, that is, the marginal value product (*MVP*). The *MVP* is calculated by multiplying price times the *MPP,* where *MPP* is the change in the quantity of beef production divided by the change in the quantity of input. For example, the change from 3.80 kg of concentrate to 4.00 kg is 0.20 kg. This quantity, divided into 0.10 = (0.60 − 0.50) kg of beef gives an *MPP* of 0.50. That physical measure multiplied by the price of beef ($1.20 per kg live weight) gives the *MVP*. The decision rule is to produce where the value of the last additional unit of product (*MVP*) just equals the product price. Assuming that the price of supplement feed is $0.13 per kg, then the optimal daily level would be 5.90 kg of feed per head (Table 8.2).

The second method (B) of determining the optimal feeding level first requires calculation of the daily supplement feed cost for one animal. This is done by multiplying the supplement unit cost ($0.13 per kg) times the daily quantity. This is called the variable cost since it is the one being varied. Then, the fixed costs (fixed in the sense that they are being held constant while the quantity of supplemental feed is being varied) are added. The cost $0.23 per kg calculated in Table 8.1 is entered in the first row of Table 8.2, which is used to describe the cost per kg of beef produced with no supplemental feeding. The calculations in Table 8.1 also show that the fixed cost would be $0.25 if there were supplemental feeding. This figure is thus entered in each row in the fixed cost column in Table 8.2. Total cost, shown in the next column, is the sum of fixed and variable costs.

Total revenue is calculated by multiplying beef price ($1.20 per kg) times quantity of beef produced (Q_y). Profit per kilogram produced, the difference between the total revenue and total cost columns, is $0.13 per head per day with no supplemental feeding. If only 3.80 kg of concentrate are fed, the additional gain does not cover feed costs and there is a loss of $0.14 per head per day. A profit begins to be realized at the level of 4.25 kg supplement feed. A maximum profit of $0.18 occurs at the 5.00 and 5.90 feeding levels.

Marginal cost is calculated by dividing the change in total cost by the change in beef output. For example, the *MC* of $1.20 is the result of dividing $0.12 ($1.02 − $0.90) by 0.10 (1.00 − 0.90). Marginal revenue of $1.20 for that same feeding level is the product of dividing the change in total revenue, which is $0.12 ($1.20 − $1.08) by change in quantity or 0.10 (1.00 − 0.90). The decision rule is to produce where *MC* = *MR,* so the conclusion, as with the previous method, is to produce using 5.90 kg of concentrate. The profit at this level is \$0.05 per kg higher than it is with no feeding.

The numbers are typical of most agricultural production situations in which profit will usually decline if only a small amount of the input is used. At the other extreme, if the input is added until output is at a maximum (8 kg of supplement in our case), profits decline to the point that they become negative.

LINEAR PROGRAMMING IN LIVESTOCK PRODUCTION ECONOMICS

In the previous example it was assumed that the optimal combination of the two inputs—supplemental feed (called X_1) and pasture, (called X_2)—had been determined, and that the resulting production function had only one variable input, X_1. Many other combinations of inputs could have been used, however. The arithmetic can quickly become tedious when the problem of finding an optimal solution involves two or more inputs in conjunction with two or more products. As a result, a quantitative method called linear programming was designed to handle this task. Popularized by Heady and Candler (1958), the method has become one of the most widely used tools in agricultural economics as well as in many other disciplines. It can be used with the budgets shown in Table 8.1, but is not a substitute for budgeting.

An important aspect of linear programming, and the reason for its name, is that the production functions are assumed to be linear. In other words, in linear programming each input is used in fixed proportions; that is, output is determined by the limiting input, because in linear programming (LP) one input cannot substitute for another one. In a LP problem, for example, machinery cannot be used in place of labor, but the two can be used in a fixed ratio to each other. Care must be taken to distinguish between the production functions just described, which result from the expansion path of two or more inputs (Figure 8.1), and the production function described in Figure 8.2. In the latter function, all inputs are held constant except one, which, if increased enough, would cause output to diminish. The combination called X, which is made up of X_1 and X_2, constitutes the one input, for in linear programming all inputs increase at a fixed rate. Output would, of course, continue to expand to unreasonable extremes as a result of fixed input-output relationships, except that restrictions are placed on the number of inputs utilized. These limitations, called constraints, along with an objective that is either maximization or minimization of something, and alternative ways to achieve the objectives, constitute the three parts of a linear programming problem.

LP concepts may be clarified by working through a simple maximization problem that is an expansion of the example presented earlier. Suppose that our farmer or cattle raiser has 100 ha, $10,000 in operating capital and 200 hours of labor (the constraints). Also suppose that this person wants to use the land only for cattle, and that the options being considered are a breeding herd, a steer fattening operation, or both.

Our objective is to determine the optimal amount of land to devote to each operation. In other words, we are now dealing with the "what to produce" part of production economics. The outputs, expected prices, labor, and operating capital requirements for the enterprises are given in Table 8.3. Note that the beef production and live-weight price are expected values. The capital requirements are based on past performance.

Table 8.3. Input and output specifications, linear programming example

Enterprise	Expected beef production per hectare	Expected price per kilogram live weight	Labor per hectare	Operating capital per hectare
	(kg)	*($)*	*(hr)*	*($)*
Breeding	175	1.00	1.8	85
Fattening	200	1.20	3.1	110

The problem can be solved geometrically by labeling hectares devoted to the fattening operation on the horizontal axis and hectares in the breeding enterprise on the vertical axis (Panel A, Figure 8.4). The second step is to plot the three constraints. In Panel A, a point is placed on 100 ha of land on both the vertical and horizontal axes and a straight line drawn between them. This line reveals all possible combinations of land use between each enterprise. The cattleman could use less than 100 ha since land is not the only restriction, but he could not use more than 100.

The next move is to add the labor constraint to Panel B, which already has the land constraint drawn in. If all 200 hours of labor were fully used on the breeding herd, then 111 ha (200 hours ÷ 1.8 hrs/ha) could be used if they were available. Only 65 ha (200 ÷ 3.1 hrs/ha) need be used if all the labor is employed in the fattening operation. As for the land constraint, the two points are plotted on the vertical and horizontal axis and a straight line drawn between them. This line shows all possible combinations of breeding herd and fattening operation that can be produced under only 200 hours of labor.

After labor, the next step is to add the capital constraint. Calculations indicate that the $10,000 of available operating capital would permit use of 117 ha ($10,000 ÷ $85/ha) in a breeding operation if that much were available. However, all the capital for a fattening operation would permit only 91 ha ($10,000 ÷ $110/ha) to be used. Each of these

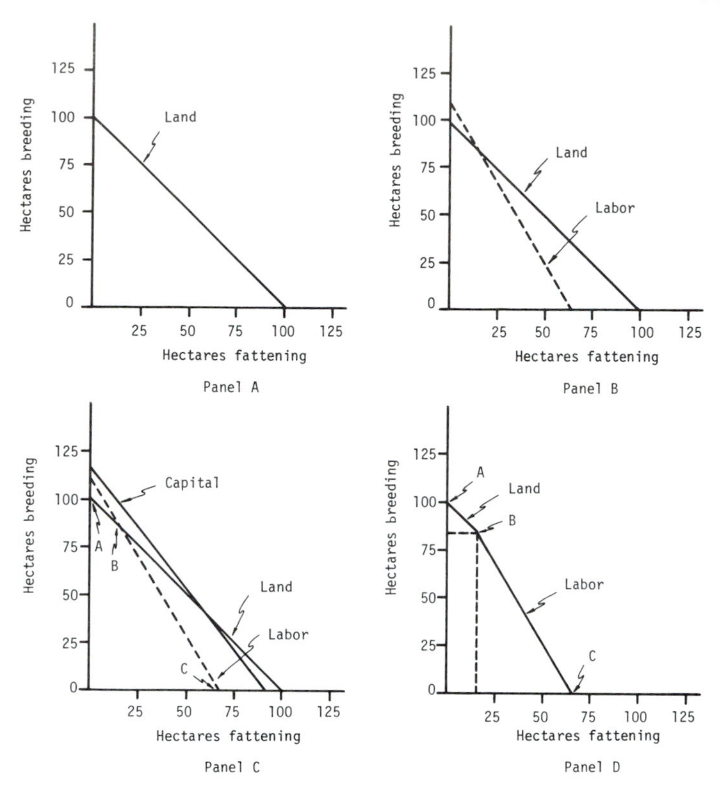

Fig. 8.4. Determination of corner solutions, linear programming example.

points is plotted on the appropriate axis in Panel C to which the land and labor constraints have previously been drawn in. The key here is that while capital is not a constraint on the breeding herd, it does limit the amount of land needed for the fattening operation to below the 100 ha limit originally set forth.

Now that all the constraints have been properly plotted, the profit maximization combination can be identified. The optimal level will always be located at a "corner" where the inside constraint lines intersect, or where the inside constraint lines cross the axis. These points are labeled *O, A, B,* and *C* in Panel D, which has been redrawn from Panel C. Land and labor are the two constraints. The optimal production combination will never fall on a straight line segment of the restriction lines.[1]

[1]This is not exactly correct since there can be a case in which the optimal solution is

One way to determine the most profitable "corner" is to calculate the income and cost at each corner, as was done in Table 8.4. In corner *O*, which is the intersection of the breeding and fattening axes, there is no production and consequently no net income. Corner *A*, which is constrained by land at 100 ha, has all resources used in the breeding operation. The 100 ha given in this activity solution are multiplied by the originally specified 175 kg of live beef production per ha, which, at $1.20 per kg, provides a total income of $21,000. The operating cost of $85 per ha results in $8,500 for the 100 ha, or a total net income of $12,500 for the entire operation.

Corner *B* is the intersection of the land and labor constraints. Eighty-five hectares are determined to be optimum for the breeding herd and 15 ha for the fattening operation. Net incomes of $10,625 and $1,950 are derived from the two operations respectively, for a total net income of $12,575. Corner *C*, the all fattening operation, provides a net income of $8,450. The conclusion is that 15 ha devoted to fattening and 85 ha devoted to breeding maximize net income. If prices of weaned calves were lower (e.g., $1.00) than prices of those sold from the fattening operation, this same combination of enterprises would still hold, except that net income would be lower ($9,600 at corner *B* versus $9,000 at corner *A*).

The optimal solutions provide the answers to the three production questions concerning what, how, and how much to produce. The "what" analysis indicates a combined breeding herd and fattening operation would maximize income. "How" is concerned with the inputs—100 ha of land, 164 hours of labor (1.8 hrs × 65 ha of land for breeding and 3.1 hrs × 15 ha for fattening), and $8,875 of operating capital ($85 × 85 ha of land for breeding and $110 × 15 ha for fattening). "How much" is concerned with the product, which in our case is 14,875 kg of beef (live weight) from the breeding herd (85 ha × 175 kg per ha) and 2,800 kg of beef from the fattening operation (15 ha × 200 kg per ha). Land has been shown to be the principal constraint. The analysis shows that some labor and capital would be left over so that if the operator has correctly specified the production function relationships, then 36 hours of labor and $1,125 could be invested in other activities. In other words, the analysis not only provides an answer to the three questions, but also provides information that can be used for planning related operations.

The preceding example of linear programming demonstrates its application to a farm level resource allocation decision problem. Most LP problems involve many more activities and thus are solved with com-

---

either *B* or *C* (indeterminant solution) when any points on the line between *B* and *C* would be optimal. However, LP algorithms will always find the corner solutions.

Table 8.4. Calculation of net returns from various corner solutions, linear programming example

Item	Breeding					Fattening					Total both enterprises
	Land	Production per ha	Operating cost per ha	Price per kg	Enterprise total	Land	Production per ha	Operating cost per ha	Price per kg	Enterprise total	
	(ha)	*(kg)*		*($)*		*(ha)*	*(kg)*		*($)*		
Corner A											
Income	100	175	...	1.20	21,000	0	0	...	0	0	21,000
Cost	100	...	85	...	8,500	0	...	0	...	0	8,500
Net	100	...	...	...	12,500	0	...	...	...	0	12,500
Corner B											
Income	85	175	...	1.20	17,850	15	200	...	1.20	3,600	21,450
Cost	85	...	85	...	7,225	15	...	110	...	1,650	8,875
Net	85	...	...	...	10,625	15	...	...	...	1,950	12,575
Corner C											
Income	0	0	0	0	0	65	200	...	1.20	15,600	15,600
Cost	0	0	0	0	0	65	...	110	...	7,150	7,150
Net	0	0	0	0	0	65	...	...	...	8,450	8,450

puters rather than by hand. Besides being applied to maximization problems, LP is commonly used in formulation of least-cost cattle rations and minimization of transportation cost. Furthermore, LP has been extended to regional and national problems for entire sectors. It is to this type of problem that we now turn.

AN APPLICATION OF LINEAR PROGRAMMING TO ARGENTINA

The beef industry in Argentina is an example of extensive-type cattle operations. There are 2.25 head of cattle per person, one of the highest ratios in the world, and beef accounts for about 15 percent of export earnings. Most beef is exported as fresh, chilled, or frozen wholesale cuts, but there is also some cooked and canned beef (mainly corned), as well as dried, salted, and smoked beef. Argentine beef exported to the United States is canned or precooked cow beef because U.S. sanitary regulations prevent imports of fresh, chilled, or frozen meats from countries where foot-and-mouth disease exists. The precooked beef is processed from the best muscle cuts of canner-type cattle and, in some instances, from similar cuts of better grade cow beef. There are two kinds of precooked beef. The first is shipped in bulk containers for further cooking before consumption. The second is precooked beef, generally quick frozen after cooking, which can be consumed without additional cooking.

Cattle are produced over much of the country, but the heaviest concentration (about 80 percent) is in the Pampa (Figure 8.5). This is a vast, originally treeless, lowland prairie with a mild climate and regular rains in the eastern part. There are no major rivers and, because of poor surface drainage, heavy rains occasionally create large shallow lakes. Most of the Pampa's soil is a deep, black, friable chernozem that is excellent for cropping. The second area of heavy concentration is the subtropical area in the northeast, which lies between the Paraná and Uruguay rivers. It is a heavily forested rolling plain containing red, leached, acid soils. The third region to produce cattle in significant numbers is the north or Chaco area, which is a vast, wooded lowland plain extending from western Paraguay and eastern Bolivia. It is subtropical and has high temperatures and abundant, even excessive, rainfall. (See Hutchison et al. 1972.)

The northeast, Chaco, and other outlying areas are characterized by cow-calf operations, which produce weaned calves or yearlings for sale to fattening ranches in the Pampa. There are some fully integrated ranches that sell fat cattle, but they are disappearing as truck transportation and feeding methods on specialized operations improve. Most of

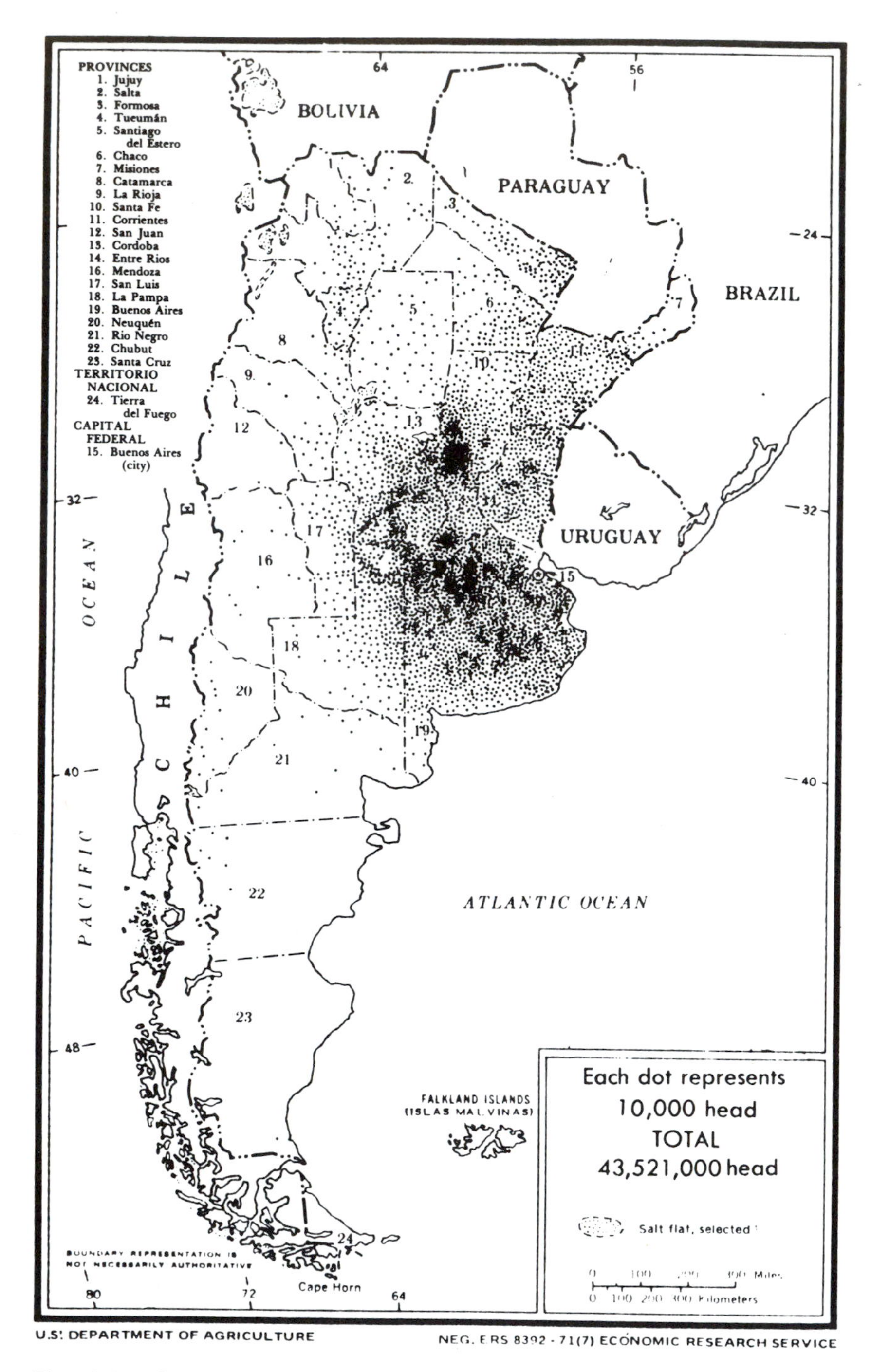

Fig. 8.5. Cattle distribution in Argentina, 1960.

Argentina's packing plants are concentrated around Buenos Aires, although there has been a move toward decentralization in recent years. The principal breeds are Aberdeen Angus, Shorthorn, and Hereford in the more temperate areas; and Santa Gertrudis, Charolais, and Zebu types in the more subtropical regions.

Estimated carrying capacity in the Pampa is 0.6 to 1.0 animal units (A.U.s) per hectare on natural or native pastures, 1.2 to 1.8 A.U.s on improved pastures, and 1.8 to 2.2 A.U.s on seeded temporary pasture (Figure 8.6). About 2.4 to 3.0 A.U.s are carried with rotation grazing, and up to 3.7 A.U.s with supplemental feeding. The current gain of about 100 to 150 kg per hectare per year could be increased to 500 kg with supplemental feeding and good pasture management (Hutchison 1972). Very few cattle are fed in confinement lots in Argentina primarily because of the low steer–corn price ratio, which only runs about 4:1 to 5:1. This ratio means that the price of live cattle per kilogram is about four or five times higher than the price per kilogram of corn. In contrast to the Argentine situation, the ratio is between 10:1 and 12:1 in the United States, a country where most cattle are fed in confinement lots prior to slaughter.

A principal characteristic of Argentine beef cattle operations is the absence of grain concentrate or supplement feeding except on a small scale to carry over essential breeding stock during severe winters. One obvious reason for the lack of supplement feeding is the mild, snow-free winters of the Pampa and northern Argentina, during which pastures

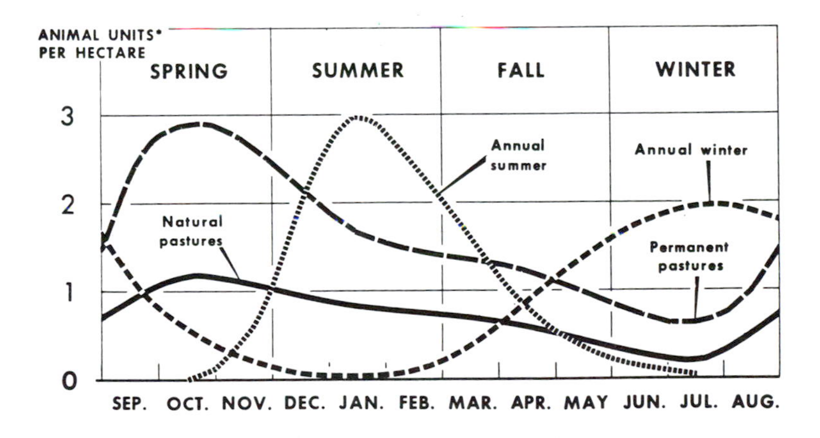

Fig. 8.6. Carrying capacity of pastures in the Pampa area of Argentina.

there remain usable although the quantity and nutritive value of forage are reduced. Cattle operations on the Pampa are usually combined with grain production, the practice being mainly one of grazing cattle during the winter and spring on fall-seeded grains (barley, oats, rye, and wheat).

It has been suggested that beef production in Argentina as well as in other areas having extensive operations could be substantially improved by: (1) increasing the carrying capacity of pastures through use of improved forage and rotational grazing; (2) using more stored and supplementary feeds during winter months and periodic droughts; (3) improving the calving rate (at birth) from a current 70 or 75 percent to more than 80 percent; (4) reducing the slaughter of young calves; and (5) leveling off the violent periodic fluctuations in productivity resulting from government policies and changes in world prices. In an effort to evaluate some of these possibilities, McGrann (1973) applied the technique of linear programming to four ecological regions of the Pampean region. In this section we summarize the method employed and the results for one zone. Our objective, like McGrann's, will be to show the economic relationships between traditional and improved methods, that is, to show a shift upward in the production function by achieving more production from the same level of inputs, or adding new production possibilities. The study by McGrann is essentially a more complicated and detailed study of the LP-type analysis we have examined.

The cattle activities specified in this 2000 ha ranch example are for the central Pampean region, which is an area of mixed crop and cattle production. The optimal enterprise organizations shown in Table 8.5 are for five cattle price levels. When both feeder steer and slaughter steer prices are low (59.5 pesos per 100 kg with an exchange rate of $1.00 (U.S.) = 3.5 pesos), the optimal solution calls for 428 ha of cultivated pasture for fattening steers, 200 ha of oats and sorghum for fattening, and 72 ha of alfalfa hay for fattening, or a total of 700 ha in pasture and hay. A total of 300 ha would be in crops.

The rational entrepreneurs in this price level would purchase 224 head of 150–450 kg steers in the fall, along with 1,019 head of 150–380 kg steers and 39 head of 150–320 kg steers. Total net beef production for the year would be 303,300 kg, or 433 kg/ha of pasture and hay produced. Net income would be 89,961 pesos.

If cattle prices increased to 137.5 pesos per 100 kg for feeder steers and 110.0 pesos for slaughter cattle, then the optimal solution calls for less crop production (259 ha) and the addition of a breeding herd consisting of 219 cows. Only 98 steers would be purchased for fattening. Beef production would increase to 311,200 kg and net income would be 170,510 pesos.

The competitive disadvantage of the steers finished to the 450 kg

Table 8.5. Optimal ranch organization and beef production for selected price levels, the central Pampean fattening area, Argentina, 1969-1970

		Price levels of feeder steers and slaughter cattle[a]				
Feeder steer price (pesos/100 kg)		59.5	85	110	136	137.5
Slaughter cattle price (pesos/100 kg)	Units	59.5	85	110	136	110.0
Net income	pesos	89,961	160,241	236,570	338,466	170,510
Operating capital	pesos	188,595	249,555	350,401	471,078	279,567
Intermediate term capital	pesos	50,783	50,813	50,028	49,531	150,573
Enterprises in optimal solutions						
Crops						
Corn	ha	300	254	155	...	241
Barley	ha	...	...	...	...	18
Total crops	ha	300	254	155	...	259
Pasture						
Cultivated pasture						
For fattening	ha	428	431	351	300	325
For breeding herd	ha	...	...	...	...	103
Annual pasture for fattening						
Oats and sorghum	ha	200	241	295	373	202
Barley I	ha	...	5	...	...	...
Pasto romano	ha	...	...	50	127	...
Annual pasture for breeding herd						
Oats and sorghum	ha	...	...	...	...	12
Barley I	ha	...	...	...	...	27
Alfalfa hay						
For fattening	ha	72	69	149	200	56
For breeding	ha	...	...	...	...	16
Total pasture and hay	ha	700	746	845	1,000	741
Cattle, using improved practices						
Breeding cows[b]	hd	...	...	...	...	219
Fat cull cows	hd	...	...	...	...	26
Ranch produced steers 180-380 kg	hd	...	...	...	...	39
Ranch produced steers 180-320 kg	hd	...	...	...	...	55
Ranch produced heifers 180-350 kg	hd	...	...	...	...	61
Purchased feeder cattle[c]						
Fall steers 150-450 kg	hd	224	253	...	...	102
Fall steers 150-380 kg	hd	1,019	1,143	1,773	2,072	96
Fall steers 150-320 kg	hd	39	...	...	...	...
Total net beef production	100 kg	3,033	3,322	4,012	4,687	3,112
Net total beef production per hectare of pasture and hay produced	kg	433	445	475	469	420

[a]These were gross ranch live-weight prices per 100 kg. The exchange rate is $1.00 (U.S.) = 3.5 pesos.

[b]This was the number of breeding cows only and does not include replacement heifers or bulls.

[c]Purchase feeder heifer activities were not included in this model.

weight, the important weight class for export, was indicated at the higher price levels. At the lower prices optimal solutions indicated a small proportion of the steers finished at 450 kg. Cattle fed to lower weights had a competitive advantage over heavier cattle because they were more efficient in feed conversion and took less time to finish out; thus the fixed costs (such as land rent or taxes) per kilogram were reduced. Lighter cattle cost less to feed, but the market price was the same for both the heavy steers (450 kg) and the light steers (380 kg). The price of heavier steers would have to increase relative to the lighter steers to compensate for their higher production cost.

Operating capital restrictions are of major concern to livestock raisers all over the world. Unlike the first example presented in this chapter, McGrann's initial analysis was made under the assumption that operating capital was unlimited and could be obtained at a real interest rate of 6 percent. By parametrically changing the level of operating capital, he was able to determine that at very restricted amounts the optimal ranch organization was an integrated cow-calf fattening system of beef production. With increased availability of capital, the optimal organization changed to a specialized fattening operation. Furthermore, total beef production is highly responsive to operating capital restrictions. For example, total beef production increased 82 percent from the low level of capital to unrestricted levels.

The linear programming study on the fattening area also showed that as price of beef increased, land used for crops was shifted to annual pasture. Most of the increase in beef production was the result of land resource adjustment—i.e., shifting of land from crops to pasture as cattle price increased—rather than of increased beef production per hectare of pasture and hay produced. The results show how a solution is determined by adding the various input costs, output prices, and constraints, and then making changes in these specifications. This solution was reached by modeling a region as though it were one enterprise.

Apart from a favorable demand for beef, factors internal to the ranch are necessary incentives for increasing its production capacity. Many ranches in the Pampean area are sufficiently large that owners have been able to satisfy their income requirements without fully exploiting ranch potential through production-increasing practices (Valdez and Nores 1979). A more intensive investment plan for growth requires increased supervision and management input; however, many of the ranch owners are not willing to put forth this increased effort, for several reasons.

One reason is that the tax system in Argentina emphasizes commodity taxation and taxation of imports and exports; these taxes are easy to collect so there is no need to develop incentives for efficient use of resources. This form of taxation places a higher tax burden on the more productive ranchers and reduces the economic incentive to increase production. The environment of high risk and difficult decisions in agriculture also reduces the incentive to invest in agriculture and diverts investment to other areas. Another constraint often overlooked is the necessity to improve the managerial input, the human resource, along with the adoption and financing of new technology.

Economic studies of ranches in Argentina have indicated that retained earnings have been the most important source of operating and investment capital (Nores 1969). If investment is limited to retained earnings, growth is likely to be restricted. The Argentine financial market has not been studied sufficiently to establish whether there is a serious credit shortage (McGrann 1973).

Many of the other inputs such as machinery and fertilizer are high in cost relative to product price in Argentina. If cost were expressed in terms of beef, a tractor would cost three times as much as in the United States. In effect, the high relative prices of inputs are due mainly to government protection of the domestic industry by means of restrictions on machinery imports and by high import duties on items like fertilizer.

SUMMARY AND CONCLUSIONS

This chapter expanded the discussion of budgeting techniques introduced in Chapter 7. We first described the three basic decisions of production economics: "how," "how much," and "what to produce" if profit maximization and economic efficiency are the decision criteria. Graphic presentations were used to illustrate the irrationality of attempting to maximize calf crop as well as to show the economic effect of underutilizing inputs.

A brief description of linear programming was given, as it has become one of the principal tools in livestock economic analysis. This technique, by which we can determine the least cost of some given practice (such as a feed ration) and profit maximization, can be used both on a limited scale—for example, to determine the optimal combination of inputs and outputs for one farm—and to study an entire region. In recent years dynamic aspects have been added so that optimal solutions over time can be determined.

One major application of linear programming has been regional and sector analysis to provide information for policymaking at the national level. An example was given for the central Pampean region of Argentina, where cattle fattening on pasture, on crop residue, and on hay is a major enterprise. The results of a study by McGrann were presented to show that production levels about twice the traditional average were possible in the fattening zone by maintaining high average weight gains for finishing cattle, production of a summer and winter annual pasture on the same land, and intensive cultivation of pasture land. The increase in production that was possible from the adoption of production-increasing practices proved to be much greater than the direct response to price increases; the optimal ranch organization in the fattening area analyzed, however, was highly sensitive to feeder cattle price increases.

Microlevel analysis helped us to identify some significant constraints: (1) lack of landowner incentive (economic or personal) to increase productive capacity; (2) price instability and inflation, which increase investment risk and debt capital use; (3) high input cost relative to product prices that leads to lower returns on investments; (4) external capital constraints and short length of loan maturity; (5) shortages of managerial ability; and (6) a traditional large capital drain to the nonagricultural sector, where there is possibly higher return and less risk.

The next chapter will explain how cattle are marketed in various parts of the world and how cost-price relationships as well as national conditions affect these markets.

REFERENCES

Heady, Earl O. 1952. *Economics of Agriculture and Resource Use.* Englewood Cliffs, N.J.: Prentice-Hall.

______ and Candler, Wilfred. 1958. *Linear Programming Methods.* Ames: Iowa State University Press.

Hutchison, John E.; Urban, Francis S.; and Dunmore, John C. 1972. *Argentina: Growth Potential of the Livestock and Grain Sectors.* USDA, FAER no. 78.

Martin, Lee R., ed. 1977. *A Survey of Agricultural Economics Literature, Vol. 2, Quantitative Methods in Agricultural Economics 1940s to 1970s.* St. Paul: University of Minnesota Press.

McGrann, James M. 1973. Microeconomic Analysis of Opportunities for Increasing Beef Production: The Pampean Area, Argentina. Ph.D. diss., Texas A & M University.

Nores, Gustavo. 1969. An Econometric Model of the Argentine Beef Cattle Economy. M.S. thesis, Purdue University.

Valdez, Alberto, and Nores, Gustavo. 1979. *Growth Potential of the Beef Sector in Latin America—Survey of Issues and Policies.* Washington, D.C.: International Food Policy Research Institute.

9
MARKETING LIVE CATTLE

The term "market" can refer to a physical location or to the result of interpersonal actions. From these possibilities it follows that marketing is a mechanism for exchange that can be defined in terms of space, time, and the form of a particular commodity. If one is asked, "How is the cattle market today?" the question relates to the price at a particular place and time for a certain type and size of cattle. But apart from the reference to a particular location such as a stockyard, the market is not a visible entity. The fact that "the market" is not a visible physical thing often leads to considerable misunderstanding by many people and is sometimes a progenitor of government price-fixing, investigations into price manipulation, and so forth.

Marketing is the performance of all business activities involved in the flow of goods and services from the points of initial production to the ultimate consumer. Cattle marketing is a specialized area, the goal of which is efficient movement and pricing from initial production to slaughter. In some cases, producers may load, haul, and deliver their cattle directly to slaughterhouses that buy on their own account. In other instances, cattle ownership might be transferred many times through numerous intermediaries involved in the marketing process. In the United States the typical practice is for feedlots to sell their cattle directly to slaughter plants; in England cattle are usually sold through local auctions.

Multiplicity of ownership between birth and slaughter is probably greatest in the United States or Canada, where in extreme cases a lightweight calf might be trucked to an auction, purchased by an order buyer, and sold to a person specializing in growing out light calves on grass. It could then be sold to a dealer, and resold to a backgrounder who would prepare the animal for a feedlot. It might then be purchased by a feeder, be placed in a feedlot until it reached slaughter weight, and finally be sold to a packing plant. During the many trips from one specialized growing or feeding area to another, the animal is loaded and unloaded at least a half dozen times. It may also travel vast distances. For example, calves born in Florida in the eastern United States sometimes spend 6 to 9 months as far west as Arizona and are shipped back east for slaughter, a trip of 3,000 or 3,500 km (Stegelin and Simpson 1980).

Do fewer intermediaries mean that the system will be more efficient? Not necessarily, as there are substantial savings available from certain types of specialization. Generally, the more developed an economy, the greater the degree of specialization. One reason for more specialization in the United States, for example, is that the mix of services changes as incomes increase. As consumers achieve higher incomes they demand better quality beef and are willing to pay for it. By quality we mean flavor, tenderness, and asepsis, as well as other subjective factors from which economists can estimate or derive retail-level demand curves. Demand and supply curves could be traced for each level of exchange, as was described in Chapter 2, but this is not normally done. The point to remember is that even though demand is ultimately set at the consumer level, prices fluctuate according to the supply of cattle. Seasonal factors, as well as acute conditions such as drought and policy announcements, determine cattle supplies, and regular long-term undulations of supply and price are due to national herd inventory changes.

WORLD CATTLE CYCLES

Cattle cycles, defined as the period from one low point in cattle numbers to the next low point, have become one of the most disrupting factors in all phases of the livestock industry in countries that are major beef importers and exporters. The United States probably has the longest history of regular cattle cycles of any country in the world in terms of known cycles, which have been traced out since the late 1800s. The lowest inventory in the last cycle (number 8) was January 1, 1979, so that 1980, the first year of increasing numbers, marked the beginning of a new cycle. Although the cattle cycle basically refers to inventory numbers, price cycles accompany the inventory changes roughly in the opposite direction of the peaks and valleys.

The first additional fed cattle of the new herd buildup in cycle number 9 will come out of U.S. feedlots and go to slaughter about 1982–1983. Per capita supply will begin to accelerate at that time. But profits will stimulate producers to keep increasing inventory. Eventually, probably around 1986 if this cycle resembles previous ones, there will be a saturation point and retail prices will begin to slip. Panic selling will begin and will cause prices to decline more rapidly, so that sales will increase and prices will decline even further. The industry will then enter another period of financial darkness.

The United States has experienced eight cattle cycles since records have been kept. The first one bottomed out in 1885. The five cycles since 1930 are shown in Figure 9.1. The shaded sections indicate periods of herd rebuilding, while the other areas depict the deceleration phase during which liquidation takes place. The cycles since World War II have run about 10 years, the last cycle of 12 years being an exception. The length of that cycle can be attributed to the ferocity with which prices increased during 1973–1974 and to the sharp, spectacular drop in 1974–1975. The upswings of U.S. cattle cycles have varied from 6 to 8 years, whereas the downturns—the liquidation of deceleration phases—on the average have lasted about 4 years.

Cattle cycles are the result of a lagged response in production to a favorable or unfavorable profit situation. Why cattle cycles occur can be understood by following the developments of a complete cycle. Consider the 1975–1977 situation in the United States—in effect, the "dark years" of the last cycle. Many producers, in response to low prices and attendant poor profits or losses, became discouraged and either reduced their inventories or left the cattle business. Other ranchers found themselves lacking the cash flow needed to meet expenses and were forced to liquidate part of their herds. In addition, nearly all cattle raisers culled cows rather heavily to eliminate breeding stock that was of marginal or low productivity. The selling of increased numbers forced prices lower and provided additional incentive to reduce herd size.

Slaughter and consumption increase during periods of low beef prices because of greater supply. In 1976, for example, per capita consumption (supply) of beef in the United States was about 59 kg (carcass weight basis). The 1980 average was 48 kg, as supply was down since producers were retaining heifer cattle for herd buildup. Thus, production of beef is up during periods of low prices simply because production of beef can be equated with slaughter. In other words, supply response relationships and high "production" should not be confused during periods of low prices.

Eventually, about 2 to 3 years after beef prices fall, cattle liquidation becomes sufficient to create a short supply. Prices are bid up slowly at first, and then more rapidly. As cattlemen become more optimistic, they hold back heifers and this action causes prices to advance even further. The initial price spurts take place in items such as ground beef that are related primarily to nonfed cattle. Prices of the lower beef grades start to rise, followed by higher priced cuts. Signals of optimism are flashed throughout the industry as slaughter cattle prices improve. Cattle feeders bid up feeder cattle assuming that fed cattle prices will advance along with nonfed beef.

Fig. 9.1. Cattle cycles since 1930, prices and inventory.

Price cycles are therefore inversely related to inventories (Figure 9.1). The production cycle lags behind the turning point in inventory numbers by about 2 years, while the turning point in cow numbers falls behind the turning point in feeder prices from 0 to 2 years. The final result is that the production cycle follows prices by as much as 4 years.

As soon as producers are confident that a new cycle has really begun in the industry, those who have resources available for new or additional livestock production hold back even more females they might have sold, or invest in breeding stock; prices then increase further and the price-inventory spiral becomes firmly established. This holding action signals the beginning of the buildup or acceleration phase. The lag between producers' decisions to increase cattle numbers and actual increase may be from about 3 to 5 years. A significant increase in cattle numbers is not forthcoming, however, until the second generation of calves reach marketable weights.

The major variable related to changing cattle numbers is a biological one. Suppose that a rancher decides to increase production by holding back ranch-raised females. There is a 9-month gestation period followed by a period of 15 to 24 months for grow-out before a heifer enters the breeding phase. After that heifer is bred, there is another period of gestation and another 15- to 36-month grow-out period before her progeny becomes beef. In areas of intensive beef production the time lapse is 4 years. In extensive operations where cattle are finished on forage, for example, in Argentina or Australia, it may be 6 years from the time a decision is made to increase herd numbers until the additional beef is ready to eat. On the down side the only way to reduce cattle numbers is to increase cattle slaughter.

A second factor to influence numbers of cattle involves short-run or seasonal relationships that make the long-run cycle more erratic and thus increase the difficulty of predicting turning points. Associated with seasonality is the impact of climatic factors such as periodic droughts or unusually good weather.

The effect of competing meat supplies is another important variable. Technological advances that have led to improved poultry production or to large, efficient hog operations have had substantial impact on the demand for beef and thus on beef prices and, of course, on cattlemen's profits and production-related decisions. In countries such as the United States there is an important interrelationship between the hog cycle and the beef cycle, even though the estimated cross-price elasticity of pork and beef is very low, 0.10 (see Chapter 2). Associated with competing meat supplies are changes in consumer tastes and preferences. It is expected that blended beef, i.e., a combination of texturized vegetable protein (tvp) and ground beef will be common by 1985. Certainly, as beef prices increase, more meatlike substitutes are used, especially in processed foods such as TV dinners (Simpson and Degner 1979).

Social, political, and economic upheavals such as war, worldwide depressions, or domestic recessions also greatly influence cattle cycles. A recent example is the so-called energy crisis related to the cartel actions initiated by the Organization of Petroleum Exporting Countries (OPEC) in 1973. At that time the beef exporting countries had increased herds substantially. The immediate action of most importing countries was to reduce imports of nonessential items like beef to conserve foreign exchange for vital petroleum imports. That action was a major factor in reducing trade and world prices.

INTERNATIONAL CATTLE CYCLES. This section demonstrates that cattle cycles in Europe, Australia, New Zealand, and several Latin American countries are strongly influenced by cycles in the United States. This relationship arises because the United States imports nearly one-third of the beef entering world trade. Thus the United States is in the position of being a price leader. The charts presented in this chapter show inventories from various countries plotted in relation to the U.S. cycles. The first graph (Figure 9.2) shows that the United Kingdom has a definite beef cycle whose high point in numbers in 1975 corresponds quite well to the U.S. peak at that time. The subsequent declines have also been parallel. Australian inventories have doubled since their low point in 1967. Another peak took place in 1976, and

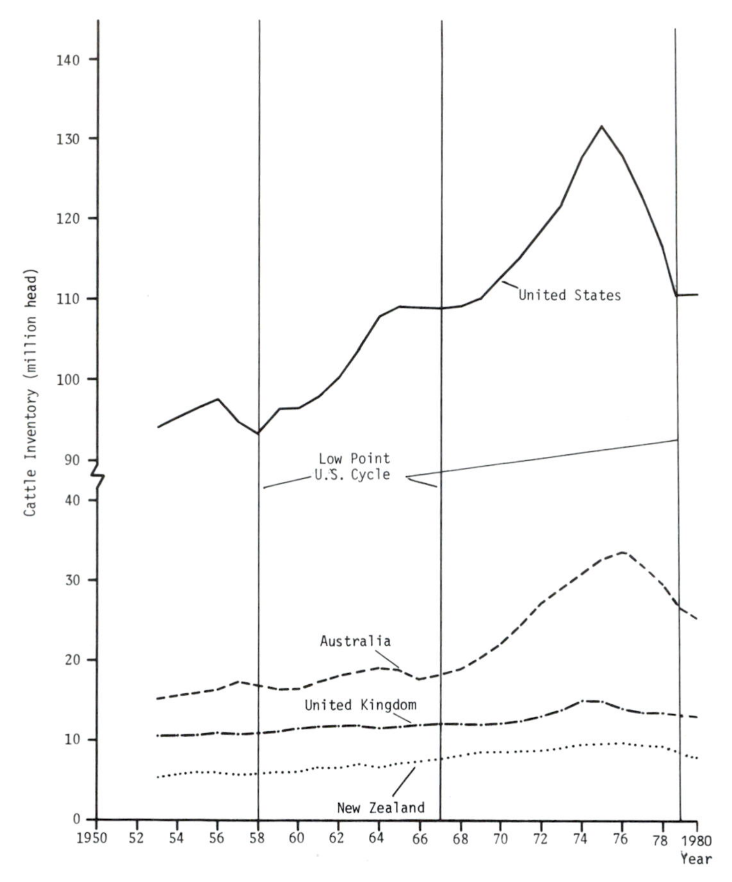

Fig. 9.2. Cattle inventory of Australia, New Zealand, United Kingdom, and United States, 1953–1980.

declines have occurred since that time. A strong relationship between the U.S. and Australian turning points is evident as far back as 1953. Although New Zealand has a strong dairy orientation, inventories there have almost doubled in the past 20 years, and most recent peaks also correspond to the U.S. cycles.

It is more difficult to trace out cycles for France and West Germany (Figure 9.3) because they are part of the nine-country European Economic Community (EEC) and thus are bound by its common agricultural policy (see Chapter 2). Also, their breeds provide both milk and meat. Less common are enterprises having a single-purpose breed. The minority are operations having two breeds of cattle, one for dairy and one for beef. In countries such as France and Italy the milk-meat production system coexists with specialized beef enterprises, and feedlots are developing. Operations in the Nordic countries, the Netherlands, and Switzerland are almost all of the milk-meat type. In Ireland, Scotland, and England much of the beef production derives from specialized beef herds. Examples of these characteristics are presented in Table 9.1, which shows the importance of dual-purpose cattle by the high ratio of milk cows to all cattle.

Beef cattle numbers in Europe as a whole have paralleled the United States quite well (Figure 9.4). Both areas had significant rises from 1971 to 1975, but have had declines ever since. The EEC has also followed the United States pattern fairly well. It may be concluded that a definite European cattle cycle exists, but that primarily countries with the most beef-type herds are affected by U.S. cycles.

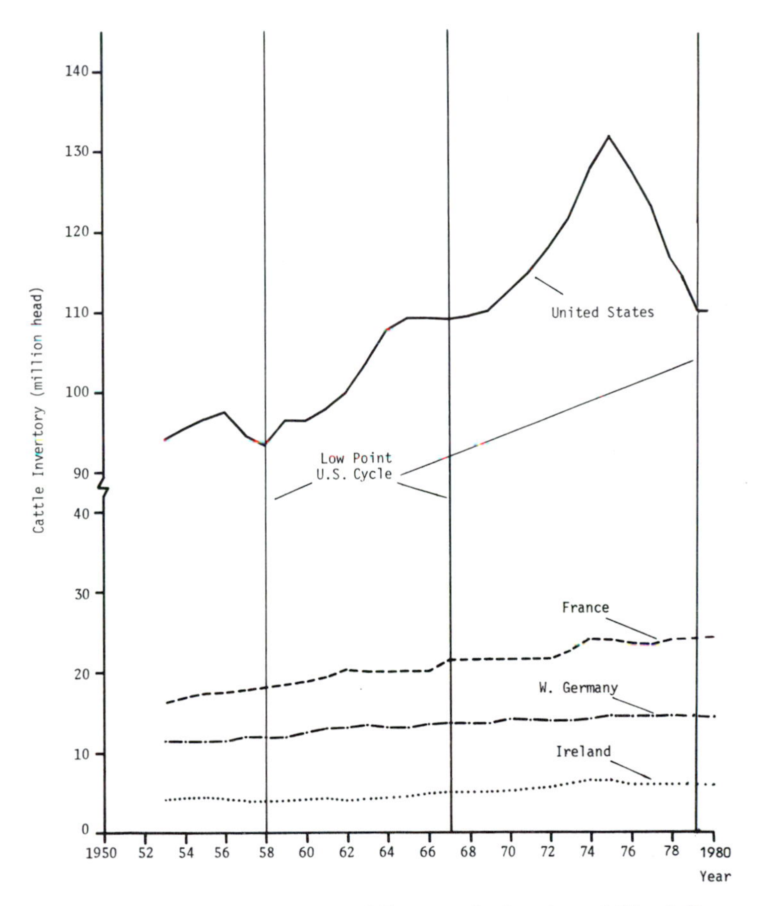

Fig. 9.3. Cattle inventory of France, Ireland, and West Germany compared with the United States, 1953–1980.

Table 9.1. Milk cows as a percentage of total cattle in selected countries, 1978

	Milk cows	Cattle	Milk cows as a percentage of cattle	Importance of dual-purpose cattle
	(1,000 head)			
France	10,230	24,133	42	High
West Germany	5,360	14,763	36	High
Ireland	1,521	7,154	21	Medium
Italy	3,660	8,500	43	High
Sweden	654	1,892	35	High
Switzerland	880	2,024	43	High
United Kingdom	3,327	13,660	24	Medium
United States	10,848	116,265	9	Low
Argentina	2,615	61,280	4	Low
Chile	760	3,492	22	Medium
Paraguay	602	5,800	10	Low
Uruguay	450	9,424	5	Low

Source: Adapted from FAO *Production Yearbook,* 1978.

There does not appear to be a beef cycle for South America as a whole (Figure 9.4). On the contrary, cattle numbers have increased sharply and constantly, from about 140 million head in 1960 to more than 210 million head in 1979. The country mainly responsible for the increase is Brazil, which has accounted for approximately two-thirds of the total increase in South America. Although there is no Brazilian cycle, definite cycles corresponding to U.S. inventory changes can be traced out for Argentina and Uruguay, the two largest South American beef exporters. The Argentine cattle cycle is largely related to weather and government policy, although inventory numbers follow patterns similar to those in the United States. Despite close market ties with the United States, long-term cycles cannot be traced out for other Latin American importing and exporting countries such as Costa Rica, Honduras, or Nicaragua, largely because procedures for estimating inventory are less than satisfactory, and because weather and internal economic conditions and policies override the external cyclical influence.

The conclusion to be drawn is that many of the larger importing and exporting countries have well-developed cycles and that their inventory changes have become more pronounced since the United States instituted beef import quotas in 1964. Given the close relationship between turning points, it appears that the United States is the leader in price making and inventory adjustment in the major beef exporting countries. This conclusion is also supported by the slight lead of U.S. prices over prices in other countries (Hinchy 1978). Cattle inventories and prices can be expected to continue fluctuating in a cyclical manner so that despite different social

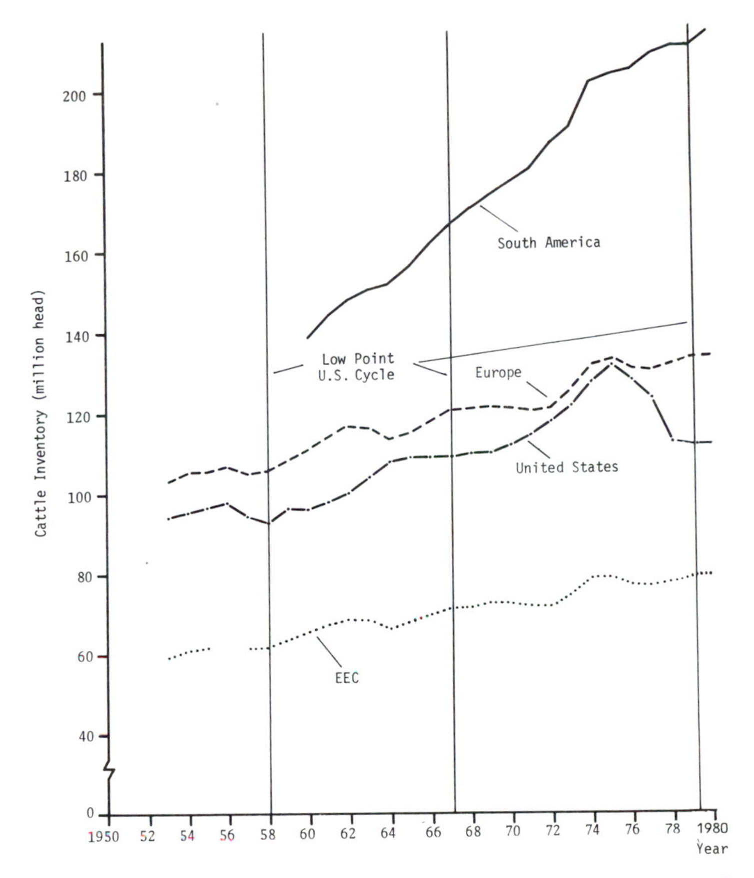

Fig. 9.4. Cattle inventory of Europe, EEC, Argentina, and South America compared with the United States, 1953–1980.

and economic systems and countercyclical import legislation by the United States and the EEC, cattle producers the world over will continue to be rewarded or penalized depending on the phase of the cycle (Simpson 1981).

WORLD BEEF PRICES AND PRICE POLICY

Beef prices are complicated by the biological constraints inherent in cattle raising. Whereas a factory producing commodities such as bottles of artificial orange juice, men's suits, or even cars can adjust production in a relatively short period of time in response to consumer demand, the beef cattle industry takes a long time to adjust production patterns; it takes years to make drastic changes. The result is that supply curves—which in effect express the relation between price and quantity produced—show little short-run response (within a year). They are more elastic in the long run; that is, there is larger change in quantity for a given change in price than in the short run. In the very short run, for example, producers who have brought stock to market are reluctant to take them home, even if prices are much lower than they expected. Thus, since supply is relatively fixed in the short run, producers are less responsive to price changes than they would be if they had more information about market conditions. Even in the short run, marketings can be speeded up or slowed down by price changes and changes in price expectations. Producer reactions are most responsive in inventory change over the longer run (Farris 1969).

Livestock supply is affected by factors other than delayed output response and cattle cycles. Production costs and alternative production possibilities affect decisions, as do seasonal variations (James and Farris 1971). Also significant is the source of beef, for example, draft or dairy animals versus specialized beef cattle. Drought, such as the one in the Sahel zone of Africa in 1972–1973, can seriously reduce the supply for several years. Weather is especially important in beef trade as most exported beef is from cattle fattened on pastures. In areas such as the tsetse fly zones of Africa, disease is a major limiting factor. Finally, response to price changes depends on the type of production system. Many migratory cattle owners count their wealth in livestock rather than in holdings of money. They may sell only enough cattle to meet their modest fixed cash requirements and, because their wealth is in cattle, they may sell less rather than more cattle as prices rise. This same phenomenon of a backward-bending supply curve occurs, of course, in the initial part of the cattle cycle even in developed countries as cattlemen hold back females to build up herds.

PRICE POLICY. Governments adopt certain price policies to promote economic development and to meet perceived welfare needs of both producers and consumers. The policies have tradeoffs and so their efficacy depends on well-defined objectives. One policy might be to control meat or prices of live cattle for the benefit of consumers and, perhaps, the regulation of butcher and trader profit margins. If prices are set too low, however, profit rates on production will decline and new investment will not be forthcoming. The general experience of meat and livestock price controls in developing countries has been unfavorable, at least in terms of stimulating economic development. Apart from restricting investment, butchers react by reducing product quality and services to customers. Black markets appear and marketing enterprises do not adopt new methods.

Another possibility in price policy is to raise producer prices. But experience attests to the difficulty of permanently and effectively guaranteeing them. Even the apparently innocuous practice of setting "floor" or minimum prices can lead to substantial problems, for there is no means of compelling independent traders to buy cattle they do not want. Official government buying organizations that attempt to regulate supply through compensatory buying schemes eventually fail unless they have abundant resources such as those in the EEC. Even the EEC's artificially generated surplus had become a problem by the late 1970s.

Despite difficulties with such policies, most lesser developed countries have found it necessary to wed the free market trading system to some type of price stabilization program. These schemes work better if the country has large processing or exporting enterprises because they will be easier to monitor. In Botswana, the Botswana Meat Commission operates a price stabilization fund that guarantees price 1 year in advance. The administration of such regulated pricing policies is costly in a modern marketing system; it is much more difficult in a traditional system.

Although there are convincing theoretical arguments for general reductions on tariff and nontariff trade barriers, in practice judicious use of export or import duties, subsidies, and quotas can effectively influence the level of domestic prices. An export duty on beef and beef animals, for example, reduces the profitability of supplying foreign markets and thus diverts more of the product to domestic consumers. Several Latin American countries have used this device to obtain government revenue and to keep down domestic meat prices. In periods of high world prices many of those countries have reversed these policies and have even restricted domestic beef supplies to encourage exports and to gain additional foreign exchange.

WORLD CATTLE PRICES. The annual price fluctuations in canned beef and chilled quarters that enter world trade from Argentina are described in Chapter 11 along with domestic wholesale beef prices. We now compare live cattle and carcass prices for a number of countries. The live cattle prices for fat stock or fed cattle, that is, animals destined for slaughter, during the period 1955–1977 are given in Table 9.2. They are graphed in Figure 9.5 for Argentina, the United States, and West Germany. The hypothesis that the United States is a price leader is supported by the close correlation between the U.S. situation and Argentine prices.

Table 9.2. Wholesale live cattle prices in selected countries, 1955-1977

Year	Argentina[a]	Canada[b]	Denmark[c]	West Germany[d]	Ireland[e]	United Kingdom[f]	United States[g]
				(cents per kg)			
1955	...	43.7	37.6		36.1	42.7	51.1
1956	...	43.6	39.7		28.9	40.4	49.2
1957	...	43.8	38.9		33.8	44.6	52.5
1958	...	52.1	39.2		36.4	43.8	60.5
1959	16.9	57.7	41.8		37.3	43.9	61.4
1960	17.9	51.4	41.3		34.4	43.1	57.8
1961	17.3	49.4	39.4		34.2	44.8	54.3
1962	15.4	53.1	37.9		35.6	45.6	61.0
1963	17.8	48.2	39.4		34.7	46.8	52.8
1964	29.6	46.3	51.0		40.9	49.0	51.0
1965	30.3	48.9	52.8		45.1	49.5	57.7
1966	26.3	52.7	47.1		42.0	50.7	58.0
1967	20.9	56.4	41.6		42.8	49.3	56.6
1968	20.4	54.9	40.7	69.1	43.5	48.6	61.2
1969	20.4	60.2	47.7	72.6	45.2	51.1	67.3
1970	27.6	64.1	50.9	75.0	48.3	55.0	66.8
1971	...	71.5	53.1	79.6	54.6	60.6	71.4
1972	37.5	79.2	81.0	104.4	67.3	79.4	78.9
1973	46.4	91.5	103.5	126.8	85.2	90.8	92.4
1974	49.4	107.1	98.1	126.9	70.2	86.4	92.4
1975	28.3	107.1	116.1	146.3	89.2	103.3	98.3
1976	40.3	105.1	122.7	145.4	99.4	101.4	86.2
1977	41.0	82.7	135.9	162.8	...	...	86.2

Source: FAO *Production Yearbook,* various issues.

[a]Fat steers, special, for export, Buenos Aires, good, medium weight.

[b]Good steers, Toronto, all weights.

[c]Steers for export, best quality, Copenhagen, not including supplement paid to producers.

[d]Steers, young, well fleshed, sixteen leading markets.

[e]Through 1964, fat cattle, 2 to 3 years old, at fairs, from 1965, bullocks, Hereford Crosses, at least 1,008 lb (2,218 kg) but less than 1,120 lb (2,464 kg), at Dublin.

[f]Certified steers and heifers, average price at live-weight auctions, including payments under Fatstock Guarantee Scheme.

[g]Steers, choice, all weights, Chicago.

Perhaps most interesting is the relationship between prices of EEC countries and those of non-EEC countries. West Germany's prices, for example increased 2½ times from 1968 to 1977, while non-EEC country prices did not even double. Furthermore, in sharp contrast to prices in Argentina and the United States, which declined after 1974, prices in West Germany continued to increase at a rapid rate. The same is true of Denmark and the other EEC countries, which set guide prices according to cost of production rather than through the forces of internal or external supply and demand.

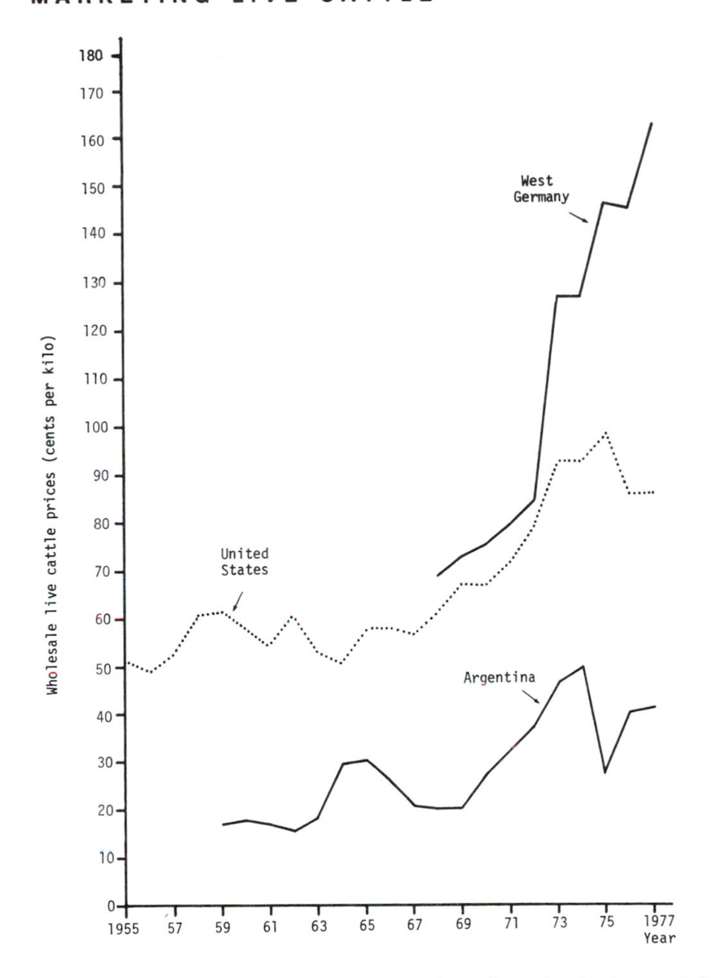

Fig. 9.5. Wholesale live cattle prices in selected countries, 1955-1977.

MARKET INTELLIGENCE AND PRICING

In most countries with more than just a rudimentary livestock marketing system, state and/or federal agencies combine to report prices, quantities, weights, and other market information from the various terminal and auction markets. In addition, reporters may contact wholesale meat markets and packers, as well as individuals known to be involved in direct sales. Producers or other interested parties in the United States and some European countries can usually obtain the latest quotes by telephoning a specialized bureau (McCoy 1980). The state and/or federal governments also organize the information and publish a series of daily, weekly, monthly, and annual reports as well as analyses of the information. In addition to the government there are various other sources of information such as breed and cattlemen's associations, magazines, universities, the cooperative extension service, and the terminal or auction markets themselves. Finally, private groups such as consulting companies and media centers produce newsletters containing information about the livestock market.

People concerned with livestock who live in countries with good market intelligence systems often are unaware of the important role that timely and reliable information plays in their daily activities. Their reliance on these systems is somewhat analogous to our reliance on the telephone or automobile; once the tool is adopted, it becomes indispensable in a business operation. Furthermore, reliable market information is vital in any country, whether prices are fixed or not, as a means of planning for a consistent supply of beef. If producers and intermediaries have sufficient market information, they will not be taken advantage of by unscrupulous traders. This concern for improving marketing practices leads to government control, regulation, and the provision of various services.

As a country develops and people become more sophisticated, various devices can be introduced to reduce marketing costs and to assist in meeting consumer preferences. One example is livestock grading. In most of Africa, each time that ownership transfer takes place, the purchaser must personally examine the animals because there is no grading system to guarantee sales made without visual inspection. Furthermore, complementary institutions such as reliable telephone service and weigh scales are not available. The excessive cost of this type of poor marketing situation can be derived by comparing marketing in Africa with that in the United States, where feedlot operators can make their purchase of feeder cattle by merely calling an order buyer and giving that person the weight, type, and grade desired. Millions of cattle are sold over the telephone each year, and in this way marketing costs are substantially reduced.

Most Western European countries have well-developed market intelligence systems that work well as long as there is sufficient information from which *the* price can be determined. As in the United States and Canada, direct marketing poses a problem, however, in that the numbers of sales reported are continually being reduced as more meat trading takes place on a formula basis. Furthermore, as the size and concentration of packers and retailers grow, there may be more opportunity to manipulate market prices. A considerable amount of money has been spent on evaluating margins throughout the world because of a widespread and persistent fear that middlemen—that is, the people responsible for all the functions between the farm gate and the consumer—are reaping undue profits from both the farmer and consumer through one or a combination of factors such as inefficiency, duplication of efforts, illegal influencing of prices, and noncompetitive buying and selling practices. Continued concern and surveillance are justified by the existence of oligopolies and oligopsonies (few sellers or few buyers, respectively) in specific localities and at different levels of the industry in every country (Wittenberg 1977). As a result, legal actions have been brought against auction markets, packers, processors, food chains, and others, mainly by groups of cattlemen.

Another problem is that although direct marketing has introduced greater efficiency, publically negotiated sales of cattle and meat have been reduced and the use of formula pricing has increased (Ward 1979). This pricing technique, in which sales are made according to a formula based on published prices in a specified source, is used for domestic pricing in much of Europe, Canada, and the United States. Furthermore, a substantial amount of meat from all over the world destined for the export trade is based on prices published in the *Yellow Sheet,* a daily market news report published by the National Provisioner, Inc. (Comision Interinstitucional de Mercadeo de Carnes 1975). If those prices do not accurately reflect competitive market conditions, sales of cattle and meat throughout the world might be biased. This would likely be a short-run condition because a price below the equilibrium level could not be sustained in the long run. Because of the complexity of the beef business, even government efforts to fix beef prices generally are only partially successful.

The concern about formula pricing led to a special study by the U.S. Department of Agriculture, which concluded that prices quoted by the *Yellow Sheet* on the average accurately reflect the sample of negotiated prices (USDA 1978). Their analytical results from price studies were inconclusive in determining differences between the average prices for sales made on a formula versus negotiated basis. A subsequent task force on meat pricing appointed by the U.S. secretary of agriculture concluded that, while formula trading should not be banned, action should be taken to enlarge the pricing base. The task force also recommended that the meat industry be encouraged to proceed with development and pilot testing of an electronic meat marketing system, but that the USDA should not own or operate this meat trading system (USDA 1979).

CATTLE MARKETING CHANNELS

The traditional marketing channel for cattle in any country has historically involved movement from the production unit to

Commercial feedlots usually sell directly to a packer buyer who offers a specific price per pound for the entire lot. The cattle are weighed, loaded on a truck, and will usually be slaughtered within 24 hours.

a local market and then to a central market. The cattle then pass through an abattoir to a central meat market and finally to a butcher shop. In more primitive systems, a butcher simply purchases an animal at a local market, kills it and sells portions of the meat from stalls. The opposite extreme is a modern integrated marketing channel in which cattle are sold directly to the abattoir, packinghouse, or processing plant and then moved directly to a supermarket.

Cattle can be sold directly from producers to slaughterhouses or, as noted, pass through a series of middlemen. Similarly, there are numerous methods for transporting cattle, depending on geographic conditions, marketing sophistication, and tradition. Trekking or trailing of cattle is common in much of Africa and Latin America but is seldom used in the more economically advanced countries, where almost all livestock are now moved by truck or trailer. Rail transport for livestock was fairly common all over the world after World War II but is now seldom used because of improved road systems. Major problems of rail transport are that cattle have to be rested, fed, and watered every 24–30 hours, there is difficulty in coordinating freight with livestock, and the volume of traffic is usually not high enough to warrant running special trains. In general, the advantages that railroads sometimes offer in lower freight rates are more than offset by indirect costs.

Water transport is used in much of the world for both local and international marketing of live cattle. There is considerable trade from Malta to Yugoslavia and from Thailand to Hong Kong. An expanding sea trade is taking place between the exporting countries of Ethiopia, Somalia, and Sudan and the importing countries of Saudi Arabia and the Persian Gulf states. Australia, Argentina, and Ireland regularly use special vessels for sea transport of cattle (Fenn 1977).

Air transport has become more important for both international shipments and domestic movements of cattle. There are now regular shipments of meat and cattle from the Bolivian lowlands to La Paz. Italy imports substantial numbers of calves for fattening by air transport, many of them from the Western Hemisphere. Several times a week whole plane loads of breeding cattle leave Miami, Florida, bound for Latin America. This volume has grown to the extent that one company specializes in setting up, disassembling, and cleaning cargo planes used for animal transport.

There are numerous types of livestock marketing institutions to facilitate transfer of cattle ownership. Some of the more common ones are:

- Terminal public markets, which are also called public stockyards or central public markets.
- Auctions, which may be owned by individuals, corporations, cooperatives, or by governments.
- Local markets and cattle assembly or concentration yards, which have fixed facilities.
- Local dealers who buy and take possession of the cattle. They primarily purchase at the production or auction level.
- Packer buyers purchasing for a packing plant.
- Packing plants and packer buyer stations.
- Order buyers who act as agents for purchasers of cattle.
- Other farmers who purchase for their own account.
- Butchers and retailers.
- Special types of auctions held infrequently, primarily for feeder calves.
- Cooperative shipping associations owned by farmers for assembling and forwarding cattle.
- Cooperative selling associations that act like cooperative shipping associations but also carry out additional functions.
- Commercial feedlot operations, which may purchase or sell cattle directly or through an agent.
- Commission men who act as agents, but do not take ownership of cattle.

• Producer bargaining associations that act as agents for their members.

These institutions may be found in nearly all countries that have primarily cow-calf types of operations, but the degree of specialization depends on the level of economic development and the location within the country. In the western part of Texas in the United States, for instance, there are few auction markets relative to cattle numbers because the ranches are large and generally sell directly to feedlots or to order buyers. In eastern Texas there is a heavy concentration of auctions because the production operations tend to be much smaller. Furthermore, because many of the livestock owners have mixed operations of livestock and crops, they may purchase stocker calves to supplement or complement their own crop production activities.

As economies develop, new functions and institutions are introduced. Nearly all industrialized countries, for example, now have some cattle feeding. The extent depends on the price of grain or other feedstuffs. If cattle prices are low relative to grain, then cattle are grown and fattened on grass or other forages. As the cattle-grain price ratio increases, at some point it becomes cheaper to fatten cattle on grain than on grass or some other forage. In the United States and Japan, as well as much of Europe, most cattle are fed intensively in confinement lots. In some areas, such as northern Mexico, Zimbabwe, and parts of the Near East, feedlots are beginning to develop. The point is that in the early stages of an economy's development there is no specialized fattening function. A large number of African producers just sell old cull animals, for example, with no thought of feeding them to increase weight and improve quality and yield.

A COMPARISON OF CATTLE MARKETING SYSTEMS

The purpose of this section is to highlight some differences in the cattle marketing systems of Europe and Africa. Other marketing systems are discussed in other chapters. As we look at the various systems it will become apparent that the type of institutions that have evolved are a product of various factors such as cattle production methods, cultural values and mores, the economic system, level of economic development, and type of cattle. Despite the diversity of variables, there are some rather predictable patterns, which we will attempt to set forth.

EUROPE. In the socialist countries of East Europe such as Bulgaria, Czechoslovakia, East Germany, Hungary, Poland, Romania, and Yugoslavia, as well as in the Soviet Union, state agencies purchase a substantial number of all cattle sold. The predictable corollary is that, in most cases, prices are fixed. Also, producers generally market directly to the processing plants or abattoirs, most of which are also state owned. State control, as well as a long history of cattle sales for specialized purposes such as fattening, has also led to the adoption of grading standards. In Czechoslovakia, the standards are based on dressing percent, while in Poland they are based on live weight.

It may seem odd that weight classifications, as in the Polish system, can be used for grading purposes, but in countries with dual-purpose cattle, calves become a residual from the milking operation and, as a consequence, quality grades are not as important as they are in countries where cattle types are more sharply divided between beef and dairy purposes. This leads us to another grouping of countries—those which have primarily dual-purpose cattle, but whose industries are based on a free enterprise system. Some of these are Belgium, France, Italy, the Netherlands, and West Germany. Most farms in these countries are small to medium operations with 10 to 50 or 75 cows. Their membership in the EEC means their prices are regulated mainly by subsidies and protection from imports to assure most producers a profit (see Chapter 11 on trade). Nearly all European countries still have weekly sales in central markets, a mechanism that is supplemented by the activities of dealers and commission-brokers, who buy directly from farmers. There are regular breed auctions in West Germany, but commercial cattle auctions are rare. In Ireland, on the other hand, both commercial and breeding cattle auctions are common.

Important changes have taken place during the past two decades in nearly all the Western European countries. The sale of cattle through the traditional livestock markets, which have generally been owned and run by municipalities in cooperation with the local agricultural organizations and traders, has declined. Meanwhile, direct sales to slaughter houses as well as contractual selling and ownership integration have increased. Direct sales of slaughter cattle to processing plants, which can be owned by a wide variety of groups such as the state, cooperatives, individuals, or corporations, vary greatly among countries as indicated below:

Country	*Direct sales (%)*
Norway	100
Sweden	95
West Germany	78
Ireland	65
France	60
United Kingdom	50
Denmark	47

These data from the mid-1970s can be compared with data for the same period from the United States and Canada, where direct sales of fed cattle are 85 and 55 percent of all fed cattle sales, respectively (OECD 1978).

Changes in live cattle marketing procedures must be analyzed with caution. It is true that direct marketing of fatstock (fed cattle) is on the increase, but auctions in the United Kingdom and municipal markets in all cattle producing countries are still extremely important in transferring ownership of calves and cull animals. The marketing of livestock in France differs from that in other EEC countries in that livestock dealers still play an important role despite the recent development of producer marketing groups. In 1972 livestock dealers handled about 82 percent of the total beef cattle.

Administered marketing systems, which are also combined with domestic production and import regulation, are found principally in countries such as Sweden or Switzerland that have high production costs. Switzerland has a parastatal cooperative (CBV) that is combined with the Swiss Federation of Livestock Producers. Each producer in this efficient system informs the federation of marketing intentions a week in advance. The federation then places a value on each animal. If cattle are not sold the CBV purchases them, provided the producer attends the sale. Buying-in prices, fixed every two weeks by the CBV, are based on production costs. Prices are allowed to fluctuate within a small range. Sweden follows a similar procedure, except that the established price is based on world prices. If the domestic price climbs above the upper limit, the import levy is lowered, and vice versa.

The use of cooperatives as a marketing tool varies considerably among countries. Cooperatives are used to a greater extent in areas that are socialist oriented, and in those having a high percentage of small farmers. Although comparative data are available for only a few countries, it is known that in the middle 1970s about 82 percent of all livestock in Sweden were marketed through cooperatives. The percentages for Norway, Denmark, and Ireland were 72, 65, and 55, respectively. West Germany and France had 34 and 20 percent, respectively. The role of cooperatives is small in Australia, Belgium, the United States, New Zealand, and the United Kingdom.

In general, cooperatives work best where input supply needs can be combined with marketing functions. Thus, farmers who use cooperatives for activities other than just livestock marketing, are more inclined to use them for marketing cattle as well. Since extensive beef operations tend to be specialized activities having few purchased inputs that a cooperative can sell more cheaply than a private organization, owners of cattle in these situations tend to use cooperatives much less than people from a mixed agricultural operation. Another factor is that the more prices fluc-

On this large ranch calves are being sorted for sale. The cows will go back to the high country pastures. Calves will be moved to high quality pasture or to a feedlot for finishing before going to slaughter.

tuate, the more likely cattle raisers are to use a marketing cooperative only when it is to their advantage. To be successful, a livestock marketing cooperative must have a continually high volume, which in turn requires a long-term commitment from its members. Such a commitment is hard to obtain unless there is a rigid pricing and marketing system and there are few alternative sale outlets.

It may be concluded that direct sales, ownership integration, and contractual arrangements will continue to expand in the European countries. As direct marketing increases, changes in market intelligence systems will have to take place. Livestock grading and classification schemes will improve with increased demands for transactions on a description basis to reduce the need for a meeting of suppliers and buyers. With increased direct purchasing there will be larger and better organized firms, all the way from producers to retailers. There will also be greater interrelatedness as large retailers, especially supermarkets, sharpen their specifications for a continuous, large, homogeneous supply of good quality meat.

AFRICA AND THE NEAR EAST. The cattle marketing systems in Africa and the Near East, rudimentary and informal as they are, have been serving the needs of traditional, subsistence level, pastoral peoples for centuries. As these regions experience changes resulting from twentieth century economic/technological progress and modernization, and as the nonagricultural community demands more, better, and less expensive meat, there is pressure to make sweeping changes in production and marketing practices. There is no one cattle marketing system for all Africa, since systems there vary widely, from relatively well-developed organizations in certain parts of Zimbabwe and South Africa, to the old market structure of Chad that in some respects dates back 200 years. In countries such as Sudan livestock marketing is almost entirely in the hands of private merchants whereas in countries like Iraq and Syria responsibility for livestock marketing has been given to cooperatives.

Marketplaces, usually in small towns or villages, remain the focal point of cattle exchange wherever livestock producers are still only marginally connected with a market economy and sell their animals whenever they need cash, rather than produce for a market economy. Few of these primary markets have more than rudimentary watering and feeding facilities or weighing equipment so that cattle are almost always sold by the head. The East African markets are often better equipped, organized, and taken care of than those in the Near East largely because markets of the Near East are generally operated by municipalities that regard the markets as a source of income and attraction of nonurban business, rather than as a development tool (FAO 1977).

Various local customs, religious traditions, and transport infrastructure (or lack of it) have created a wide range of sales practices. In some parts of Africa cattle are trailed in and several hundred head usually change hands within a few weeks, whereas in Somalia livestock marketing takes place at all main centers practically every day so that trading is usually restricted to a few hundred head or less. Secondary or wholesale markets (terminal markets) are important only in a few countries such as Sudan. In that country about 50 percent of all the animals are marketed through the Omdurman facilities (Fenn 1977).

Trekking is still the most common method of transport in much of Chad, Mali, Niger, and Sudan, where cattle will often be driven more than 500 km (Ariza-Nino et al. 1980). Although cattle can actually gain weight during trekking if grazing is good, cattle have to be fully mature to withstand the transport; they also run the risk of contracting and spreading disease. Also, this method is clearly unsuitable for cattle that have been fattened under intensive conditions and that carry good flesh. Finally, trekking has little flexibility as a marketing system for it is limited to periods when forage is available.

Ethiopia, a country in which about 70 percent of the offtake is consumed in rural areas, provides a useful case study of the type of marketing problems that arise when trekking is the dominant system of moving cattle. The poor distribution of large slaughterhouses means that cattle have to be trekked great distances without the benefit of watering, feeding, and resting facilities, or quarantine control. As a consequence, animal weight losses are high and, once the cattle do arrive in the poorly organized markets, which are little more than open pastures or rough holding corrals, drovers are frequently exploited (World Bank and FAO 1977).

The Ethiopian case contrasts with the movement that takes place from Fada N'Gourma in Upper Volta to the capital city of Lomé, in the country of Togo. Both countries are in northwest Africa, just west of Nigeria. On the basis of a detailed analysis of the transport cost, Josserand and Sullivan (1979) recently concluded that traders involved in the 765–800 km movement do not make undue profits, since return on investment is about 8–12 percent on an annual basis (Table 9.3). They also compared the cost of truck transport versus trekking and discovered that trekking is the cheaper method even though the trader's capital is tied up for 45–60 days as opposed to 13 days (Table 9.4).

Table 9.3. Margins of cattle trekked and trucked from Fada N'Gourma in Upper Volta to Lomé, Togo, April 1978

	Trekking		Trucking	
	Mature steer (300 kg live weight)	Cow (250 kg)	Mature steer (300 kg)	Cow (250 kg)
		(U.S. $)[a]		
Purchase price in Fada	195.45	159.09	195.45	159.09
Marketing costs[b]	56.07	56.07	64.86	64.86
Total cost delivered Lomé	251.53	215.16	260.31	223.95
Price at Lomé stockyard	281.82	240.91	281.82	240.91
Merchant's margin	30.29	25.75	21.50	16.96
As percent of total investment	12	12	8.3	7.6
Price Lomé stockyard per kg, carcass weight[c]	1.88	1.93	1.88	1.93

Source: Adapted from Josserand and Sullivan, 1979.
[a]Exchange rate of U.S. $1.00 = 2.20 CFA.
[b]See Table 9.4.
[c]Assuming a 50 percent dressing percentage.

Table 9.4. Cost of trekking cattle versus mixed transportation from Fada N'Gourma, Upper Volta to Lomé, Togo, April 1978

	Trekking[a]		Mixed Transport[a]	
	Cost/ head[b]	Total herd cost	Cost/ head[b]	Total herd cost
	(U.S. $)[c]			
Purchase-related costs				
Intermediary's commission	1.14	51.14	1.14	51.14
Identification mark	0.05	2.05	0.05	2.05
Gardiennage	0.23	10.23	0.23	10.23
Taxes (Voltaic)				
Merchant's import license	0.36	16.36	0.36	16.36
Veterinary inspection	0.68	30.68	0.68	30.68
Export tax	29.55	1,329.55	29.55	1,329.55
Export permit	...	0.91	...	0.91
Vaccination	0.23	10.23	0.23	10.23
Taxes (Togolese)				
Import tax	5.45	245.45	5.45	245.45
Laissez-passer sanitaires	...	0.45	...	0.45
Transportation				
Salary, 3 herders	...	163.64	...	54.55
Food	...	143.18	...	40.91
Salary and food, seller/agent	...	90.91	...	34.09
Return transport, herders and agent	...	90.91	...	90.91
Trucking of cattle, Dapango-Lomé	...	...	...	675.00
Commissions and other costs				
Intermediary's commission at border	...	45.45	...	45.45
Damage to crops, waterholes, wells, etc. or to trucks	...	34.09	...	22.73
Para-official taxes	...	22.73	...	22.73
Mortalities, losses of cattle, forced sales	...	90.91	...	90.91
Market tax	...	22.73	...	22.73
Dillalis commission	...	51.14	...	51.14
Gardiennage at Togblékopé or stockyard	...	70.45	...	70.45
Total	56.07	2,523.19	64.86	2,918.65

Source: Adapted from Josserand and Sullivan, 1979.

	Trekking	Mixed
[a]Size of herd (head)	45	45
Distance trekked (km)	765–800	165
Distance trucked (km)	...	560
Duration trekking (days)	45–60	12
Duration trucking (days)	20	1

[b]Expenses calculated on a per head basis. The total per head is derived by dividing the total cost by 45 head.

[c]Exchange rate of U.S. $1.00 = 220 CFA.

Another interesting finding was that while cattle trekked over distances of several hundred kilometers are often reported to lose considerable weight during the journey, it was discovered that weight loss is very slight if the animals are walked at a normal pace (about 20 km / day) and taken care of properly. Josserand and Sullivan found that in some seasons cattle actually put on weight. The conclusion one reaches is that in-depth research using partial budgeting is useful prior to making policy recommendations about market improvements.

A major cattle marketing difficulty, the wide variation in prices for cattle in market economies, can be viewed as both a problem and an opportunity. To the manager, it is a problem to anticipate seasonal and cyclical price variations and to adjust the production system to anticipated changes. It is also an opportunity to make extra money from a good marketing strategy. This type of market illustrates the strong incentives placed on an industry to produce what consumers want—at a time, place and form desired. On the other hand, the depressed prices during drought signals to consumers the incentive to increase purchases.

The material in this chapter has provided a sketch of the diversity of marketing methods, organizations, and facilities throughout the world. Just as in production, cattle marketing improvements in many parts of the world have lagged behind many other agricultural industries. There are a variety of reasons for this associated with the nature of the business, the solutions of which require the cooperation of a group or groups of producers. Marketing decisions are complex and must be made by keeping fundamental economic and social factors in mind.

Governments in the industrialized countries, and in virtually all of the Western Hemisphere, have been able to reach current levels of marketing efficiency because of improved methods of collecting and disseminating market information and because of infrastructure development, such as the construction of roads, communication services, and marketing facilities. It is ironic that governments in much of Africa are still struggling to devise a means of adequate market reporting by radio, while electronic markets based on the principle of a centralized trading system—via some form of telecommunication such as a telephone with computerized record keeping capabilities—are now being tested in the United States (Davis and Sporleder 1978).

Whereas cattle are still bartered by the individual head in Somalia, tele-auctions are commonly used in Canada and in the United States (Engleman et al. 1979). In that electronic system, in which potential buyers are connected with the auctioneer and each other by telephone or teletype, the animals are inspected, marked, and graded prior to the auction. Lots are organized on paper and bids are made across long distances.

Telephone auctions are not common but teletype auctions are widely used, especially in Canada.

A new concept in auctions is to use video tapes of cattle taken on the production units so that potential buyers can view the animals rather than make a personal inspection. The cattle are then auctioned off from a giant TV screen. This method exposes consigned cattle to many more buyers than an individual market possibly can. The system is used in several areas of the western United States, and has proved to be popular where cattle are of a fairly uniform grade and quality.

SUMMARY AND CONCLUSIONS

Many variations of live cattle marketing exist within some countries and throughout the world. All of these, however, must deal with trying to create greater value by improving place, form, and time utility. Good planning and herd management and good communication among sellers and buyers are essential to effective marketing. Marketing systems that facilitate moving feeder cattle to available feed and moving slaughter cattle to the slaughter plant within the shortest time and distance consistent with good pricing practices are those that rate best for market performance.

Clearly the greatest problem, and the most difficult problem to solve, is supply and price instability. Development of supplemental and confinement feeding has improved this situation, and modern transportation and communication has made remarkable contributions; but maintaining an orderly flow of cattle to markets is still a major problem.

Perhaps the major conclusion one can draw about world cattle marketing systems and methods is that human beings and the organization of a system are much more important variables than are technological factors. Furthermore, approaches to cattle marketing are governed by the social and production system of a country, and any plans for market improvement or change must take place within a rather narrow range of possibilities and as part of an overall advance in the country's standard of living.

Improvements in marketing efficiency depend on attitudes, related infrastructure, and cost-benefit relationships. For this reason, trekking is a more efficient method of transport than trucking in some areas. It also explains why some marketing systems that appear to be inefficient fail when certain "obvious" changes are introduced. We now turn to a discussion of beef processing and beef marketing in an effort to add one more piece to the puzzle of why the world's beef business operates as it does.

REFERENCES

Ariza-Nino, Edgar J.; Manly, D. W.; Shapiro, Kenneth H. 1980. *Livestock and Meat Marketing in West Africa*. Vol. 5. Ann Arbor: University of Michigan. Center for Research on Economic Development.

Comision Interinstitucional de Mercadeo de Carnes. 1975. *Estudio de Mercado de Ganado Vacuno y de la Carne de Res en Costa Rica*. San Jose, Costa Rica: Ministerio de Agricultura y Ganaderia.

Davis, Ernest E., and Sporleder, Thomas L. 1978. *Computers for Feeder Cattle Marketing*. Texas Agric. Ext. Serv. L-1714. College Station: Texas A & M University.

Engleman, Gerald; Holder, David L.; Paul, Allen B. 1979. *The Feasibility of Electronic Marketing for the Wholesale Meat Trade*. USDA, AMS-583.

Farris, John. 1969. *Factors Affecting Cattle Prices*. NCRE Publ. no. 25. East Lansing: Michigan State University.

Fenn, M. G. 1977. *Marketing Livestock and Meat*. 2nd ed. Rome: United Nations, Food and Agriculture Organization.

Hinchy, Mike. 1978. The Relationship between Beef Prices in Export Markets and Australian Saleyard Prices. *Quart. Rev. Agric. Econ*. 31:83–105.

James, John B., and Farris, D. E. 1971. *Factors Affecting Price Differences of Cattle in the Southwest*. Texas Agric. Exp. Sta. B-1106. College Station: Texas A & M University.

Josserand, Henri P., and Sullivan, Greg. 1979. *Livestock and Meat Marketing in West Africa*. Vol. 2. Ann Arbor: University of Michigan. Center for Research on Economic Development.

McCoy, John H. 1980. *Livestock and Meat Marketing*. 2nd ed. Westport, Conn.: AVI Publishing Co.

Organization for Economic Cooperation and Development. 1978. *Structure, Performance and Prospects of the Beef Chain*. Paris.

Simpson, James R. 1978. *Cattle Cycles: A Guide for Cattlemen*. Food Resour. Econ. Dep. Staff Pap. 100. Gainesville: University of Florida.

______. 1981. *An Assessment of the United States Countercyclical Meat Import Act of 1979*. Food Resource Econ. Dep. Econ. Info. Rpt. 152. Gainesville: Univ. of Florida.

Simpson, James R., and Degner, Robert L. 1979. Blended Beef: Big Opportunity for Central Processors? *Meat Process*. 18(2):36, 38, 43.

Stegelin, F. E., and Simpson, J. R. 1980. *An Economic Analysis of the Effect of Increasing Transportation Costs on Florida's Cattle Feeding Industry*. Food Resour. Econ. Dep. Staff Pap. 161r. Gainesville: University of Florida.

U.S. Department of Agriculture. 1978. *Beef Pricing Report*. AMS, Packers and Stockyards Program. December.

______. 1979. *Report of the Secretary's Meat Pricing Task Force*. June.

Ward, Clement E. 1979. *Meat Marketing and Pricing—An Evaluation of Current and Proposed Systems*. WP-34. Stillwater, Okla.: Oklahoma State University.

Wittenberg, J. 1977. *A Regional Analysis of Beef Cattle Prices*. Dep. Agric. Econ. Manage. Misc. Study no. 63. Reading, England: University of Reading.

World Bank and United Nations Food and Agriculture Organization. 1977. *The Outlook for Meat Production and Trade in the Near East and East Africa*. 2 vols. Washington, D.C., and Rome.

10
BEEF PROCESSING AND MARKETING

The processing and marketing of beef is a complex business owing to technological factors, the large number of products, the transformation that takes place in the form of the product, and the continual changes in input and output prices. Furthermore, the methods vary greatly, from sophisticated capital intensive operations in which microbial count is taken regularly, to the most primitive situations in which butchers kill an animal, hang it from a tree, and use axes to chop off pieces of meat that are then sold directly to consumers without being weighed, cleaned, or packaged. Naturally, the quantity and quality of services affect the marketing margins. For example, a cow might be slaughtered, cut up, and cooked over an open flame on a ranch in Australia, in which case the marketing margins are quite small, or that same cow could be shipped to a slaughter plant, slaughtered, the carcass deboned, and the meat frozen. Then, it could be boxed as subprimals, shipped to Los Angeles, California, transported inland, thawed, and eventually eaten as hamburger, or possibly processed further into canned beef stew. In this case the marketing margin would account for most of the value.

In developed countries, beef slaughter and processing plants have generally increased in number since World War II, as a result of improved refrigerated transportation. There has been a general decentralization as new plants have been built in the more concentrated production areas. This has perhaps been most striking in the United States as the number of meatpacking firms for all livestock increased from 1,999 in 1947 to 2,833 in 1963. The number of establishments grew from 2,154 to 2,992 (National Commission on Food Marketing). New plants were concentrated in the western corn belt and the Great Plains of the United States rather than in major cities, where they were located prior to World War II. The new plants were generally one-story, high-volume, single-species plants. These plants typically were designed to slaughter and chill carcasses 12 to 24 hours and load for a 2- to 3-day trip on refrigerated truck or 3- to 4-day trip by train to the East or West Coast. The beef was normally moved to retail stores in carcass form and sold at retail 1 week to 10 days after slaughter.

Since the mid-1960s, more and more of the beef is moving to retail in boxes in the form of primals and subprimals. This allows mechanized handling and reduces storage and transportation costs. Some of the larger plants have a weekly slaughter capacity of over 20,000 head. Small plants with lower wage rates have been able to compete with these large plants, however, as labor costs remain from two-thirds to three-fourths of the plant total operating costs above the cattle costs. Economies-of-size relate to better by-product utilization and use of laborsaving equipment. Being a high volume business, profits for those not involved in boxed beef were generally less than 1 percent of sales whereas profits of plants shipping boxed beef were nearly 2 percent of sales. Costs vary considerably, depending on the sophistication of the marketing system.

WHOLESALE MEAT MARKETS

In traditional meat marketing channels, beef passes directly from the slaughterhouses to a central meat market where butchers purchase their supplies from wholesalers in the form of carcasses, quarters, or primal cuts[1] Butchers in areas such as Chile go directly to slaughterhouses, where sales are carried out by traders who have purchased live cattle and have paid to have them slaughtered in hopes of making an economic return from this marketing function and possibly from price changes. In some more traditional areas such as the African countries, retail butchers often buy their own livestock and have them slaughtered.

As cities grow in size and sophistication, the brokerage and actual handling of meat is taken over by specialized jobbers. Urban growth brings about fairly rapid changes as old-fashioned public slaughterhouses are replaced by modern facilities that are more impersonal and geared to larger clients, to a greater extent. In most developed countries, retail grocery chains have taken the place of butcher shops except in a few specialized locations. The chains have developed greater market power and along with it have gradually adopted technological advances in refrigeration, transportation, ordering, and processing, with the result that many of them now purchase carcasses from packinghouses, process them in their own facilities, and distribute beef via their own fleet of trucks. In an effort to gain back the market power, packing plants have begun to exert themselves more aggressively in distribution by taking advantage of boxed beef, a system of cutting, packaging, and distributing

[1]Much of the material in this and the next two sections is drawn from Fenn 1977.

subprimal cuts in boxes (Dietrich and Farris 1976). In all events, this continual struggle for market power has led to substantial efficiencies and cost savings for consumers.

RETAIL MEAT MARKETING

Although the methods of retailing beef vary greatly throughout the world, in all countries meat retailing is a highly specialized profession. In some Muslim countries butchers have become almost a distinct caste, dressing in a certain way and controlling their businesses through a tight system of apprenticeships and inheritance. When meat is scarce, black markets—that is, clandestine sales—develop and the butcher takes on a special importance in people's lives. Another reason for the butcher's traditional prominence is that meat is a highly perishable commodity whose preparation and sale require both management and technical skills. In most developing countries, for instance, there is no refrigeration and the stocks must be completely sold the same day the animal is slaughtered.

Low income consumers may not be able to afford more than crude

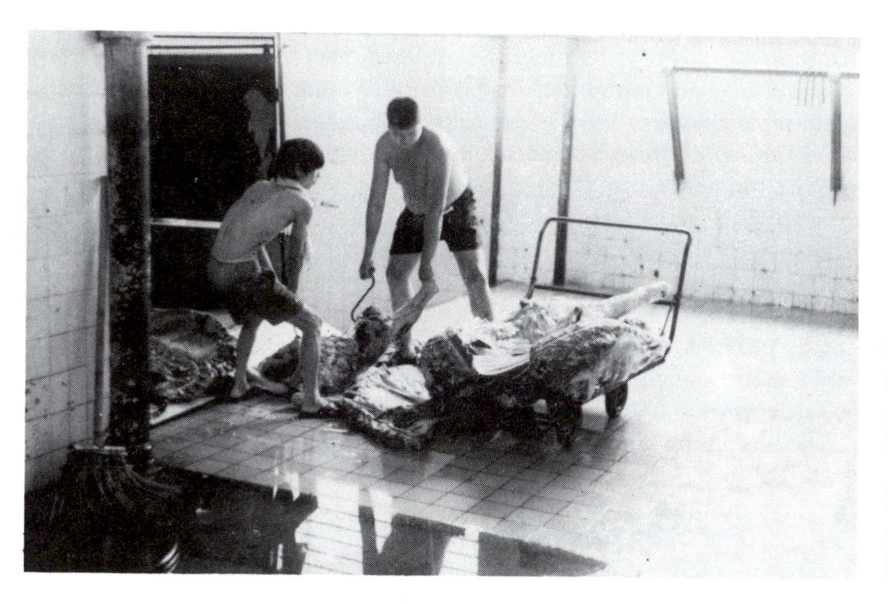

U.S. Meat Export Federation Photo

Despite the tremendous advances in beef processing and marketing, there are still many beef handling operations more primitive than this one in Southeast Asia.

In many developed countries, packaged beef in refrigerated self-service counters has the highest sales volume of any item in the supermarket.

cuts of meat and are not bothered by public stalls or by minimum standards of sanitation, but higher income clients demand higher quality meat, more services, and assurance that the product is handled in a hygienic manner. Consequently, specialized butcher shops have developed, only to be replaced by the impersonal retail counter of the supermarket. The change is a result of substituting capital for labor as wage rates increase. It is unlikely that supermarket retailing of beef will gain much prominence in Asia and Africa in the near future, but this form of retailing has become quite important in Latin America.

Despite a variety of innovations, meat retailing is a costly, labor intensive function. Self-service retailing of meat in supermarkets was an innovation with so many obvious advantages in the United States that it quickly spread throughout the country following World War II. It is now common in most developed nations. It allows consumers to take their time in selection, eliminates waiting, and reduces labor costs compared with service meat counters.

The rather inefficient processing of fresh meat in the back room of retail stores is the next problem to solve. Attempts at central processing of the fresh items—much of it beef—has not been completely satisfactory. A few companies have tried it and some have discontinued central processing of fresh meat. Quality control and maintaining an attractive package throughout shipping and handling are the main problems. Labor costs can be reduced by eliminating processing in the back room and it is expected that systems and technologies will soon be developed to make this an acceptable practice.

CATTLE SLAUGHTER AND PROCESSING PLANTS

The livestock slaughterhouse, or abattoir, has a strong influence on both the livestock and retail trade as it is a focal point in determining the manner in which meat is handled throughout the meat and livestock marketing process. For example, sloppy, inefficient, poorly managed, unsanitary conditions will be reflected along the livestock and meat marketing channels, just as rigid veterinary inspection of live cattle and carcasses with appropriate price discounts will lead to improved herd management. A strict grading system will encourage producers to specialize because they will be rewarded for producing higher quality cattle. The abattoir also has a great influence on meat marketing, for minimum sanitary levels are set there. In view of the drastic change in the form of beef that takes place in slaughter plants, it is little wonder that they are the object of considerable attention by government officials, planners, and others related to the world's meat business.

Slaughter facilities, as we have said, vary considerably. The most primitive unit is a simple slaughter slab, which may or may not have a roof, found in low income country villages or alongside small public markets. The next level of facility is a small slaughterhouse provided by the local government. It will have rudimentary slaughter facilities; its major equipment will be a hoist, and it will probably have one or two livestock pens and a water supply. The larger metropolitan areas have more extensive, although quite primitive public facilities in which livestock owners slaughter and dress their own animals. These are common throughout the developing countries.

Another type of system is the "industrial" abattoir. Here the operations are carried out by the plant's staff rather than by the livestock owners. In developing countries and all the centrally planned economies, these plants are owned by a public authority. Some of these industrial plants process their own cattle while others operate on a custom basis. Still others do both. This category includes plants of every design and scale, both privately and publicly owned.

Abattoirs in most parts of the world have usually been owned and run by public authorities rather than by private companies because of the need for disease control. And the offensive odors, the noise, the nature of the business, and the waste disposal problems have led to government intervention. Perhaps most important, abattoirs require large in-

Modern beef slaughter plants are generally one story, single species plants with facilities to vacuum pack boxed beef and have it ready to transport in less than 48 hours.

vestments, are quite conspicuous, usually have monopolistic powers, and have been located near sales outlets owing to a lack of refrigeration facilities. Since World War II, government ownership of abattoirs has come under more criticism as the management and operations of abattoirs have become increasingly complicated and technically different from the usual services provided by local government authorities. Furthermore, government management of abattoirs has the natural effect of dampening innovation and cost effectiveness measures in the abattoirs' operations. This reaction affects the entire marketing chain. As a consequence, the private sector has taken over an increasingly important role in beef processing.

In most of the market economies, beef slaughter is among the most competitive and efficient food processing industries. After tax, profits on sales average about 1 percent. This is partly because the ratio of gross sales to capital investment is low. Although cost of packing plant construction is a substantial investment, the investment per annual head of capacity is almost insignificant compared with the investment per annual head of capacity in cattle production. For this reason, it is "good industry economics" to maintain sufficient slaughter capacity to accommodate the wide fluctuations in marketing due to drought and the cattle cycle.

Just as the live cattle marketing system starts the process of assembling larger lots and sorting them into more uniform groups, the slaughter plant continues this process. The by-products must be removed, processed, and packaged; the carcasses cleaned, sorted, and chilled. In some cases, processing into primals and subprimals begins in the same plant, but this is a separate stage.

In the United States, the innovation of vacuum-packing primals and subprimals of the beef carcass is rapidly being adopted. Started in the 1960s by food retailing corporate chains, it moved into the meat packing industry in the 1970s and accounted for 50 to 75 percent of all beef slaughtered by 1980. This practice lengthens the storage life of beef and facilitates packing in boxes and handling with a forklift. It saves space in trucks and in expensive refrigerated storage. For these reasons, by 1990 the authors expect this practice will be fully adopted as a standard practice of the United States beef industry.

PRESERVATION, STORAGE, AND TRANSPORTATION

A major concern of processors, health officials, and of course consumers is to insure that the meat that is retailed and ultimately consumed is hygienically acceptable. As such, processors and distributors attempt to prevent physical deterioration of the product, a change that takes three forms: microbial spoilage or the excess growth of microorganisms on the meat surfaces; discoloration; and rancidity that arises from chemical changes in the meat's fat. The technologies for control of meat spoilage are well known, and depend mainly on appropriate storage conditions and strict cleanliness. Management is invariably the limiting factor in proper meat handling.

There are various methods of preserving meat, the principal one of which is chilling by refrigeration to a temperature between $-1°C$ and $+4°C$ directly after slaughter. The carcasses are usually hung at the slaughter plant or are in refrigerated transport for 1 to 4 days as this allows the meat to improve in flavor and texture. Sale of chilled meat is now an accepted practice in most countries of the world, although some low income consumers in every developing country are still accustomed to buying the meat "hot" soon after slaughter. In addition to these traditional tastes and preferences, there are many areas, especially in small towns or rural regions, where price-cost-volume relationships are such that refrigeration cannot be economically justified.

Another technique in wide use, especially in international trade by oceangoing vessels, is the freezing of carcasses, but this method frequently causes changes in texture and appearance, and some loss of flavor. An improvement is quick-freezing, but costs are prohibitive except for specialized sale of small cuts for the retail trade. Still another variation is freeze-drying whereby the meat is placed in an airtight drying chamber until sublimation occurs at freezing temperatures. This method works very well, but high labor and energy requirements prevent it from being used extensively even in countries with high beef prices.

A substantial amount of the beef now entering world trade and do-

These ready-to-cook frozen ground beef patties are an example of the many further processed products available from modern packers and provisioners.

mestic markets in many developed countries is vacuum packed in airtight plastic containers. This method facilitates product handling, retards spoilage, and reduces costs. Along with this innovation has come boxed beef, a system in which carcasses are broken down into primal or retail cuts in a centralized processing location and then distributed (Sporleder 1972). The processing of boxed beef requires considerable management skills since the meat must always be maintained at low temperatures (Stafford 1974). This system has eliminated the need for some of the butchers in many retail outlets and is a cost-effective means of handling meat since there is less bulk and more compactness in handling the beef in boxes rather than in hanging it. Labor unions resisted the technique, however, because much of the handling can be done by conveyor and forklift and thus the amount of labor required for processing is reduced.

It is technically feasible to store frozen carcass meat for several months, but meat storage costs are high in relation to the product. The investment in facilities and operating capital tied up in meat add up quickly when inventories are allowed to accumulate. For this reason meat processors attempt to turn over the product as quickly as possible. For the same reason, endeavors to even out seasonal fluctuations in live cattle marketings will probably be more practical than long-term meat storage.

Air transport of beef has become a reality in many areas, primarily in developing countries where poor road infrastructures make truck transport prohibitive. An advantage of air transport is that with careful handling refrigeration can be avoided, even in the hot climates, as the atmospheric temperature is quite low, about 8°C at a normal flight altitude of 3,000 m. Several Latin American countries such as Costa Rica and Bolivia have considerable experience in the use of aircraft for beef transport; elsewhere beef has been shipped by air from Kenya and Sudan to Libya and the Near East. Despite the potential cost advantages of air transport, its intracountry use can be expected to decline as improved roads are constructed. At the international level only minimal use will probably be made of air transport owing to the added complications of coordinating inspection, slaughtering, and flight schedules, all of which normally take place in different locations.

PLANT LOCATION

Changes in livestock slaughter have taken place rather rapidly since World War II in most countries of the world. The major noneconomic factor is the demand for more hygienic conditions, which has led to more centralization and eventual closing of many small slaughterhouses that could not easily be brought up to required health or economic standards. A related aspect is greater trade and consequently pressure from abroad to meet stricter health standards. Consequently, numerous new facilities have been constructed in the 1960s and 1970s, and considerable expansion will probably continue to take place through the turn of the century in both developed and developing countries.

The major factor in determining the location of an abattoir or packing plant is the relative advantage of transporting meat rather than moving live animals. Related to this is the availability of refrigeration at distribution and retail points, the availability of labor, current industry structure, extent to which cattle are fed concentrates, and the availability of grain. Because consumers in many areas still have a preference for "hot" meat from freshly killed cattle, they can be expected to exhibit a certain reluctance to accept a chilled product. But the experience in countries such as in Chile, Ivory Coast, Nigeria, and Zaire demonstrates that customers quickly become accustomed to the chilled product when traditional "hot" meat is no longer available.

Modernization of the meat processing industry has long-term advantages for all parties, but there is a natural resistance to change in the marketing channel as it means making new investments in facilities and skills. At the national level, however, these costs can be offset through direct marketing of cattle to packing plants by owners, who can be paid on a grade and yield basis rather than by the head. Other advantages of the modern techniques are better health control, the possibility for improved scheduling of deliveries, and reduced processing costs. The result in developed countries is a trend away from slaughtering livestock in city abattoirs or even near big cities and toward constructing abbatoirs in producing areas. An extensive study by Mittendorf (1978) concludes that in the developing countries, however, the establishment of slaughterhouses in consuming areas is generally most economic because the transport of meat is usually more expensive than that of livestock. Also, the costs of a cold chain—that is, refrigeration operation from a slaughterhouse situated in a producing area to an area of consumption—are considerably higher in developing countries than the costs of comparable operations in industrialized countries. As illustrated in Chapter 9, trekking of animals is also a relatively low-cost factor in certain areas and must be considered in any feasibility study for developing countries.

PLANT USE AND ECONOMIES OF SIZE

Unit costs, size, and feasibility of locating cattle slaughter and processing plants is a source of concern to governments and private investors. Many factors can influence a decision to establish or refurbish a plant in a specific location, but the major elements are: (1) plant size, (2) use of plant capacity, (3) type of facilities, (4) throughput or efficiency, and (5) input use. The more important input costs are those of cattle assembly, labor, and distribution of the final product. In fact, labor costs are of such magnitude that a small plant can often compete on favorable terms with a large plant if it has a labor cost advantage. A large plant may also have an assembly cost disadvantage if it must reach out long distances to obtain sufficient supply to operate near capacity. Thus, when considering construction of a new plant one needs to consider the above factors and design the plant for the local conditions. Large plants are only more efficient if labor costs are kept in line and if an abundant supply of cattle is available locally.

The size distribution of meat packing plants in the United States in 1972 shows that of 2,475 plants in operation that year, 79 percent were small, having less than 50 employees; these small-size plants accounted for 11 percent of total sales. While total sales from small-size plants is not high, it does illustrate that many small plants have been able to compete with larger plants (Table 10.1). This is an indication that plants have been designed to serve the specific density of livestock in their area, or the size of their market. Most new plants built in the United States since 1972 have been large-volume plants located in areas where large-scale feedlots can provide large numbers of cattle weekly, but it must be kept in mind that inplant economies are not the only consideration when decisions are being made about plant size.

Table 10.1. Meat packing plants by number of employees and value of shipments, United States, 1967 and 1977

Distribution of number of employees	Number of plants[a]		Value of shipment[b]	
	1967	1977	1967	1977
		(%)		
1-4	43.4	49.0	0.7	1.0
5-19	21.2	22.5	2.6	2.7
20-49	15.6	9.9	7.9	4.7
50-99	8.2	7.4	11.1	10.2
100-249	6.3	5.8	18.2	18.4
250-499	3.1	3.1	21.8	21.6
500-999	1.1	1.5	11.8	18.9
1,000-2,499	0.8	0.9	14.4	16.8
2,500 or more	0.3	0.2	11.5	5.7
Total	100.0	100.0	100.0	100.0

Source: U.S. Department of Commerce, 1972.

[a]In 1967 there were 2,697 establishments (864 with 20 or more employees), and in 1977 there were 2,590 establishments (737 with 20 or more employees).

Packing plant cost relations are illustrated by Figure 10.1, in which all costs are declining except assembly costs (distrbution cost is not shown). An important point here is that as a plant reaches out to procure adequate supplies, assembly costs eventually override the economies of plant size. Thus, in this illustration, a plant handling more than volume *X* would be increasing costs, especially if distribution costs were added to the illustration. Texas beef slaughterers, for example, purchased 92 percent of their steers and heifers within the state of Texas in 1974, and purchased 90 percent directly from feedlots. This method of procurement kept assembly costs to a minimum (Dietrich and Farris 1976).

Once a plant is in operation, plant costs are minimized when it is used at its practical capacity, that is, for a specific number of animals per day. If it can break even with a 4-day week, then operating 9 hours per day 5 days a week will probably turn a good profit. Generally it is not economical to operate much less than a full day. The cost of cleaning is the same and often labor must be paid for a full day if hourly employees report to work. Seldom is a plant profitable if it has chronic labor problems.

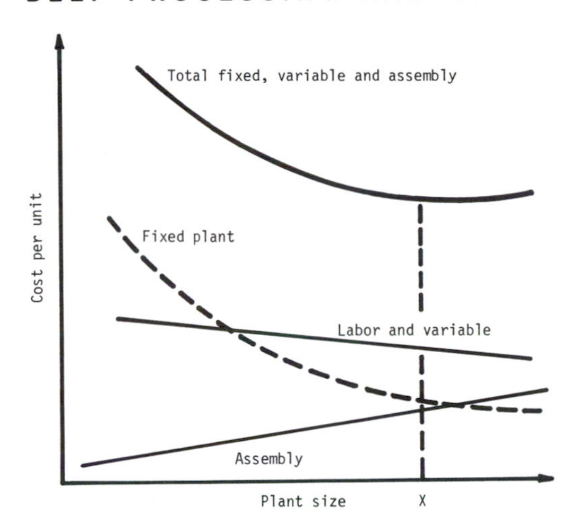

Fig. 10.1. Packing plant relationships affecting cost per unit processed.

The cost and revenue relationships in packing plants are now described on the basis of a 1979 study of beef slaughtering and processing plants in West Virginia, in the United States (Durst and Kuehn 1979). This study was chosen because the plant sizes are small and thus should have considerable relevance to both developing and developed countries.

Plant investment costs and operating expenses are based on engineering studies that represent typical operations. Three sizes are evaluated: 6 head per day (Plant A); 20 head per day (Plant B); and 25 head per day (Plant C). Total investment costs in 1979 U.S. dollars for the three sizes are about $95,000, $237,000, and $258,000, respectively (Table 10.2). The output varies from 1,059 head annually for the small plant operating at 70 percent of capacity, to 7,560 head for Plant C operating at 120 percent of capacity.

The components of total annual costs including investment, payroll, miscellaneous operating expenses, and utilities for the three plants at five levels of capacity utilization are presented in Table 10.3. The labor requirements are 8, 17, and 25 employees, respectively. Total costs for Plant A vary from $97,000 annually when the plant is used at 70 percent of capacity to $146,000 when it is operated at 120 percent of rated capacity. Annual costs for the 25 head per day unit vary from $265,000 to $423,000 for the same capacity ranges.

The cost per animal slaughtered for the three plant sizes and five capacity utilization levels is presented in Table 10.4. The considerable difference both within a plant and between plants indicates the importance of constructing the proper size in relation to volume of animals in the area and of utilizing the plant at its highest level of capacity. The infor-

Table 10.2. Investment cost and annual output as a percent of capacity for three sizes of West Virginia beef slaughter plants

	Plant A	Plant B	Plant C
	6 head/day	20 head/day	25 head/day
		($)	
Investment			
Building	61,472	157,420	172,400
Equipment	33,277	78,617	82,779
Land	669	1,414	1,908
	95,418	237,451	257,587
		(head)	
Annual output as a percent of capacity			
120	1,814	6,048	7,560
100	1,512	5,040	6,300
90	1,361	4,536	5,670
80	1,210	4,032	5,040
70	1,059	3,528	4,410

Source: Adapted from Durst and Kuehn, 1979.

Table 10.3. The components of total annual costs (investment, payroll, miscellaneous, operating, and utilities) by level of capacity utilization

Plant	Capacity utilization	Investment	Payroll	Miscellaneous operating	Utilities	Total
	(%)			*($)*		
	120	11,869	89,213	42,528	2,622	146,232
	100	11,869	72,388	36,710	2,385	123,335
A[a]	90	11,869	66,664	33,807	2,270	114,610
	80	11,869	60,940	30,904	2,141	105,854
	70	11,869	55,215	28,001	2,021	97,106
	120	29,012	208,803	98,522	5,611	341,948
	100	29,012	165,212	83,784	4,982	282,990
B[b]	90	29,012	150,709	76,429	4,661	260,811
	80	29,012	136,206	69,075	4,348	238,641
	70	29,012	121,705	61,720	4,033	216,470
	120	32,056	260,962	123,034	7,159	423,211
	100	32,056	205,898	104,417	6,351	348,722
C[c]	90	32,056	187,578	95,127	5,951	320,722
	80	32,056	169,263	85,838	5,549	292,706
	70	32,056	150,944	76,549	5,150	264,699

Source: Durst and Kuehn, 1979.
[a]Plant A—100% capacity = 1,512 head per year.
[b]Plant B—100% capacity = 5,040 head per year.
[c]Plant C—100% capacity = 6,300 head per year.

mation in Table 10.4 can be used to determine economies of size. This aspect is presented in Figure 10.2, which shows the cost per head plotted against annual capacity.

If each plant operated at its rated capacity, the costs per head would be about $81, $56, and $55. This demonstrates the great advantage of moving to the two larger plants. Economies of size are also derived when the plant is operated at 70 percent of capacity, although the unit costs are higher. The relationship between utilization and size shown by the dashed lines in Figure 10.2 indicates that the cost per head would be the same whether the 25 head per day plant (size "C") were operated at 70 percent of capacity or whether a plant with about 3,900 head rated capacity (one somewhat smaller than "B") were operated at 100 percent of capacity.

Efficiencies of each size of plant, and information on strategies of managing them, can be derived from Figure 10.3, which is also based on data from Table 10.4. In this figure the total cost per head is plotted against capacity utilization in percentage terms rather than plant size, as was the case in Figure 10.2. The results indicate that the smallest plant can achieve considerable economies by operating at the highest levels of capacity, for the cost drops about $10, from $92 per head when it is operated at 70 percent of capacity, to $81 when operating at 120 percent of capacity (Figure 10.3). The larger plants do not reap as much benefit

Table 10.4. The components of total cost and total cost per animal processed (average cost) by level of capacity utilization

Plant	Capacity utilization	Investment	Payroll	Miscellaneous operating	Utilities	Total
	(%)			*($)*		
A	120	6.54	49.18	23.44	1.45	80.61
	100	7.73	47.88	24.28	1.58	81.47
	90	8.72	48.98	24.84	1.67	84.21
	80	9.81	50.36	25.54	1.77	87.48
	70	11.21	52.14	26.44	1.91	91.67
B	120	4.80	34.52	16.29	0.93	56.54
	100	5.67	32.78	16.62	0.99	56.15
	90	6.40	33.23	16.85	1.03	57.51
	80	7.20	33.78	17.13	1.08	59.19
	70	8.22	34.50	17.49	1.14	61.35
C	120	4.24	34.51	16.27	0.95	55.97
	100	5.09	32.68	16.57	1.01	55.35
	90	5.65	33.08	16.78	1.05	56.56
	80	6.36	33.58	17.03	1.10	58.07
	70	7.27	34.23	17.36	1.17	60.03

Source: Durst and Kuehn, 1979.

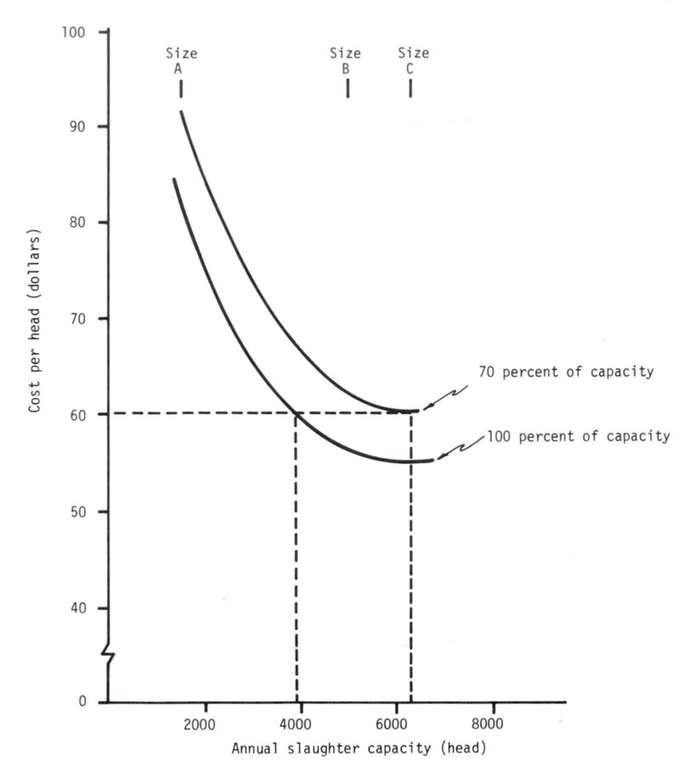

Fig. 10.2. Economies of size for West Virginia slaughter plants operating at 70 and 90 percent capacity.

from operating at a higher level of capacity because there is only a spread of about $5, depending on the use. In fact, costs actually increase slightly when the plants are used above 100 percent of capacity because employees must be paid higher wages for working overtime.

Revenue, and consequently profit, from the plants depends on the charge per head. If there are several plants in competition with each other, then management must determine what effect different fee levels will have on total revenue. Also, net revenue is directly related to capacity and size. In the West Virginia study three different revenue levels called low, medium, and high were examined. A base slaughtering fee of $5.00, $10.00, and $15.00 per head for each level was assumed, to which a processing charge of $0.045, $0.055, and $0.064 per kg for the three sizes, respectively, was added. An additional revenue of $4.40 per 100 kg of live weight was added for sale of hide and offal, i.e., the "drop."

Net revenue per animal processed for the three different fee levels

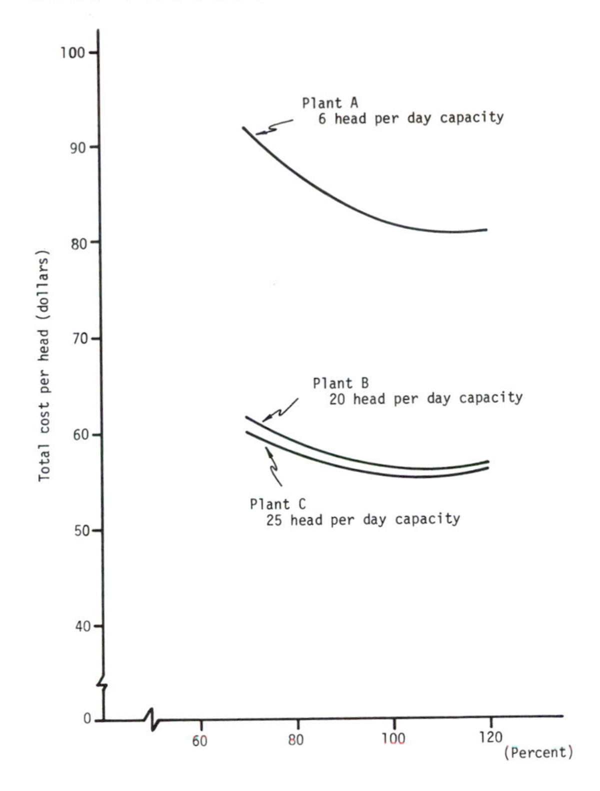

Fig. 10.3. Total cost per animal slaughtered for three sizes of plants (adapted from Durst and Kuehn 1979).

and five capacity utilization levels is presented in Table 10.5. There is little difference between Plants B and C for any one revenue level because only small economies of size are achieved by moving to the higher level of capacity. There are considerable differences, however, between the two larger plants and the six head per day unit, as net revenue for the latter is considerably lower and is even negative when the unit is operating at only 70 percent of capacity.

The importance of appropriately determining slaughtering charges is also shown in Table 10.5. In this case, Plant C would be much better off to charge the highest fee even though the plant would be used only at lower capacities. If the high fee were charged and the plant operated at 70 percent, for example, then a net revenue of $46.32 would be derived. The best that could be obtained from operating at the medium fee charge level would be $35.49 when operating at 100 percent of capacity.

Table 10.5. Net revenue per animal processed by level of capacity utilization at high, medium, and low revenue levels

Plant	Capacity utilization	Annual output	Net revenue per animal by fee level		
			High	Medium	Low
	(%)	*(head)*		*($)*	
	120	1,814	25.72	10.22	−5.27
	100	1,512	24.77	9.63	−6.23
A	90	1,361	22.13	6.63	−8.87
	80	1,210	18.86	3.36	−12.14
	70	1,059	14.64	−0.85	−16.36
	120	6,048	49.80	34.30	18.80
	100	5,040	50.19	34.69	19.19
B	90	4,536	48.84	33.34	17.84
	80	4,032	47.15	31.65	16.15
	70	3,528	44.98	29.48	13.98
	120	7,560	50.35	34.86	19.36
	100	6,300	50.99	35.49	19.99
C	90	5,670	49.78	34.28	18.78
	80	5,040	48.26	32.76	17.26
	70	4,410	46.32	30.82	15.32

Source: Durst and Kuehn, 1979.

MEASURING THE BENEFITS FROM PACKING PLANTS

A major goal of developing countries is to increase the amount of economic activity in that country because this leads to higher incomes and greater investment, which in turn stimulate further economic activity. Because beef and cattle exports make up a substantial percentage of export earnings in some countries, it is useful to measure the amount of economic activity provided by the different types of export products. In this section we provide measurements on the direct and indirect benefits from five different types of beef exports: live cows, bone-in beef quarters, frozen boneless manufacturing beef, cooked/frozen beef, and canned beef. The results are presented for three South American countries—Argentina, Brazil, and Uruguay—but the findings have relevance for all beef exporting countries.

A flow diagram of beef processing in a fully integrated South American beef packing plant is presented in Figure 10.4. As is evident, some operations, such as killing and subsequent cooling, are common to all beef products. On leaving the cooler, carcasses for all products except bone-in beef quarters require deboning and trimming. Only meat surrounding large bones is exported in the form of boneless manufacturing beef (used by importers in further processed products such as frankfurters, hamburger, and canned products containing beef). Small pieces of meat that can only be trimmed with a relatively high amount of labor are transferred to the cooking operation for processing into other types of product or are sold to other packers. After manufacturing meat is deboned, it is wrapped in plastic and placed in cartons that weigh about 20–25 kg and then frozen.

The meat for cooked/frozen beef is also first deboned from the carcasses, although less care is taken in the process than for manufacturing beef since it is cut into small pieces. Other bones such as ribs are also trimmed. The meat is then dumped into a horizontal cooker through which it slowly passes, and the cooked meat as well as the broth, called extract, is expelled at the opposite end. The extract is passed through a filtering process to remove excess fat and particles. Then it is packed and exported for further processing into items such as bouillon cubes. The meat is pressed into plastic tubes and frozen. Cooked/frozen beef is used

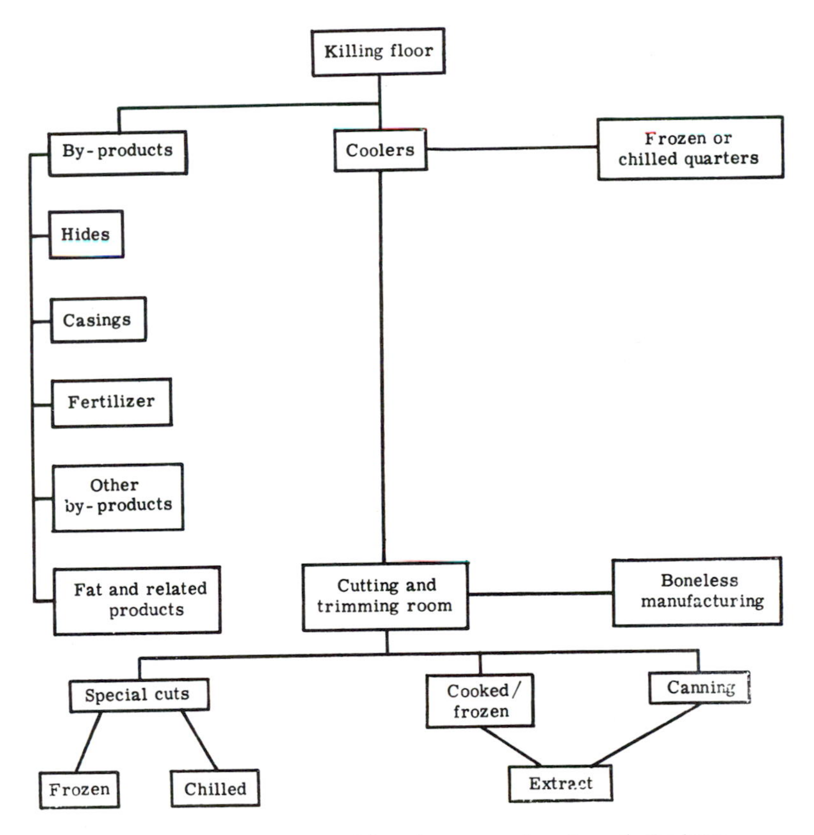

Fig. 10.4. Flow diagram of beef processing in a fully integrated South American beef packing plant.

This Brazilian packing plant is old but has been renovated to meet European sanitary requirements for export to Western Europe.

mainly by importers for further processing in products such as chili con carne or other products requiring cooked meat.

Cooked/frozen beef was developed in 1951 by Deltec International Ltd., a U.S. corporation which at that time had extensive holdings in South America. The purpose in developing the product was to provide an alternative to canned beef for the U.S. market that would also comply with U.S. sanitation standards resulting from South America's endemic foot-and-mouth disease. There are no published sources on world trade in cooked/frozen beef, but an indication of the volume can be obtained by examining Argentina's exports of the product since, apart from Brazil, Argentina has been the only exporter of any consequence. Argentine exports increased from about 3,000 tons in 1964 to 25,000 in 1977. At an average value of $1,972 per ton in 1977, the value of that year's exports was $49.3 million. The United States receives the bulk of these exports and most of the remainder goes to the European Economic Community.

As shown in Figure 10.4, the process for canned beef is the same as for cooked/frozen meat except that after leaving the cooker it is mixed with condiments (Continental Can Co., Inc. n.d.). About 20 percent more fat is usually added for corned beef, while for other canned products such as beef gravy or beef cubes little or no fat is added. The mixture is then poured into a large hopper and automatically portioned out to the cans. The small cans are automatically vacuum sealed while the large 6-lb institutional size (2.7 kg) cans are frequently closed by hand. All cans are placed in a retort and the contents cooked again, after which they are stored and then checked for defects.

The selection of an economic sector for development stimulation can be based to some extent on direct and indirect impacts on output (gross national product) and income. To demonstrate how the results of the disaggregation process can be applied, we will now consider the effects of a $1 million change in final demand on the five subsectors analyzed, using multipliers from the 1963 Argentine input-output study as a base.

The first step in determining the multipliers is setting up cost models.[2] From these budgets it was determined that the processor's greatest costs are for the raw product, varying from 62 percent of the total for canned beef, where the product is quite concentrated, to 81 percent for bone-in quarters (Simpson and Farris 1975). These percentages are inversely related to labor use, which varies from 4 to 15 percent of the total costs (Table 10.6). The budgets were inserted in the various input-output studies and run on a computer from which the multipliers were derived. The output multipliers (type II) for the 1963 Argentine study, as well as for the other Latin American studies, are given in Table 10.7. The interpretation is that for each $1.00 worth of exports there are *$X* worth of direct, indirect, and induced effects generated in the economy. In all cases the product ordering was the same and very little difference is observed in the absolute size of the multipliers.

The effect that a $1 million change in export sales would have on the base economy is presented in Table 10.8. The total change is much lower for live cow exports than for any other product. Cooked/frozen beef is highest, with $4.87 million output and $1.16 million income.[3] Canned beef is second, with $4.41 million output and $940,000 income. Quarters

---

[2]From extensive interviews with two large international packers in Paraguay, three in Argentina, three in Uruguay, and one in Brazil.

[3]The multipliers have been developed for exports of cow beef, but the results could be adapted with little modification to exports of steers. In all likelihood there would be no important difference in the absolute size; that is, live steer exports can be expected to have the lowest multipliers, followed by quarters. The various subprimal cuts from steers probably have a multiplier slightly larger than that presented for frozen boneless manufacturing beef, but smaller than that for canned or cooked/frozen beef. Only the lower quality steers and the less valuable portions of the higher quality steers would be considered to be the same class of raw product as cow beef.

Table 10.6. Estimated pretax costs for bone-in beef quarters, frozen boneless manufacturing beef, cooked/frozen beef, canned beef, and live cow exports, live-weight basis, 1973

Item	Bone-in quarters	Frozen boneless manufacturing	Cooked/ frozen	Canned	Live cow exports
			($/metric ton)		
Livestock	302.11	302.11	302.11	302.11	302.11
Fuel and electricity	7.89	8.41	6.25	4.92	
Foods				1.26	
Meats				5.89	
Textiles	4.17				
Ready-made wearing apparel	0.49	1.84	2.13	4.38	
Wood	0.40	0.35	0.28	0.63	
Paper and cartons		6.86	3.43	6.79	
Printing and publications				0.63	
Chemical products				0.04	
Metals				2.00	
Vehicles and machinery	1.53	1.75	2.09	3.13	
Other industries		1.40	6.56	0.95	
Commerce	6.58	9.31	11.99	11.99	
Transportation and storage	5.56	6.09	5.13	2.86	23.87
Other services	8.97	9.60	9.68	10.95	2.90
Labor	17.83	48.35	56.08	73.72	13.09
Depreciation	11.58	9.53	11.98	20.00	
Defective cans				0.00	
Imports	5.82	6.12	8.91	32.75	
Total	372.93	411.72	426.62	485.00	341.97

Source: Simpson and Farris, 1975.

Table 10.7. Output multipliers for export of live cows, bone-in beef quarters, frozen boneless manufacturing beef, cooked/frozen beef, and canned beef from four different disaggregated input-output studies

Input-output study	Live cows	Bone-in quarters	Frozen boneless manufacturing	Cooked/ frozen	Canned
1963 Argentina	3.33	4.04	4.01	4.87	4.41
1970 Argentina		4.05	4.03	4.85	4.41
1959 Brazil	3.19	3.77	3.68	4.76	4.18
1961 Uruguay	3.03	3.47	3.44	4.18	3.79

Source: Simpson and Farris, 1975.
Note: Type II multipliers include direct, indirect, and induced effects.

and manufacturing beef are about the same and tie for third place. Assuming that economic development is the goal, the estimates indicate that cooked/frozen or canned beef should receive the greatest emphasis, followed by either manufacturing beef or quarters. We can conclude that even though canned beef is from lower quality animals, in terms of economic development it can be an important product.

Table 10.8. Effects of a $1 million change in final demand on income and output (GNP) for five types of South American beef exports

Exports	Change in output	Direct change in household income	Direct, indirect, and induced income change	Direct, indirect, and induced output (GNP) change[a]
			($)	
Live cows	1,000,000	27,860	568,001	3,331,400
Bone-in quarters	1,000,000	131,580	822,875	4,040,800
Frozen boneless manufacturing	1,000,000	58,170	762,068	4,007,400
Cooked/frozen	1,000,000	321,130	1,163,390	4,872,200
Canned	1,000,000	173,600	943,013	4,411,200

Source: Simpson and Farris, 1975.

[a]The change in output multiplied by the type II output multiplier for the disaggregated 1963 Argentina input-output study (see Table 10.7).

SUMMARY AND CONCLUSIONS

Processing and marketing are essential links between the producer and consumer. Considerable change has taken place in all marketing channels in almost every country of the world, yet great disparities still exist in sanitation practices, product forms, and efficiency. There are many lessons to be learned from studying the various systems, but perhaps it is most important to realize that proper legislation is fundamental to improving every step in the marketing chain. Industries cannot be forced to change; rather they must be stimulated into recognizing that adoption of innovations is in their best interests. Furthermore, the legislation must be appropriate to the time, place, and level of economic development in the area. One of the greatest threats to effective implementation of change in meat marketing programs is inability and/or failure to monitor and enforce regulations.

A case study presented in this chapter shows the great economies of size available from moving to larger meat processing plants. The analysis demonstrates the need to correctly determine the number of cattle in an area so that the proper size can be built to avoid problems of excess capacity. The data also indicate that potential investors in smaller plants, such as those examined in the study and found in many parts of both the developed and developing world, are better off economically to take advantage of economies of size and to operate at a lower percentage of capacity. The larger the plants become, however, the smaller the advantages from economies of size. Eventually, a point is reached at which capacity utilization becomes more important in cost reduction than economies of size.

Now that we have an overall view of the world's beef business and a better understanding of pricing, live cattle marketing, and production and processing relationships, we turn our attention to the how and why of the international trade in beef.

REFERENCES

Continental Can Co., Inc. n.d. *The Canning of Corned Beef.* Canning memo, rev. 6.

Dietrich, R. A., and Farris, D. E. 1976. *The Texas Meat Packing Industry, Structure, Operational Characteristics and Competitive Practices.* Bull. B-1164. College Station: Texas Agric. Exp. Sta.

Durst, Ronnie L., and Kuehn, John P. 1979. *The Feasibility of Establishing Low Volume Beef Slaughtering-Processing Plants in West Virginia.* Bull. 666. Morgantown: West Virginia University.

Fenn, M. G., ed. 1977. *Marketing Livestock and Meat.* Rome: United Nations, Food and Agriculture Organization.

Mittendorf, Hans. 1978. Factors Affecting the Location of Slaughterhouses in Developing Countries. *World Anim. Rev.* 25:13–17.

National Commission on Food Marketing. 1966. *Organization and Competition in the Livestock and Meat Industry.* Technical Study No. 1, Washington, D.C.

Simpson, J. R., and Farris, D. E. 1975. The Benefits for Economic Development from Selected South American Beef Exports. *World Anim. Rev.* 13:9–15.

Sporleder, Thomas L. 1972. *Primary Packaging Cost Analysis for Fresh Beef from Packer to Retail Distribution Center: A Case Study.* Texas Agric. Mark. Res. Dev. Cent. MRC-72-7. College Station: Texas A & M University.

Stafford, T. H. 1974. *Methods and Costs of Distributing Beef to the Food Service Industry.* N.Y. Food Life Sci. Bull. no. 36. Ithaca, N.Y.: Cornell University.

U.S. Department of Commerce. 1972. *Census of Manufacturers.*

11 WORLD TRADE

The international market for beef is diverse and large. Its market structure is characterized by a few large exporting and importing countries, with a large fringe of smaller trading countries. Trade in beef represents about 8 to 10 percent of world production (OECD 1980). The developed economies are the key markets for beef, and the U.S. and the European Economic Community (EEC) each accounted for almost one-third of the total imports in 1978. Two-thirds of the EEC trade, however, is among EEC members. The major exporting areas are Oceania and the Pampas area of South America.

Prior to the 1960s, the British dominated international trade in beef. The United Kingdom was by far the largest importing country and had played a major role in developing the beef industry in the major exporting countries. The British had no significant trade restrictions on meat except for health and sanitation. Italy was also a substantial importer, and the combined western European countries that now comprise the EEC were clearly the largest beef-deficit and -importing countries of the world. In the early 1960s, the United States became the leading importer owing to rapidly rising per capita income and the growing "fast food" industry, coupled with few trade restrictions. In 1964 the United States enacted an import quota and in 1979 revised the law to make the quota countercyclical. It still remains the largest single import market, importing mostly "manufacturing beef," while it is beginning to export a substantial volume of high-quality beef.

Areas of the world such as southern South America and Oceania have a comparative advantage in beef production because of relatively low population pressures and natural conditions that favor cattle raising. Other areas, such as the United Kingdom and Japan, have a comparative disadvantage in beef production, with the result that they are beef importers. Canada and the United States are both major importers and major livestock product exporters. This chapter describes why certain trade patterns have developed and explains how the mechanisms operate.

WHY NATIONS TRADE

Absolute differences in the prices of goods between two countries are the major basis for international trade. When the price differences are greater than the transport and handling costs, it is profitable for the country with high prices to import from the country with low prices. Long-run prices are lower in one country than another because long-run costs are lower. Costs differ for an unexpected reason—because pretrade cost ratios within each country are dissimilar. For example, assume that there are only two countries, Australia and the United States, and, in isolation, only two commodities are produced by each country, beef and tractor parts. Further assume that the cost ratios differ as shown below:

Commodity	United States	Australia
	(U.S.$)	*(A$)*
Tractor parts	1.00	3.00
Beef	2.00	1.00

In this example, the cost ratio of beef to tractor parts in the United States is 2:1 whereas it is 1:3 in Australia. These ratios indicate that the United States has a comparative advantage in producing tractor parts and a comparative disadvantage in producing beef. Australia, on the other hand, has a comparative advantage in beef production and a comparative disadvantage in the production of tractor parts. Even without adding exchange rates, it is apparent that a basis exists for trade since the ratios are different. If the exchange rates were U.S.$1 = A$1, trade would definitely be profitable.

There is a range of exchange rates for which trade is profitable, and the limits are determined by the rates that are necessary to make the product prices equal. In this example, one limit on the exchange rates is that which would make the price of tractor parts the same in both countries, i.e., U.S.$1 = A$3. In other words, if a person had three Australian dollars it would be possible to buy one unit of tractor parts in either country. The other limit of exchange rates relates to the other commodity. In this example, it is U.S.$1 = A$0.50 or alternatively, U.S.$2 = A$1, as this is the amount that would make beef the same price in each country. Trade is possible at either limit or at any exchange rate lying between them. International exchange rates and product unit cost ratios do fluctuate, thus changing the comparative advantage and disadvantage of countries in different commodities.

Countries specialize in the production of certain commodities mainly because of their particular resources. Patterns of beef production and trade are also influenced by population densities, tastes and preferences,

This plant in Paraguay produces canned beef for export. It operates only about six months a year, mostly during the dry season.

political factors, and historical patterns. The modern era of beef trade, for example, started when Europeans began exchanging smoked, cured, and salted products. By the mid-1800s substantial herds in the Western Hemisphere were supplying hides and salted meat to Europe. Meat concentrate and canned meat came into general use in England by 1850 (Hanson 1938). The canned product, which came to be known as corned beef, provided a means of storing beef both during shipment and later in warehouses and retail outlets. Canned beef was a major food for soldiers during both world wars, so that by the end of World War II Argentina and Uruguay had substantial balance of payment surpluses as a result of their canned and preserved beef exports.

Major advances in refrigeration technology after World War II led Argentina and Uruguay to become heavily involved in shipping beef quarters to Europe. The United States imported almost no chilled or frozen beef until the 1950s because hoof-and-mouth disease (or foot-

and-mouth disease as it is also known) was endemic in almost all of the Latin American countries. A successful campaign to eliminate the disease as far south as the Panama Canal was undertaken just after World War II. Once it was established that the disease was indeed eradicated in those countries, chilled and frozen beef as well as live animals (from Mexico) were permitted to enter the United States. There was a lag before countries could respond to the export potential, however, so that substantial exports were not forthcoming until the mid-1960s.

Eradication of hoof-and-mouth disease (aftosa) has had a substantial impact on the economies of the Central American countries, which now depend to a large extent on beef exports for foreign exchange. In 1979, for example, about 60 percent of the beef produced in Honduras and Costa Rica was exported (Table 11.1). In the same year the U.S. allocation level amounted to 59 percent of production in Honduras, 54 percent of that in Costa Rica, and 22 percent of that in Guatemala. About 90 percent of beef exports from these countries went to the United States.

Despite the successful campaign in Mexico and Central America, aftosa continues to be a major factor influencing trade flows. Because of an outbreak of hoof-and-mouth disease in the United Kingdom, beef shipped to Europe is now deboned. The South American countries cannot ship their fresh, frozen, or chilled beef to the United States and thus depend almost exclusively on Europe and the USSR. As will be shown, slow expansion of these markets puts a great strain on countries like Argentina, Brazil, Paraguay, and Uruguay, which could produce much more beef if market outlets were available and if international prices remained at profitable levels.

TRADE FLOWS

World beef trade doubled from the early 1960s to the late 1970s. During these two decades total world exports of the forty-seven major beef trading countries (about 95 percent of total exports) increased from 1.9 million metric tons to 4.3 million tons (see Appendix Table A.5). There are many more importing countries than the forty-seven shown in Table A.5, so that the import totals of 1.8 and 3.6 million metric tons during the same two periods are estimated to account for only about 85 percent of the world's imports.

Unfortunately, discrepancies exist between statistics sources. The two major sources, the United Nations Food and Agriculture Organization and the U.S. Department of Agriculture, are particularly at odds. Table 11.2 compares data on exports and imports of beef for 3 years. An

Table 11.1. U.S. beef import allocation compared with total exports and production for selected countries, 1979

Selected country	Production	Exports	Exports as % of production	U.S. import allocation: product weight	U.S. import allocation: carcass weight[a]	Country allocation as a % of total allocation[b]	Exports as % of production	U.S. actual imports: product weight	U.S. actual imports: carcass weight[a]	U.S. allocation as % of production[b]	U.S. actual imports as % of exports
	(1,000 metric tons)		(%)	(1,000 metric tons)		(%)		(1,000 metric tons)		(%)	
Australia	1,793.8	1,050.0	59	399.8	547.7	56	59	399.5	547.3	31	52
Canada	995.0	50.0	5	35.2	48.2	5	5	35.3	48.4	5	97
Costa Rica	81.4	47.9	59	32.3	44.3	5	59	30.4	41.6	54	87
Dominican Republic	40.0	3.7	9	1.6	2.2	0	9	1.4	1.9	6	51
El Salvador	37.6	7.7	20	5.1	7.0	1	20	4.5	6.2	19	81
Guatemala	94.6	22.6	23	15.0	20.6	2	24	14.6	20.0	22	88
Honduras	53.0	31.7	60	22.7	31.1	3	60	22.7	31.1	59	98
Mexico	1,037.0	4.8	0	3.8	5.2	1	0	2.4	3.3	1	69
New Zealand	490.6	327.7	67	164.2	225.0	23	67	162.4	222.5	46	68
Nicaragua	76.3	43.8	45	31.1	42.6	4	57	30.8	42.2	56	96
Panama	41.0	1.4	3	0.5	0.7	0	3	0.4	0.6	2	43
Others	...	...	...	1.5	2.1	0	...	1.0	1.4	...	...
Total	...	...	...	712.8	976.7	100	...	705.4	966.5	...	...

Source: Production and export data from USDA, *Foreign Agriculture Circular: Livestock and Meat,* FLM 2-80. Import allocation, product weight basis estimates by personnel of USDA, Dairy, Livestock and Poultry Division, FAS.

[a]Production and export data are based on carcass weight. The import allocation is on a product weight basis as are actual imports. FAS experts estimate that 85 percent of U.S. imports are boneless, 15 percent are bone-in. The conversion factors to carcass weight basis are 1.43 and 1.00, respectively. Thus a conversion factor is a weighted average of the two, or 1.37.

[b]Carcass weight basis.

Table 11.2. A comparison of FAO and USDA data on beef exports and imports, 1974-1976

	Exports			Imports		
Year	FAO estimates all countries	USDA 46[a] countries	Ratio FAO/USDA	FAO estimates all countries	USDA 46[a] countries	Ratio FAO/USDA
	(1,000 metric tons)		*(%)*	*(1,000 metric tons)*		*(%)*
1974	2,269	2,701	0.84	2,317	2,602	0.89
1975	2,374	3,280	0.72	2,468	2,837	0.87
1976	2,700	3,833	0.70	2,578	3,158	0.82

Source: FAO *Trade Yearbook,* 1976, and USDA, 1978.
[a]Data from only 46 countries available for all three years.

apparent enigma exists; the USDA statistics for fresh, chilled, and frozen beef are higher than FAO data for "fresh" beef, even though fewer countries are listed. Also, there are considerable discrepancies between years. Table 11.3 highlights the statistical differences. Because USDA data are published about 2 years in advance of FAO data, USDA data are used in this book.

Exports of beef from the forty-seven major countries amounted to about 8 percent of their production in 1961 and about 10 percent in 1979. In other words, although production increased more than 65 percent from the early 1960s to the late 1970s and trade more than doubled, there was only a slight increase in trade as a percent of production.

The largest beef importer is, by far, the United States, a country that has traditionally accounted for nearly one-third of total world imports (Figure 11.1). Because imports in other countries have grown faster, however, the U.S. share of world beef imports declined from 35 percent in 1961–1963 to 28 percent in 1977–1979 (Table 11.4). Several other significant changes in share of world imports have also taken place. France's imports have grown from 1 percent to 7 percent of total imports. West Germany's share increased from 5 to 7 percent, and the USSR's share went from 2 to 5 percent. Imports by the United Kingdom have declined over the past 15 years, from 28 to 16 percent of the total.

Table 11.3. A comparison of FAO and USDA data on beef and veal exports for three major beef exporting countries, 1976

	Beef exports		
Country	FAO	USDA	Ratio FAO/USDA
	(1,000 metric tons)		
Australia	551	860	0.64
New Zealand	228	383	0.59
Argentina	227	534	0.43

Source: FAO *Trade Yearbook,* 1976, and USDA, 1978.

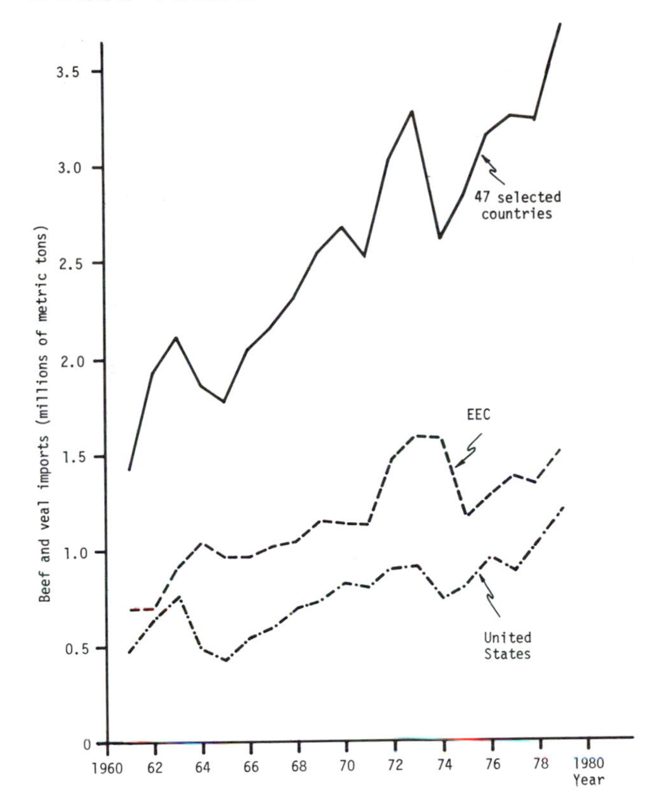

Fig. 11.1. Beef and veal imports, EEC, United States, and forty-seven selected countries, 1961–1979.

More than 70 percent of world beef imports are accounted for by just six countries, France, West Germany, Italy, the United Kingdom, the United States, and the USSR.

More than half of the fresh, chilled, and frozen beef imports into the United States in 1977 were from Australia (Figure 11.2). Most of this beef was destined for manufacturing purposes. The other major country shipping beef to the United States is New Zealand. Nearly three-fifths of U.S. canned beef comes from Argentina, and about one-third from Brazil. These are also the major countries shipping canned beef to the nine EEC countries.

A number of countries are both substantial importers and exporters of beef. When trade is calculated on a net basis, both France and West Germany lose their position in the top six countries, replaced by Greece and Japan. The latter country now accounts for 6 percent of net world imports while Greece has 4 percent. The United States has remained con-

Table 11.4. Share of beef and veal trade by major importing and exporting regions and countries

Region and country	Average by year 1961–1963	1970–1974	1976–1978
		(%)	
Exporting country			
Argentina	29	18	16
Australia	19	21	25
Brazil	2	5	4
Ireland	5	6	7
New Zealand	9	9	8
Uruguay	4	4	2
Other	32	37	38
Total	100	100	100
Importing country			
France	1	5	7
Federal Republic of Germany	5	9	7
Italy	8	12	10
United Kingdom	28	15	16
United States	35	30	28
USSR	2	3	5
Other	21	26	27
Total	100	100	100
Net importing country			
Greece	1	2	4
Italy	8	15	13
Japan	0	4	6
United Kingdom	31	17	19
United States	38	40	39
USSR	0	3	7
Other	22	19	12
Total	100	100	100
Net importing region			
EEC	45	41	33
Western Europe, other	6	8	10
Eastern Europe	8	3	0
North America	38	38	41
Other	3	10	16
Total	100	100	100

Source: Appendix Table A.5.

Note: The 18 for Argentina means, however, that Argentine exports in the 1970–1974 period (548,800 metric tons) accounted for 18 percent of the exports of the 47 selected countries during the period (2,990,000 metric tons). The 41 for EEC as a net importing region for 1970–1974 means that EEC countries accounted for 41 percent of the net imports from among the 47 selected countries.

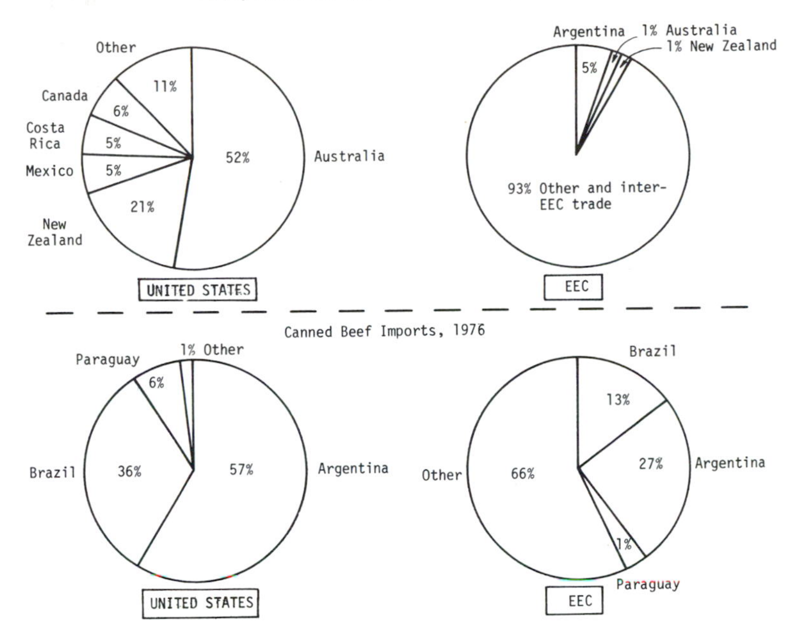

Fig. 11.2. Fresh, frozen, chilled, and canned beef imports by the United States and EEC countries, on a percentage basis from the country of origin, 1977 (USDA 1979).

stant at about 40 percent of net world imports over the past 15 years, while the USSR has increased from almost no imports to 7 percent of net world imports in the period 1961–1963 to 1976–1979.

Considerable change has taken place in net beef imports when analyzed by region. The EEC as a whole dropped from 45 percent in the early 1960s to 33 percent in the late 1970s, primarily because the United Kingdom's share has declined substantially. In the meantime, the share of the other non-EEC countries of western Europe increased slightly, from 6 percent to 10 percent. Eastern Europe's share has declined dramatically, from 8 percent to becoming a net exporter. The net share for other, nontraditional areas such as Japan and the Middle East has correspondingly increased.

Export country ranking has also undergone change. Argentina's once dominant position has steadily eroded from nearly 30 percent of all world exports in the early 1960s to 16 percent by the late 1970s (Table 11.4 and Figure 11.3). In the meantime, Australia increased its portion from 19 to 25 percent. Brazil's share has increased from 2 to 4 percent of

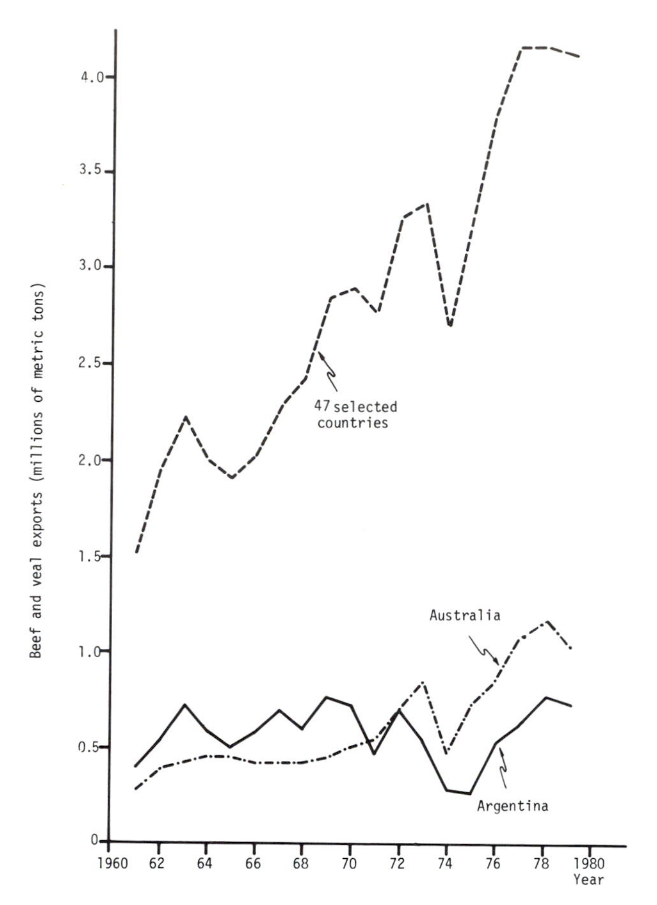

Fig. 11.3. Beef and veal exports, Argentina, Australia, and total for forty-seven selected coutnries, 1961–1979.

the world total, and the three other leading exporting countries, Ireland, New Zealand, and Uruguay, have just maintained their positions. These six countries accounted for about two-thirds of world exports in the late 1970s.

About 10 percent of the beef entering world trade is in canned form. Argentina has traditionally been the leading exporter, but this position is now being challenged by Brazil (Table 11.5). About 58 percent of world exports are from the Western Hemisphere, 27 percent originate in Europe, and only about 10 percent are from Oceania. Africa plays a minor role.

Europe accounts for about 70 percent of world canned beef imports followed by the Western Hemisphere with 28 percent (Simpson and Mi-

The Smithfield meat market in London was the international price barometer for beef until Britain entered the European Common Market in the 1970s. It still plays a major international role, but its ties to exporters outside the EEC are sharply reduced.

rowsky 1979). The other 2 percent is about equally divided between Asia and the Middle East. The United Kingdom is the leading canned beef importing country, accounting for about 32 percent of the total. The United States and West Germany each share about 22–23 percent (Table 11.5).

Partly because of hoof-and-mouth disease, Argentina's importance as a beef exporting country has languished, whereas Australia and New Zealand, both free of the disease, have prospered. The effect that these sanitary restrictions have had on trade flows is graphically portrayed in Figure 11.4, which shows fresh, frozen, and chilled beef exports for 1977 for Argentina and Australia. In that year Argentina exported about 280 thousand metric tons of fresh, frozen, and chilled beef on a product weight basis of which the largest share, 30 percent, went to the EEC and 20 percent to other European countries. Just two countries, Spain and the USSR took 22 percent more. Africa accounted for an additional 16 percent, with the result that other countries were just 12 percent of Argentina's total exports.

Australia, with no disease barriers to bar trade in other parts of the

Table 11.5. Exports of canned beef as a percentage of total world exports by selected regions and countries, 1962, 1970, and 1976

Region or country	Year		
	1962	1970	1976
		(%)	
		Exports	
Region			
Africa	9	6	5
Asia and Oceania	8	10	10
Europe	27	27	27
Western Hemisphere	56	57	58
Total	100	100	100
Selected country			
Argentina	34	41	26
Australia and New Zealand	8	10	9
Brazil	6	8	25
France	5	6	6
Paraguay	8	6	4
United States	1	2	2
Uruguay	7	1	1
Yugoslavia	2	4	3
Other	29	22	24
Total	100	100	100
		Imports	
Region			
Africa	3	2	0
Asia and Oceania	8	3	1
Europe	62	59	70
Middle East	1	1	1
Western Hemisphere	26	35	28
Total	100	100	100
Selected country			
Federal Republic of Germany	3	12	23
United Kingdom	48	31	32
United States	23	30	23
Other	26	27	22
Total	100	100	100

Source: Simpson and Mirowsky, 1979.

world, shipped 39 percent of her 645 thousand tons in 1977 to the United States and only 1 percent to the EEC (Figure 11.4). Japan, the USSR, and the Middle East accounted for 29 percent of shipments. Australia's flexibility is demonstrated by the fact that 31 percent of its imports were destined for other countries. Considerable advances are being made in Australia's cattle production techniques, and it appears that substantial expansion can be expected (DeBoer 1979).

Because of relatively secure access to almost all fresh, chilled, and frozen beef markets, Australia has gradually phased out exports of

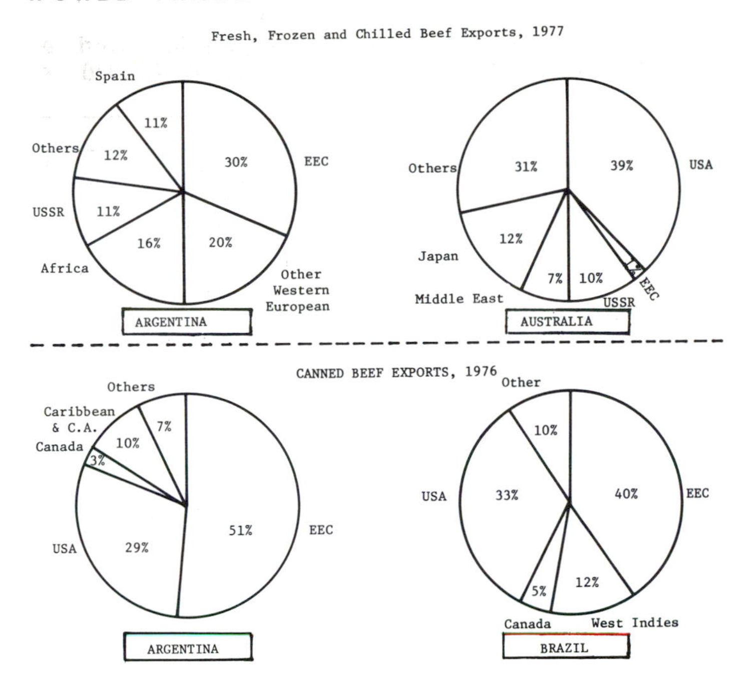

Fig. 11.4. Fresh, frozen, and chilled beef exports from Argentina and Australia and canned beef exports from Argentina and Brazil on a percentage basis by country of destination, 1977 (Ministerio de Economia 1978; DeBoer 1979; USDA 1978, 1979; Simpson and Mirowsky 1979).

canned beef. Argentina, on the other hand, is forced to maintain this export item. In 1977 nearly 50 thousand tons of beef were exported in canned form (product weight basis), of which more than half went to the EEC and 29 percent to the United States (Figure 11.4). Brazil also exported most of its canned beef (73 percent) to just two areas, the EEC and the United States.

There is no one international price of beef. Rather, there are prices for each of the different products—for example, chilled loins, bone-in quarters, and frozen boneless manufacturing beef. Furthermore, these prices fluctuate and are not well known to the public, as most of the sales are in direct two-party contracts. However, an indication of average annual prices can be derived by dividing total f.o.b. value of exports by quantity. This has been done for Argentine chilled quarters and canned beef, and the resultant average prices are plotted against total beef exports from 1964–1978 in Figure 11.5. For the 7 years 1964–1970 the quantity exported fluctuated considerably but with no trend, while prices remained stable for canned beef at $800 per ton and for quarters at about

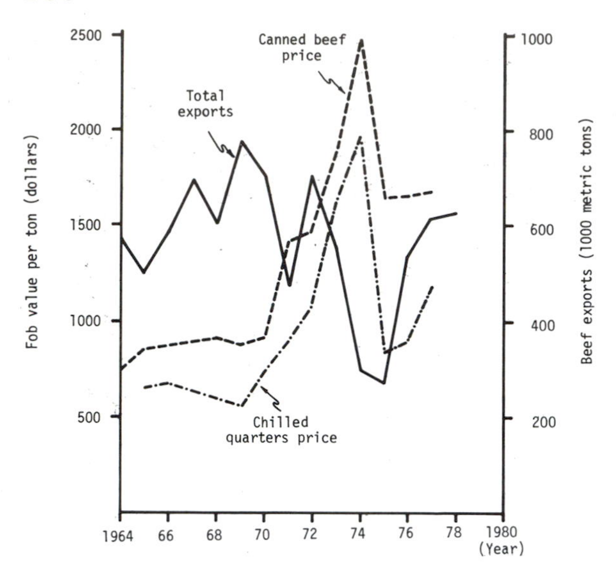

Fig. 11.5. Argentine exports of beef and veal compared with f.o.b. prices of chilled quarters and canned beef, 1964–1978 (Ministerio de Economia, *Sintesis Estadistica,* various issues).

$600. In 1971, prices began to skyrocket, and canned beef and chilled quarters reached an average of $2,481 and $1,970, respectively, in 1974. Argentine exports declined substantially beginning in 1973 as production declined in response to cattlemen holding back breeding animals for building up herds and to a reduced demand from many of Argentina's trading partners (Figure 11.5). In 1974 and 1975 the EEC countries substantially reduced their imports and another market outlet was eroded. International prices also fell dramatically that year, chilled quarters dropping to half the previous year's price. Cattlemen responded to lower live cattle prices by massive liquidation of herd numbers so that while exports increased at the low prices, canned beef constituted a greater proportion of the total.

A new subpattern of trade is now developing as numerous countries around the world import more grain-fed beef. Much of this beef in countries like the Philippines and Hong Kong is destined for the high-class restaurant trade. In countries like Japan retail sales of fed beef are growing despite extremely high prices. The United States has aggressively promoted beef exports to Europe, Japan, Iran, and Indonesia,[1] and as a

[1]Considerable expense goes into the efforts of entities charged with promotion of beef exports such as the Australian Meat Board, the New Zealand Meat Board, the Junta Nacional de Carnes in Argentina, and the Meat Export Federation in the United States.

result, shipments increased from 15 thousand metric tons in the early 1960s to 75 thousand tons in 1979. The increase took place almost entirely in the late 1970s, primarily as a result of increased trade with Japan. By 1979 the United States exported $227.2 million of fresh or frozen beef and veal, or about 13 percent of the $1.8 billion in beef imports (see Appendix Table A.6).

BY-PRODUCT TRADE

The major product of the world's beef business other than meat and edible portions of the animal is hides, which, as a function of cattle production, are tied to cattle inventories. Hide production, which is measured in metric tons rather than numbers of hides, increased from an average of 3.0 million tons during 1960–1964 to about 4.5 million metric tons by 1978 for forty-nine selected beef producing countries (see Appendix Table A.7). These countries account for about two-thirds of world hide production. In terms of the forty-nine countries, the United States produced about 25 percent of all hides, followed by Argentina with 8 percent, Brazil with 8 percent, and Australia with 4 percent.

Approximately 640 thousand metric tons of hides were exported annually during 1959–1963 compared with 1.5 million tons in 1978 (see Appendix Table A.8). Trade in hides, as a percentage of production, has increased continuously over the past 2 decades. By the early 1970s the proportion had increased to 31 percent, and by 1978 to about 35 percent of hide production in the forty-nine countries.

The United States is the world's leading hide exporter, with 45 percent of the selected countries' hide exports in 1978. The percentage has increased steadily since the early 1960s, when the ratio was 26 percent. The major reason for the change is specialization in the manufacturing of leather goods in other countries. Australian exports, as a percentage of all hide trade, have increased, but Argentina's share and absolute amount have fallen sharply in response to increased specialization domestically in leather goods manufacture.

The countries of destination for hides have also changed over the past 2 decades. Italy has increased her share of imports from 11 percent to 22 percent, and Japan from 15 to 19 percent (see Appendix Table A.9). Mexico, once a net hide exporter, has now become a major importer.

By-products of all types are an important consideration in determining a country's net position with respect to the beef cattle industry as a whole. A good case in point is the United States, which accounts for about 40 percent of net world trade in beef and veal but which achieved a

Some of the important economies of size in beef plants are related to by-product processing. This is a hide curing and tanning operation. The U.S. is the leading cattle hide exporter.

positive balance of payments in 1978 of $62 million for beef and cattle products as a whole, although by 1979 there was a deficit of $121 million (see Appendix Table A.6). In 1979, over a half billion dollars of tallow and greases were exported along with $874 million in cattle hides. The balance for all meat products and livestock was a negative $183 million in that year.

POLICIES, AND TARIFF AND NONTARIFF TRADE BARRIERS

Foreign economic policy embraces the varied activities of national governments that bear, directly or indirectly, on the composition, direction, and magnitude of international trade and factor flows. It can be defined as the activities of governments that are primarily intended to regulate, restrict, promote, or otherwise influence the conduct of trade and investment. Government policies should not be considered exclusively domestic or foreign, but rather should be classified according to their effects on each other since they are interdependent.

Governments have many goals in foreign economic policy, of which several stand out as being especially important to considerations of tariff and nontariff barriers on beef. One natural tendency is for countries to preserve national self-sufficiency and promote protectionism. Vested private interests, such as cattlemen's associations or leather goods manufacturers, lobby with legislators and government agencies to raise trade barriers for their commodity even though they may simultaneously exert pressure to reduce barriers on commodities for which they are consumers. Governments, in attempting to improve the general economic welfare, recognize that although increased trade provides cheaper goods for more people, imports may create a disadvantage (or seem to) for certain groups. Trade is also viewed as a source of disturbance to a policy of full employment (Root 1978). An example of the dramatic impact a change in trade policy can have on the domestic beef market is the change in domestic wholesale beef prices in England after joining the EEC. Prior to 1972 England had lower wholesale beef prices than the United States. By 1976 prices were considerably above those in the United States and since have gone higher.

The pros and cons are further complicated by concern about balance of payments and trade as a development device. It can be argued that imported beef has few positive effects on domestic development except to the extent that it releases land for other uses. Apart from the substantial benefit of imported beef in satisfying consumer needs and wants, the importing of beef, especially products that require further processing, should be recognized as a tool in assisting other countries to develop (Adams and Simpson 1977). In brief, trade policy is a weighing of various considerations that make up a country's national and international goals and responsibilities.

Tariffs (or customs duties) are taxes imposed by a government on physical goods as they move into or out of an area for the purpose of revenue, protection, or balance of payments. They are usually designed by lobby and pressure groups rather than by legislators working for the overall good of a country. There are two basic kinds of customs duties: ad valorem, which is a duty stated as a percentage value of an article; and specified duties, which are expressed in terms of an amount of money per quantity of goods. Compound duties are a combination of the ad valorem and specific types. Compound duties are the tariffs most commonly applied to imports of beef and live cattle.

Nontariff barriers such as quotas or unusual sanitation restrictions are a means of restricting quantities moving into a country. Import quotas are the most widely used barrier, but nations must take care in administering them as unilateral quotas can lead to antagonism and retaliation. One type of quota, known as a tariff quota, permits a product to enter duty free or, up to a certain point, at a given rate of duty. The EEC has incorporated tariff quotas into its beef import policy. Another type of nontariff barrier is the voluntary export quota, according to which a country agrees to voluntarily restrict its exports to a certain negotiated level. The United States has built its beef import legislation around this type of trade barrier.

The export controls established in beef exporting countries usually are intended to accomplish one or more goals. One objective is to assure that a significant proportion of certain products in short supply be retained for the home market. Another objective is to control surpluses on a national or international basis in order to achieve production or price stability. A conflicting goal is the use of beef exports for generating foreign exchange. Objectives are difficult to meet as prices are subject to cyclical fluctuations. Also, quantitative restrictions by importing countries are sometimes enacted at most inconvenient and unexpected times.

Health restrictions, a common type of nontariff barrier, are difficult to administer because of two main problems. First, it may not be possible to apply the desired principle of nondiscrimination, a major provision in the General Agreement on Tariffs and Trade (GATT). Second, implementation of the nondiscrimination principle may not be the only precondition required for an international meat trade agreement because of sanitation and health concerns. This second point is salient to veterinary objectives.

Specific health restrictions are difficult to formulate because infections of animal diseases do not always produce symptoms in the individual animal that would be identified at antemortem and postmortem inspection in the slaughterhouse. Thus it becomes necessary to prevent the shipping of all meat from infected areas to noninfected areas. As might be expected, another troublesome aspect is defining the area that is infected. The result is that a major controversy surrounds proposed definitions for size of the area considered to be potentially infected, and duration of the restrictions. The general practice is to consider a whole country infected even though only a small part may be involved. For this reason the establishment of officially certified specific disease-free areas appears to be a most appropriate and promising approach (Griffiths 1972).

Another means of insuring that disease is not used as a nontariff barrier is by the worldwide adoption of uniform regulations for the control of animal diseases. It is possible to agree on certain basic principles, but the method of applying the principles depends on livestock production conditions, the immediate disease situation, patterns of trade movement, feasibility of control measures, and many other factors that vary from country to country and region to region (Konigshofer 1979). Thus, although it is generally agreed that a multilateral veterinary agreement is desirable, there is as yet no mandate from governments to draft any specific proposal. The conclusion one may draw is that bilateral health regulations will become increasingly stringent and, as such, will become a stronger deterrent to international beef trade.

BEEF IMPORT POLICY OF THE UNITED STATES

The amount and origin of U.S. beef imports were regulated from 1964 through 1979 by Public Law 88–481. The quota, which amounted to about 6 to 7 percent of domestic production, was based on a market sharing principle. Compliance with the quota was voluntary but a trigger point for application of restrictions was set at 110 percent of the negotiated level. A major drawback of the law was that imports increased with production.

As a result, beginning in 1978 various legislative proposals were set forth for a law based on a countercyclical formula according to which imports and production would vary inversely. Finally, the Meat Import Act of 1979 was signed into law the last day of that year. In the new law an updated formula for determining annual quotas is provided as follows:

$$\text{Annual quota} = \text{Average annual imports (1968–1977)} \times \frac{\text{3-yr moving average of domestic production}}{\text{10-yr average of domestic production (1968–1977)}} \times \frac{\text{5-yr moving average of domestic cow beef production}}{\text{2-yr moving average of domestic cow beef production}}$$

The portion of the formula provided by the 5-year moving average of domestic cow beef production divided by the 2-year average establishes the countercyclical part of the formula. This annual quota has a lower limit of about 1.2 billion pounds with an adjustment to increase this as the U.S. market grows.

BEEF IMPORT POLICY OF THE EUROPEAN ECONOMIC COMMUNITY

The EEC accounts for about one-third of the net beef imports in the forty-seven countries listed in Appendix Table A.5. The EEC's beef policies are of utmost importance to countries such as Argentina, Botswana, and Uruguay that do not have access to the United

States for sales of fresh, chilled, or frozen beef. The community arrangements are especially important to the large number of beef, veal, and milk producers throughout the EEC. In fact, milk and beef account for more than one-third of the value of EEC agricultural output, and beef is the major item in the household budget.

Because of the rapid expansion of beef trade occurring in the early 1960s, the first steps were taken in 1964 to regulate beef and veal marketing on a common market basis. The basic objectives of the common agricultural policy (CAP) have been to develop production in order to guarantee the security of supply for consumers by ensuring a satisfactory income for producers, and to stabilize prices and consequently the cyclical movement of production. Several important changes have been made since regulations for the original six member states went into effect on July 29, 1968. A system of permanent intervention was agreed upon at the end of 1972, for example, and other changes were brought about in early 1977. Our discussion centers on regulations in effect since 1978.

The basic regulation provides for a system of price supports that attempts to keep community market prices as close as possible to an agreed common price level. The main price mechanisms are internal market measures such as support buying and private storage, regulations affecting imports from outside the EEC, and refunds on EEC exports to third world countries. A key point is that EEC countries in general have a comparative disadvantage in the production of beef, with the result that prices have to be set at rather high levels to encourage production.

Community members hold annual meetings at which time a guide price is set at a level considered desirable under current and future market conditions. The guide price was $1.60 (79.89 pence) per kilogram ($0.73 per pound) for the 1978–1979 marketing year in the United Kingdom. The price varies slightly among EEC countries. A compulsory intervention price set at 90 percent of the guide price is used to determine the buying-in prices for EEC support buying, a tactic that is used when domestic prices fall to an unacceptable level. The intervention price in the 1978–1979 marketing year was $1.44 per kilogram (71.90 pence) in the United Kingdom. Since 1972 there has been permanent intervention, which means that a guaranteed price has existed.

Reference prices, a weighted average of actual cattle prices recorded in representative markets and abattoirs in the nine member countries are established for each nation according to the relative contribution each cattle class makes to total beef production. Cow prices, for example, are weighted at 60 percent of the Dutch reference price, but only 23 percent of the U.K. reference price. Calculations of the reference price made on a weekly basis also take into account the size of cattle production. Upper and lower buying-in prices tied to the intervention price are set for each country. The buying and storage are carried out by both government and private traders via a common pool of funds.

Cattle and meat imports from non-EEC countries are covered by fixed and permanent customs duties, but the more important factor is a variable levy calculated on the difference between the EEC guide price and the value of the imported product or cattle. The general effect of the levy is to bring imported beef and cattle prices plus the import tax up to EEC prices. The actual import levy is dependent on the weekly relationship between the EEC reference price and the EEC guide price. On October 2, 1978, for example, the U.K. levy on frozen boneless cuts was $1.83 per kilogram (91.407 pence). From April 1977 through 1978 import levies were almost 100 percent of the purchase price in world markets. The high prices resulting from these duties have the effect of substantially reducing the quantity consumed.

The EEC has insured that imports are maintained within planned limits by establishing quotas in addition to the levy. In 1977 almost all cattle entered the EEC under concessionary agreements, and almost four-fifths of chilled or frozen beef and veal imports came in under special arrangements. Manufacturing beef for use in further processed products, such as canned beef, is subject to quota but has little or no variable levy. Under the home convention, which is an agreement between the EEC and more than fifty developing countries in Africa, the Caribbean, and the Pacific, a certain amount of beef can be imported into the EEC duty free from the four African countries of Botswana, Kenya, Madagascar, and Swaziland. The allocation in 1978 was about 34 thousand tons (approximately 3 percent of EEC imports), of which more than 60 percent was awarded to Botswana.

Several other controls are designed to protect EEC producers. Export refunds (subsidies) are paid to EEC exporters to enable them to compete in world markets. Furthermore, special subsidies are paid to farmers who have high production costs, such as those living in hilly areas. Safeguard clauses that allow for greater protection in the event of severe disturbances within the EEC countries are also a part of the scheme. Action under the safeguard clause was operative in the period from July 1974 to March 1977; at this time import licenses for most categories of cattle, calves, beef, and veal were suspended, largely in response to foreign exchange spending arising from the rapid increase in petroleum prices, and because of rapid herd buildups in some of the EEC member countries. The result was that imports by EEC countries (including those countries who entered the community later) dropped from 849 thousand tons in 1973 to 159 thousand tons in 1979. Net imports were equivalent to 14 percent of production in 1973, but had declined to just 2 percent of production by 1979 (Table 11.6). This vast reduction in im-

Table 11.6. Beef imports and exports as a percentage of beef production in the EEC, 1973 and 1979

Country	1973 Production	1973 Net trade*	1973 Proportion	1979 Production	1979 Net trade*	1979 Proportion
	(1,000 metric tons)		*(%)*	*(1,000 metric tons)*		*(%)*
Belgium	270.0	(8.2)	(3)	290.6	5.0	2
Denmark	191.6	104.7	55	245.0	180.6	74
France	1,565.2	(19.4)	1	1,791.0	8.0	0
Federal Republic of Germany	1,356.5	(250.5)	(19)	1,520.0	70.0	5
Ireland	216.3	151.5	70	378.0	297.5	79
Italy	1,077.0	(454.0)	(42)	1,065.0	(335.0)	(31)
Netherlands	326.2	28.3	9	342.0	70.0	20
United Kingdom	948.1	(401.3)	(42)	1,010.0	(455.0)	45
Total	5,950.9	(848.9)	(14)	6,641.6	(158.9)	(2)

Source: USDA, 1978 and 1980.
*Net imports shown in parentheses.

ports is a prime example of the accomplishments made possible by coordinated production and trade policies.

Wholesale beef prices for two EEC countries, France and the United Kingdom, as well as for Australia, New Zealand, South Africa, and the United States are given in Table 11.7 for the period 1955–1977. This table and Figure 11.6 point to considerable differences among the countries, especially since 1968 when the EEC instituted its pricing system. At the same time that Australia suffered an amazing price decline in domestic prices (from $1.29 per kilogram in 1973 to $0.51 in 1975), the domestic price in France increased from $2.41 per kilogram to $2.82. Furthermore, the price in France continued to increase substantially through 1977 (up to $2.91) while the price in Australia increased only five cents per kilogram. By 1977 the wholesale price in France was nearly six times as high as it was in Australia.

SUMMARY AND CONCLUSIONS

World trade in beef and veal doubled from the early 1960s to the late 1970s, but it has remained at about 10 percent of production by the major beef exporting countries. The United States accounts for about one-third of total world imports, which is just slightly less than total imports in the nine EEC countries. Australia has taken over as the major beef exporting country with about 25 percent of world exports, leaving the once dominant Argentina with about 15 percent of the total.

An analysis of trade flows and world beef trade structure leads to

Table 11.7. Wholesale beef prices in selected countries, 1955-1977

Year	Australia[a]	France[b]	New Zealand[c]	South Africa[d]	United Kingdom[e]	United States[f]
			(U.S. cents per kg)			
1955	31.6	73.4	34.0	40.8	71.3	85.5
1956	30.2	85.1	21.6	42.0	56.7	83.5
1957	32.8	84.6	24.7	42.8	59.8	86.8
1958	38.5	94.2	40.1	41.0	68.5	99.3
1959	47.7	79.8	41.7	43.0	72.7	100.0
1960	50.6	84.7	43.2	41.1	68.8	97.4
1961	43.8	84.7	35.3	41.2	59.8	91.1
1962	38.6	92.4	41.4	42.5	66.9	98.8
1963	43.2	99.0	38.3	41.3	63.9	90.7
1964	48.2	114.0	46.0	48.5	80.1	86.8
1965	54.0	116.7	46.0	56.2	84.5	93.7
1966	60.3	118.1	45.5	56.9	80.0	94.9
1967	63.1	116.5	43.6	64.0	77.5	90.8
1968	65.1	119.5	47.4	66.1	...	96.3
1969	65.9	126.5	55.8	66.6	...	104.9
1970	70.9	142.8	59.1	66.1	86.2	104.4
1971	74.4	147.3	64.7	71.8	104.4	115.7
1972	80.6	196.1	88.3	71.3	121.9	123.0
1973	129.2	241.0	84.6	107.7	154.5	149.7
1974	78.4	222.0	53.2	145.0	146.4	146.4
1975	50.6	282.0	78.5	132.4	161.1	160.1
1976	62.1	272.2	62.0	118.1	172.8	134.5
1977	55.9	291.0	63.9	119.4	181.6	...

Source: FAO *Production Yearbook,* various issues.
[a]Steers, first and second export quality, 650-700 lb wholesale price, Brisbane.
[b]Steers, first quality, wholesale price, excluding tax, Paris.
[c]Steers, quarter beef, good average quality, opening schedule price for meat operators and exporters, North Island.
[d]Prime beef A, cold dressed weight, auction price, Johannesburg.
[e]Wholesale price for English longsides, top quotation, Smithfield Market.
[f]Steer carcasses, choice, 500-600 lb wholesale price, Chicago.

the conclusion that beef trade is governed to a much greater extent by government policies than by demand and other economic considerations. One major nontariff barrier is sanitation regulations. Although these rules are necessary to protect both consumers and producers, they may create distinct noneconomic advantages for certain countries, such as Australia and those in Central America, while effectively discriminating against a whole bloc of countries elsewhere, as in South America and in most of Africa.

Policies of importing countries are designed to protect producers and simultaneously achieve a high degree of self-sufficiency. These policies are especially critical since there is considerable evidence that in the foreseeable future beef surpluses will be a greater problem throughout the world than beef deficits. As we have demonstrated at several

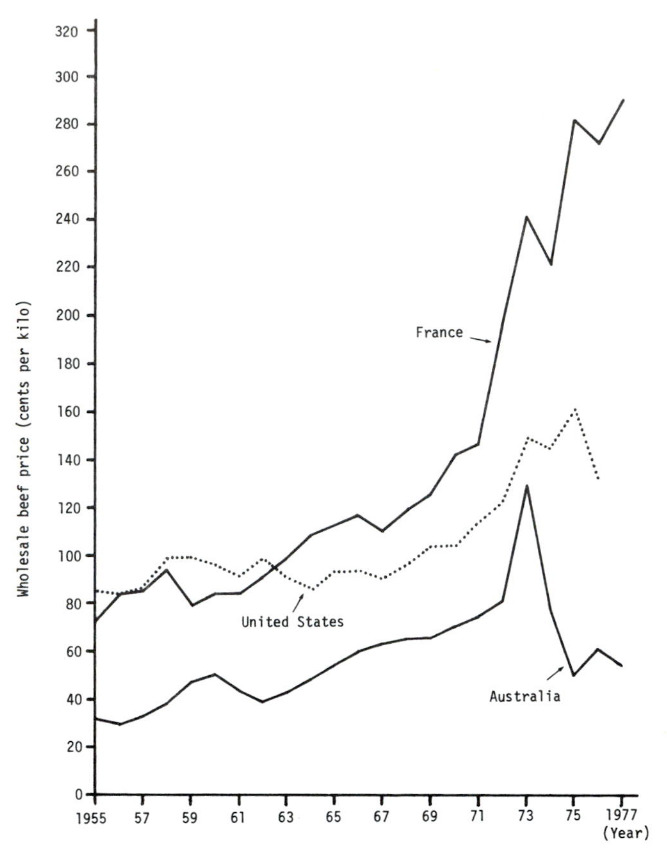

Fig. 11.6. Wholesale (carcass) beef prices in selected countries, 1975–1977.

points in this book, the potential productive capacity is high in many areas, so that the principal problem for exporting countries will be to cope with the cyclical nature of world beef prices and to develop export markets.

The United States instigated a meat import law in 1964 which, while effective as a quota device, proved to have a destabilizing effect on national and international prices. The EEC's policy has succeded in improving production and internal market structure, but has introduced a high social cost because member country prices had to be fixed considerably higher than international prices. This policy has led to overproduction, has caused exports of complementary dairy products to be subsidized, and has created considerable economic hardship for a number of beef exporting countries. The CAP has also prevented beef consumption from rising because artificially high prices are passed on to consumers.

World trade in beef and beef cattle is interrelated with processing, marketing structure, live cattle prices, and other marketing aspects. These topics, discussed in previous chapters, have provided additional insights into the how and why of beef trade. Policy recommendations and projections of trade, as well as forecasts of beef consumption and production, are presented in the next chapter.

REFERENCES

Adams, J. W., and Simpson, J. R. 1977. Estimates of South American Beef Multipliers. *Growth and Change* 8(4):28–32.

DeBoer, A. J. 1979. *The Short Run and Long Run Position of Australian Beef Supplies and the Competitiveness of Australian Beef in International Trade.* Ann Arbor: University of Michigan, Center for Research on Economic Development.

European Economic Information Service. 1978. *The Common Agricultural Policy: Beef and Veal.* London.

Griffiths, R. B. 1972. Disease Free Zones and Beef Export. *World Anim. Rev.* Vol. 3, 19–25.

Hanson, Simon G. 1938. *Argentine Meat and the British Market.* Stanford, Calif.: Stanford University Press.

Konigshofer, H. O. 1979. International Trade in Meat: Veterinary Aspects and the Role of FAO. *World Anim. Rev.* 30:2–7.

Ministerio de Economia, Junta Nacional de Carnes. 1978. *Sintesis Estadistica 1977,* Republica Argentina.

Organization for Economic Cooperation and Development. 1980a. *Trade by Commodities.* Series C.

______. 1980b. *The Instability of Agricultural Commodity Markets.* Agricultural Products Market Series 5180031. Paris.

Root, Franklin R. 1978. *International Trade and Investment.* 4th ed. Cincinnati, Ohio: South-Western Publishing Co.

Simpson, James R. 1978. *Projections of Production and Consumption of Beef in Latin America for 1985.* Food Resour. Econ. Dep. Staff Pap. 88. Gainesville: University of Florida.

Simpson, James R., and Mirowsky, Jill. 1979. *World Trade in Canned Beef, 1962–76.* Cent. for Trop. Agric. Rep. no. 2. Gainesville: University of Florida.

U.S. Department of Agriculture. 1978a. *Foreign Agriculture Circular: Livestock and Meat.* FLM 9-78. Washington, D.C.

______. 1978b. *Foreign Agriculture Circular: Livestock and Meat.* FLM 10-78. Washington, D.C.

______. 1979. *Foreign Agriculture Circular: Livestock and Meat.* FLM 7-79. Washington, D.C.

______. 1980. *Foreign Agriculture Circular: Livestock and Meat.* FLM 2-80. Washington, D.C.

Valdez, Alberto, and Nores, Gustavo. 1979. *Growth Potential of the Beef Sector in Latin America—Survey of Issues and Policies.* Washington, D.C.: International Food Policy Research Institute.

12
PROJECTIONS AND POLICY CONSIDERATIONS

This chapter first discusses projections of world supply and demand for beef, then ties these forecasts to feed grain projections, and finally relates the two to policy considerations. The main objective in setting forth the projections is to determine the extent to which global beef production should be stimulated and the areas of importance. Policy and technical considerations required to bring about desired changes in the world's beef business are related to recommendations in the last part of the chapter.

PROJECTIONS

Projections of per capita beef and veal consumption, total consumption and production, as well as deficits and surpluses in 1985 and 1990 for forty-eight selected countries are presented in Table 12.1.[1] The data for 1990 are also given to provide a reference point. Additional historical data for 1961, 1971, and 1978 appear in Appendix Tables A.2, A.3, and A.4.

Per capita consumption in ten countries is projected to decline in the 20-year period from 1970 to 1990, while a decrease is projected for twelve of the forty-eight countries in the 5-year period from 1985 to 1990. Only a few countries such as Australia, Canada, Ireland, New Zealand, the USSR, the United States, and Uruguay have any significant absolute in-

[1]The results are derived from a computer simulation model in which projections of consumption and production for each country are first totaled. Then imports for each country are divided by population to calculate imports per capita. The difference between production and consumption is then calculated for each country and deficit countries' net imports are allowed to increase only at the rate of 5 percent annually above the highest historic imports per capita. In this way, each country's consumption is adjusted to demand. Historically, consumption has lagged behind production by about 50,000 metric tons annually (out of 25–30 million metric tons annual production or consumption) for the forty-eight selected countries. Total consumption for all countries is then reduced to the 50,000-ton difference by an interactive procedure.

Table 12.1. Per capita beef and veal consumption, total beef and veal consumption and production, and deficit, 48 selected countries, 1970, with projections to 1985 and 1990

Country	Per capita consumption			Total consumption			Production			Deficit			Surplus		
	1970	1985	1990	1970	1985	1990	1970	1985	1990	1970	1985	1990	1970	1985	1990
	(kg)			(1,000 metric tons)											
Argentina	81.6	84.7	86.5	1,956.0	2,429.0	2,611.3	2,624.0	2,839.4	2,972.4	...	...	...	668.0	410.4	361.1
Australia	41.7	78.0	86.9	510.1	1,286.2	1,546.5	1,002.8	2,328.4	2,690.9	...	...	...	492.7	1,042.2	1,144.4
Austria	22.7	29.3	31.0	166.6	226.6	243.3	153.8	213.7	231.0	12.8	12.9	12.3	...	...	...
Belgium and Luxembourg	27.7	32.5	34.4	270.9	340.1	365.5	270.0	344.6	372.6	0.9	...	...	...	4.5	7.1
Brazil	18.6	20.7	21.2	1,721.9	3,000.5	3,506.9	1,845.0	2,902.3	3,249.1	...	98.2	257.8	123.1	...	...
Bulgaria	11.3	13.3	14.1	95.6	124.0	134.7	93.2	126.1	137.2	2.4	...	...	...	2.1	2.5
Canada	42.6	53.8	55.9	891.9	1,425.8	1,584.3	850.6	1,302.2	1,433.3	41.3	123.6	151.0	...	...	...
Chile	21.3	16.6	16.1	224.3	204.2	215.4	182.3	209.1	224.9	42.0	...	...	...	4.9	9.5
Colombia	23.6	17.8	17.1	417.6	623.0	690.3	425.4	578.2	627.5	...	44.8	62.8	7.8	...	...
Costa Rica	9.5	11.9	11.9	20.9	31.1	35.2	46.3	87.3	101.9	...	...	...	25.4	56.2	66.7
Czechoslovakia	19.1	27.5	29.7	387.9	427.8	474.2	299.2	424.3	471.2	88.7	3.5	3.0	...	...	...
Denmark	20.9	18.3	18.3	103.1	94.6	95.9	191.6	250.0	260.0	...	...	...	88.5	155.4	164.1
Dominican Republic	6.4	6.6	6.5	27.2	47.3	55.2	31.5	49.0	54.5	...	...	0.7	4.3	1.7	...
El Salvador	5.9	5.8	5.4	20.5	32.9	35.8	20.3	32.6	35.5	0.2	0.3	0.3	...	...	...
Finland	20.4	25.2	26.6	96.2	118.6	125.9	106.0	130.3	140.5	...	...	...	9.8	11.7	14.6
France	29.5	31.9	32.7	1,520.2	1,820.0	1,923.3	1,565.2	1,850.1	1,941.9	...	...	...	45.4	30.1	18.6
Federal Republic of Germany	24.9	25.7	26.4	1,526.8	1,613.7	1,697.6	1,356.5	1,532.2	1,618.3	170.3	81.5	79.3	...	...	...
Greece	18.1	25.3	27.0	158.9	233.0	253.4	90.1	156.1	180.6	68.8	76.9	72.8	...	...	...
Guatemala	7.7	8.6	8.8	40.1	70.6	83.2	57.1	99.2	112.6	...	...	...	17.0	28.6	29.5
Honduras	5.0	8.7	9.2	13.3	36.9	46.0	30.0	64.9	76.1	...	...	...	16.7	28.0	30.1
Hungary	9.1	10.9	11.3	96.1	118.2	123.2	117.5	152.0	162.4	...	...	...	21.4	33.8	39.2
Iran	2.3	2.8	2.7	51.8	123.6	141.3	51.6	86.0	97.3	0.2	37.6	44.0	...	...	...
Ireland	17.7	31.2	35.2	56.5	108.5	128.8	216.3	483.2	566.1	...	...	...	159.8	374.7	437.3
Israel	18.6	14.2	13.8	56.8	61.9	66.2	19.4	29.5	33.0	37.4	32.4	33.2	...	...	...
Italy	20.4	28.6	30.9	1,366.5	1,644.5	1,815.4	1,077.0	1,283.0	1,432.5	289.5	361.5	382.9	...	...	...
Japan	2.7	5.2	5.5	311.2	640.4	696.6	278.0	439.0	503.2	33.2	201.4	193.4	...	...	...
Mexico	10.0	14.8	14.5	538.8	1,227.4	1,412.6	590.0	1,226.6	1,411.8	...	0.8	0.8	51.2	...	...
Netherlands	20.0	20.2	20.3	256.9	295.2	306.9	326.2	375.0	385.0	...	...	...	69.3	79.8	78.1
New Zealand	48.1	59.0	61.9	112.6	206.9	232.3	392.8	603.9	682.8	...	...	...	280.2	397.0	450.5
Nicaragua	13.6	15.9	16.6	29.6	51.2	62.7	62.6	94.7	108.7	...	...	...	33.0	43.5	46.0

Source: Simpson, 1979.

Table 12.1. *(continued)*

Country	Per capita consumption			Total consumption			Production			Deficit			Surplus		
	1970	1985	1990	1970	1985	1990	1970	1985	1990	1970	1985	1990	1970	1985	1990
	(kg)			*(1,000 metric tons)*											
Norway	14.1	19.6	20.6	56.0	82.8	88.7	57.2	73.3	77.6	...	9.5	11.1	1.2	...	...
Panama	22.7	29.4	30.2	32.7	65.1	76.4	35.3	64.7	74.0	...	0.4	2.4	2.6	...	...
Paraguay	40.8	29.0	28.6	98.7	102.7	116.6	125.7	112.4	111.2	...	...	5.4	27.0	9.7	...
Peru	7.3	4.5	3.8	119.7	91.8	90.1	109.3	78.3	75.6	10.4	13.5	14.5	...	...	...
Philippines	2.3	2.9	2.9	84.9	174.9	200.1	75.9	166.0	188.9	9.0	8.9	11.2	...	...	...
Poland	15.0	25.0	27.9	525.0	917.1	1,055.3	549.4	950.7	1,080.0	...	...	...	24.4	33.6	24.7
Portugal	9.5	16.3	17.9	91.8	150.4	169.5	87.2	110.6	124.2	4.6	39.8	45.3	...	...	...
South Africa	21.8	19.7	18.9	439.6	650.3	714.9	443.8	600.4	638.1	...	49.9	76.8	4.2	...	...
Spain	12.2	17.0	18.8	410.7	663.7	772.4	308.2	592.9	688.9	102.5	70.8	83.5	...	...	...
Sweden	18.6	17.0	16.7	149.7	149.1	150.2	161.1	138.9	135.2	...	10.2	15.0	11.4	...	...
Switzerland	26.8	27.0	27.5	163.4	185.9	194.5	136.6	167.9	180.6	26.8	18.0	13.9	...	...	...
Turkey	5.0	4.5	4.3	183.6	234.4	254.4	183.6	234.4	254.4	0.0	0.0	0.0	...	...	...
USSR	20.9	30.2	32.6	5,028.0	8,513.3	9,571.8	5,015.5	8,135.7	9,204.7	12.5	377.6	367.1	...	...	...
United Kingdom	24.9	21.2	20.3	1,385.2	1,242.3	1,218.7	948.1	1,000.0	1,032.3	437.1	242.3	186.4	...	...	...
United States	53.1	61.4	63.3	10,915.9	14,463.2	15,621.8	10,103.0	13,441.2	14,625.4	812.9	1,022.0	996.4	...	...	...
Uruguay	59.0	80.0	80.0	211.7	274.0	286.6	378.8	374.0	395.2	...	...	...	167.1	100.0	108.6
Venezuela	19.5	24.2	25.3	200.9	395.8	473.4	200.9	390.5	447.5	0.0	5.3	25.9	...	...	...
Yugoslavia	11.3	17.4	18.1	180.7	404.8	435.7	233.0	381.4	412.0	...	23.4	23.7	52.3	...	...
Total	...	...	...	33,324.8	47,356.3	52,110.8	33,524.9	47,306.2	52,060.6	2,203.5	2,967.0	3,172.9	2,403.8	2,916.9	3,032.6

crease in per capita consumption from 1970 to 1985 and 1990, and most of that will have taken place in the 1970s. Some countries exhibit spectacular increases in percentage consumed, but the base is so low that the absolute amount in per capita terms is rather small. Japan, for example, is expected to increase beef consumption 100 percent, but this amount represents an increase of only 2.7 to 5.6 kg. Because Japan's population is so large, however, total consumption is calculated to increase from 311 thousand tons in 1970 to 701 thousand tons in 1990.

Projections of total consumption suggest that almost every country in the world should register increases in total beef and veal consumption, even though their per capita consumption might decline. The United States, with 15.7 million metric tons in 1990, is projected to be the largest beef consuming nation at that time. This total is up about 30 percent from the 10.9 million tons consumed 20 years earlier in 1970. The USSR is projected to be in second place in 1990 with 9.6 million tons. These two countries are followed by Brazil, Argentina, and France, with 3.5, 2.6, and 1.9 million metric tons, respectively.

The largest beef and veal producing country by 1990 is also projected to be the United States, with 14.6 million tons annually. The USSR follows with 9.2 million tons. There is considerable difference between these two countries and those that rank behind them. The third-place country, Brazil, for example, is forecast to produce about 3.2, Argentina 3.0, and Australia 2.7 million tons.

Total world production and consumption, which are approximately equal, increased from the 23 million metric tons recorded in the 1960s to 33 million in 1970 and 40 million in 1979. They are projected to 47 million in 1985 and 52 million in 1990 (Figure 12.1). The periods of greatest growth in production and consumption, 1966–1969 and 1974–1976, correspond with the deceleration phase of the cattle cycles found in most of the major beef exporting and importing countries, as it is the period of greatest slaughter.

The statistics in Table 12.1 that might influence policy making and recommendations most are those on the deficit and surplus of individual countries. These statistics are the equivalent of net imports or exports. Thus, even though they are not trade projections per se—since many countries such as France and the United States are both importers and exporters—they do indicate potential trade patterns. Statistics on trade for the forty-eight countries, including net trade, are presented in Appendix Table A.5 for 1960–1978.

The projections given in Table 12.1 indicate that if past trends continue Brazil will shift from being a net exporter to importing about 260 thousand tons of beef by 1990. Canada's net imports are projected to grow from the 40 thousand tons recorded in 1970 to 150 thousand tons in

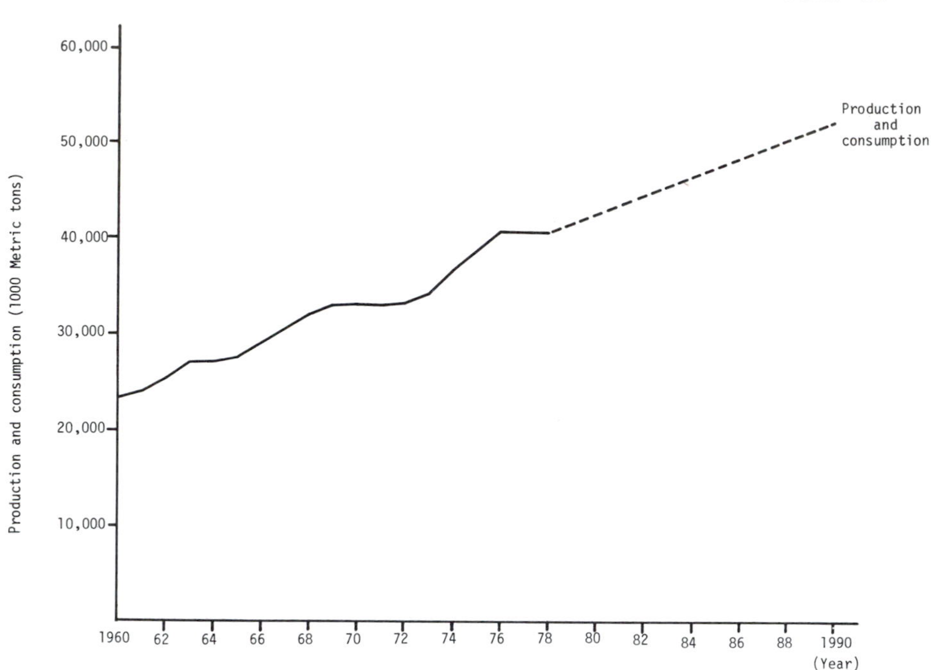

Fig. 12.1. Beef and veal production and consumption for forty-eight selected countries, 1960–1978, with projections for 1985 and 1990.

1990, and Italy's to grow from 290 to 380 thousand tons over the same 20-year period. Japan is also expected to increase imports significantly, its imports shooting up from 33 thousand tons to 193 thousand tons by 1990. South Africa, Sweden, Venezuela, and Yugoslavia shift from being surplus countries to having a deficit. The United States continues to be the major importing country, its deficits growing from 813 thousand tons in 1970 to about 1 million tons in 1990.

These projections, which closely agree with those made by the United Nations, indicate that considerable changes will take place in the surplus beef producing countries if past trends continue (FAO 1979a). Argentina, for example, slips from first position with 668 thousand tons in 1970 to fourth in 1990 with a projected 360 thousand tons (Table 12.2). In the meantime, the output of Australia and New Zealand is expected to double, thus placing them in the first and second positions, respectively. Ireland's surplus is expected to increase from 160 to nearly 440 thousand tons by 1990. Denmark and the Netherlands, both European Economic Community (EEC) countries, continue to maintain their positions. Uruguay, whose surplus is projected to decline from 167 thousand tons in 1970, to less than 110 thousand tons in 1990, moves from fourth to sixth place.

Table 12.2. Beef and veal surplus and ranking by country, 1970, with projections for 1990

Country	Surplus 1970	Surplus 1990	Ranking 1970	Ranking 1990
	(1,000 metric tons)			
Argentina	668	361	1	4
Australia	493	1,144	2	1
Costa Rica	25	67	8	8
Denmark	89	164	6	5
Ireland	160	437	5	3
Netherlands	69	78	7	7
New Zealand	280	450	3	2
Uruguay	167	109	4	6

Source: Table 12.1.

The shifts in ranking of countries with surpluses, as well as the absolute quantities of surpluses in those countries are more a result of political and associated factors than technical developments. For example, both Argentina and Uruguay have an extensive capability to increase production well above that projected, but their traditional export markets are expected to decline rather than grow unless there is a substantial change in policy. Their major market has traditionally been the EEC countries, which had a deficit of 535 thousand tons of beef and veal in 1970, but are expected to have a surplus of nearly 60 thousand tons in 1990 (Table 12.3). This means there will be fierce competition for the remaining traditional markets to which the countries with foot-and-mouth disease, such as those in South America, have access. But Table 12.1 indicates these nations will require almost no additional imports beyond 1970 levels.

There will be some growth in nontraditional markets but competition in these markets will be as severe as in the traditional markets.

Table 12.3. Deficit and surplus in beef and veal in the EEC countries, 1970, with projections for 1990

Country	1970 Deficit	1970 Surplus	1990 Deficit	1990 Surplus
	(1,000 metric tons)			
Belgium-Luxembourg	0.9	. . .	. . .	7.1
Denmark	. . .	88.5	. . .	164.1
France	. . .	45.4	. . .	18.6
Federal Republic of Germany	170.3	. . .	79.3	. . .
Ireland	. . .	159.8	. . .	437.3
Italy	289.5	. . .	382.9	. . .
Netherlands	. . .	69.3	. . .	78.1
United Kingdom	437.1	. . .	186.4	. . .
Total	897.8	363.0	648.6	705.2
Net	534.8	. . .	. . .	56.6

Source: Table 12.1.

Japan's deficit, for example, is expected to grow from 33 thousand tons in 1970 to 193 thousand tons by 1990, but much of its demand will be for high-quality fed beef of the type produced by the United States. Also, Australia and New Zealand have a great comparative advantage with respect to Japan because of their location. The USSR's deficit is expected to grow from 13 thousand tons in 1970 to 370 thousand tons by 1990, but this market will be unstable and possibly much less vigorous if the USSR continues to promote the import of feed grains as a means of improving meat consumption.

The conclusion that can be drawn from these projections is that although trade will continue to increase substantially during the 1980s, much of it will be interregional and there will be little opening for new beef exporters except in special cases. Over the long run, the traditional beef exporting countries in Latin America and Oceania will continue to have severe problems in marketing beef available for export. The United States initiated a countercyclical beef import bill in 1980 but it may have little effect on trade flows as beef quotas are well fixed. The EEC countries have an extremely tight control on production and trade that is not likely to be broken, and they will probably continue to find that paying producers heavy subsidies to reach self-sufficiency is much more palatable as a trade-off than using scarce foreign exchange to import beef.

We can conclude that on a world level there will be an adequate supply of beef and veal up through 1990 in the forty-eight major beef producing and consuming countries. Selected countries, however, will suffer actual declines in per capita consumption. In those where population growth rates are high, growth will take place mainly in lower income classes that are not traditionally beef consumers. Higher-income countries that are projected to have declines in per capita consumption are mainly those in which beef policy since 1970 has been aimed at national self-sufficiency. These policies have resulted in higher beef prices, and thus lower consumption. Countries not covered in the projections account for only about 15 percent of world beef consumption, but there is a concerted effort in these areas to increase beef production in order to reduce protein deficiencies. One method of increasing production is to increase efficiency of livestock, for example, through feed grains, which are being effectively used in the developed countries.

UTILIZATION OF GRAIN IN LIVESTOCK FEEDING

Feed grains are, by far, the major type of concentrate feeds used in all developed regions as well as in Latin America and in the Near East. In other developing regions, grain feeding of animals is a recent practice. Pastures and fodders still predominate as the world's most important animal feedstuffs, but progress has been slow in expanding the yield of these types of feeds. Grain output has increased much faster, the use of feed grains for livestock advancing in both total and concentrate feeds.

The major factor stimulating increased use of feed grains, and concentrate feeds in particular, has been the growing demand for livestock products, which is the result of population increase and real per capita income improvements. On the production side, major breakthroughs have been made in production technology. Only modest productivity increases are expected during the 1980s and 1990s, but by the year 2000 other major advances will be realized (FAO 1979b). One innovation is photosynthesis enhancement, a technique that will aid in boosting the growth rates of crops by improving the natural process through which plants form carbohydrates and absorb nitrogen for protein synthesis. In addition, several other important technological advancements loom on the horizon, such as improved crop hybrids, new pest control strategies, better fertilizer and water management systems, and the genetic development of plants that can thrive on saltwater. As for livestock, significant advances can be expected in selective breeding, livestock management, intensive feeding methods, and the shortening of fattening cycles. Induced twinning in beef cattle is also likely to be practical in special situations. These and other factors have led, and will continue to lead, to increased dependence on concentrates.

There is a positive correlation between the percentage of total consumption of feed grains and a country's level of economic development (Figure 12.2). For example, whereas 60 percent or more of all grains consumed in developed countries are fed to animals, the proportion in developing countries is 13 percent.[2] Developing countries also only account for a minor amount of feed grain use throughout the world, and despite great caloric deficits use of feed grains rose only from 15 percent in 1962–1964 to 17 percent in 1977. The United States continues to be the largest user of feed grains, but future world trends will depend on developments in the USSR (the second largest user), as well as trends in eastern and southern Europe, in Japan, and in the developing countries where the use of feed grains is growing faster than in North America or Western Europe.

FAO projections indicate that the rate of growth in demand for feed grains will decrease in developed countries, but will increase between 4.0 and 5.3 percent annually in the developing countries (FAO 1979c). This

---

[2]Background data for this section is primarily from FAO 1979*c*.

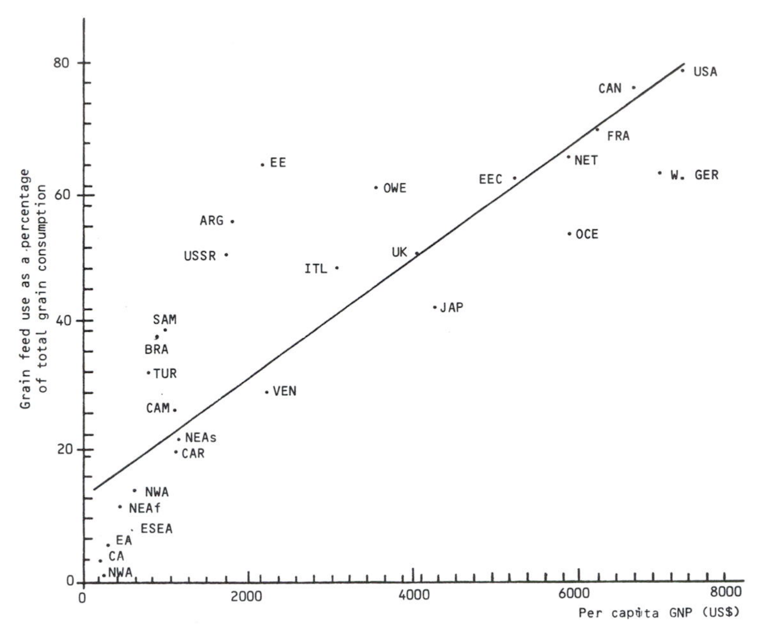

Fig. 12.2. Relationship between per capita GNP and share of grain used for feed, 1972-1974 (FAO 1979a—see Chap. 2 for key to countries and regions).

rate is considerably higher than in the past and also higher than that projected for the developed countries. The problem is that developing countries have shortages of land, capital, and modern technology for feed grain production. Also, governments generally place low priorities on the stimulation of feed grain production. The result is that the production of coarse grains in developing countries is projected to grow only from 2.1 to 2.4 percent annually, so that these countries will probably become increasingly dependent on imported feedstuffs. Naturally, a worldwide effort is being made to improve productivity, but despite this work livestock productivity will probably fail to increase to any extent in countries that do not adopt grain feeding.

Data on feed grain consumption by various classes of animals are available for only a small number of developed countries, and only fragmentary information is available for a few developing countries. Nevertheless, a profile is presented in Table 12.4, although it provides only a tentative pattern of world feed grain consumption by livestock.

Projections of feed grain use set forth in Table 12.5 clearly show the declining growth rates of feed grain use in developed countries (1.9 percent annually) compared with developing countries, where the growth is expected to average 3.9 percent annually. Both figures, as well as the world annual projected growth rate in feed grain use of 2.2 percent, are

Table 12.4. Estimates of the relative consumption of feed grains by type of livestock, 1972–1974

Region	Feed grain consumption by type of livestock			
	Cattle	Swine	Poultry	Total[a]
	(%)			
World	46	31	23	100
Developing countries	30	42	27	100
Developed countries	48	28	23	100

Source: FAO, 1979.
[a]Rounding errors prevent some rows from adding to 100.

below the earlier period of 1962–1964 to 1972–1974 when the world growth rate was 4.6 annually. Despite the change in growth rates, developing countries will still use only about one-fifth as much feed grains as the developed countries.

Table 12.5. Projections of feed grain use to 1985

Regions and grain types	1962–1964 average	1972–1974 average	1977 preliminary	1985 projection	Annual growth rate: 1962–1964 to 1972–1974	Annual growth rate: 1972–1974 to 1985
	(million metric tons)				(%)	
All grains						
World	309	483	508	628	4.6	2.2
Developing countries	47	71	84	113	4.2	3.9
Developed countries	262	412	424	515	4.6	1.9
Wheat						
World	27	72	71	98	10.3	2.6
Developing countries	2	4	6	6	4.3	3.9
Developed countries	25	68	65	92	10.8	2.6
Coarse grains						
World	283	411	437	530	3.8	2.1
Developing countries	45	67	78	107	4.2	3.9
Developed countries	238	344	359	423	3.8	1.7
Maize						
World	143	203	220	265	3.5	2.3
Developing countries	21	32	39	58	4.4	5.1
Developed countries	122	171	181	207	3.4	1.6
Barley						
World	60	109	123	153	6.1	2.9
Developing countries	12	7	19	25	3.6	3.1
Developed countries	48	92	104	128	6.7	2.6
Sorghum and millet						
World	24	39	38	51	4.8	2.2
Developing countries	9	16	18	21	5.6	2.4
Developed countries	15	23	20	30	4.4	2.0

Source: FAO, 1979c.

An integral part of the beef and grain projections is that even though beef production and feed grain production will continue to increase dramatically in developing countries, beef cattle productivity probably will not increase to any great degree. This conclusion is reinforced by a comparison of increases in productivity per head of inventory in different regions (Jasiorowski 1973). Europe's productivity increased 38 percent from 1960–1977, while productivity in the United States and Australia grew 26 and 60 percent, respectively (Table 12.6). But South America' productivity grew only 14 percent. Africa, a region whose production per head is only one-sixth that of the United States and one-fourth that of Australia, did not register any improvement over the 17-year period. Consequently, unless there are drastic changes in domestic and foreign trade policies that will serve to expand the use of feed grains even further in that region, as well as to stimulate the adoption of other technological innovations, the gap in productivity per head between the developed and the developing countries will continue to grow.

Another aspect of the productivity problem is that to a certain extent a vicious circle prevails as most consumers in low income countries cannot afford to buy more than minimal amounts of beef even with low prices. But relatively higher beef prices are necessary to bring about intensive feeding practices and to improve the grain-price ratio to the extent that grain feeding is economically justifiable. Thus greater productivity per animal is not possible without higher prices, but increased prices will bring about a decline in demand for the product. Balancing the factors in a development situation is thus of major importance in setting forth policy.

IMPROVING CATTLE AND BEEF MARKETING

Problems and opportunities in cattle marketing can be classified in a variety of ways, but the following should aid in conveying the scope of the subject.

1. Problems that relate to the general level of cattle prices: These problems range from supply and demand on a local market for beef to the national and international markets. Furthermore, problems in this category involve the variability of the general price level and gearing production to the market.
2. Problems that relate to pricing individual cattle and individual lots of cattle in relation to the general price level: This involves the accuracy and extent of timely dissemination of market information as well as accurate identification of the value of individual animals.

Table 12.6. Increase in production per head of inventory, by region, 1960–1977

Region	1960			1970			1977			Increase, production/head		
	Production[a]	Cattle inventory	Production/head[a]	Production[a]	Cattle inventory	Production/head[a]	Production[a]	Cattle inventory	Production/head[a]	1960–1970	1960–1977	1970–1977
	(1,000 metric tons)	*(millions)*	*(kg)*	*(1,000 metric tons)*	*(millions)*	*(kg)*	*(1,000 metric tons)*	*(millions)*	*(kg)*		*(%)*	
Africa	1,800	114	15.8	2,490	154	16.2	2,549	162	15.7	3	0	−3
North and Central America (other than United States)	1,497	47	31.9	1,925	57	33.8	2,410	63	38.3	6	20	13
United States	7,183	96	74.8	10,006	112	89.3	11,600	123	94.3	6	26	1
South America	4,360	160	27.3	6,041	198	30.5	6,857	220	31.2	12	14	2
Asia	1,330	249	5.3	4,048	288	14.1	4,672	358	13.1	166	147	−7
Europe	6,320	113	55.9	8,896	124	71.7	10,291	134	76.8	28	38	7
Australia	643	17	37.8	1,039	22	47.2	1,940	32	60.6	25	60	28
USSR	...	74	...	5,381	95	56.6	6,672	110	60.7	...	...	7
Total world[b]	28,000	920	30.4	40,245	1,125	35.8	48,519	1,213	40.0	18	32	12

Source: Production and inventory data compiled from FAO *Production Yearbook*, 1962, 1971, and 1977 issues.

[a]Indigenous production only, i.e., does not include imported animals.

[b]Contains an estimate of missing countries.

3. Problems that relate to market operations: These include physical handling, assembly, sorting, groupings, transfer of ownership, transporting, financing, and regulation of marketing practices.

Finding solutions to these problems will require action by (1) individuals or firms, (2) local or regional groups, (3) the national cattle industry, primarily through its various associations, and (4) government and university researchers (Farris and Dietrich 1975). One solution for individuals or firms, for example, is to spread risk of price and income variability by diversifying enterprises through planned programs. In the United States this includes, for example, the practice of hedging price risk on the futures market.

Integrating backward into feed production or integrating forward into marketing activities is another possibility because increased efficiencies as well as greater income can result from business diversification. Group action in cooperation or joint ventures are other possibilities (Carpenter 1973). Gains can also be made by reducing bruising, injury, shrink, and death loss during transit. More research needs to be done on incentives and penalties.

In some cases more direct marketing by producers directly to feedlots and/or slaughter facilities can reduce costs. Marketing associations that bring together large lots of cattle in a central location to attract widely dispersed buyers are providing good results in some areas where there are numerous small entrepreneurs. Also, individuals and firms can make better use of market information both in their daily and long-term planning by understanding how the available information can be used in a decision making framework. In addition, economists have begun to develop models for evaluating risk.

Although group marketing organizations are used in only a few areas, their success in some European countries shows that they have considerable potential, especially for forward integration. In the United States local affiliations of the National Cattlemen's Association who provide commission sales, order buying, and credit services, are examples of successful group actions. Cooperative feeder calf sales and group feeding of cattle are other examples (Johnson 1971).

Groups can also make outstanding contributions by providing market information for special needs. Two examples of market information services are those provided by the Texas Cattle Feeders Association and the National Cattlemen's Association (NCA). It is of utmost importance that cattle groups of all types develop a market orientation to deal with problems before a crisis develops or before opportunities disappear.

The greatest single problem for the world's cattle industry as a whole, as well as for many countries, is the lack of a coordinated program for orderly marketing by which supplies can be adjusted to avoid unnecessary buildups of beef. There is no problem in the industry that is harder to solve or about which less is being done. Other agricultural groups have made substantial progress in managing their marketing and supply problems to reduce the impact of adjustments, however, and it appears that much progress could be made in the beef business.

In the United States a national program of orderly marketing could be accomplished on a voluntary basis through better supply and demand information, with the cooperation of the federal government or by the industry itself. Prior to joining the EEC, England approached the problem of orderly marketing through joint efforts of a meat board and the government. A study indicates that in Australia a buffer fund scheme would provide better results for smoothing cattle prices than the alternatives of supply management, two-price schemes, or a single price schedule (Australia 1979).

The cattle industry in the United States could have separate programs for fed cattle, feeders and stockers, and cull cows. For example, projected economic outcome of feeding cattle to extra weights could be regularly published for the feedlot industry by regions. Similarly, economic guides to profitable heifer replacement and cow culling strategies would assist the decision making of individual producers. The short-run strategies would also help solve longer run problems of cattle cycles, as they are mainly a product of accumulated overreaction to anticipated market conditions by large numbers of producers. Marketing orders and agreements are other possibilities for achieving more orderly marketing of beef cattle.

One serious stumbling block is the general lack of knowledge about livestock and meat marketing, for inadequate marketing can be a critical obstacle to development in many situations. The marketing system could become a dynamic force in rural development if higher priority were given to the development of the human skills in marketing organization and management (Mittendorf 1978).

Most of the proposals for improving marketing systems also apply to the developing countries. The marketing systems in the developing countries are less efficient than in the developed countries, along both agency and functional lines, in large part because a developing country as a whole has a low level of marketing efficiency. Perhaps the best guide is the experience of other countries, which through trial and error have developed efficient beef marketing systems. Regardless of the country, the evidence shows that the most successful development programs are those in which government has accurately determined beef industry needs and has provided appropriate legislation, information, assistance, and regulation. In almost every case the approach has been one of assistance rather than extensive government involvement in production and marketing functions.

DEFINING IMPROVEMENT IN RELATION TO GOALS

The first step in determining how the beef business can be improved is to establish a vision of what is "good" or "desirable" for society as a whole with respect to the beef business. Then, objectives and goals can be set for specific programs. Within that framework, program policies can be determined. Overall, our belief is that a "good society" is one that emphasizes personal freedom for both consumers and producers. Consequently, we view the role of government as primarily that of a provider of services and a watchdog against unfair competition, rather than that of an owner or manager of production processes. We also believe that improvements can best be attained by "pulling" or stimulating economic development rather than by "pushing" it through the creation of large government bureaucracies.

We define "improvement" or "progress" in the beef business as one or more of the following: (1) an increase in per capita consumption of beef, (2) better distribution as well as increased beef consumption in a country, (3) reduction in the long-term real price of beef to the consumer, and (4) increased income to cattle producers and to other people engaged in the beef chain. Furthermore, we view production, marketing, and their various facets as facilitating functions used to meet the needs and desires of consumers. In other words increased trade, greater use of fertilizer, or crossbreeding is a means rather than the final goal. This distinction is highly important, for it shows that some commonly accepted measurements of progress, such as increased production, are really goals for a specific program rather than for the overall vision of what is good or desirable.

There are many worthwhile overall or national goals and objectives, such as reducing death loss, eradicating disease, or reducing the cost of production and marketing. Once these goals and objectives are set, specific program goals or objectives as well as targets and policies can be posited. These are set forth subject to certain constraints or obstacles that may require economic or policy trade-offs. A government intent on improving the balance of payments (overall national goal), for example, may design a specific program aimed at increasing production of export type cattle (the program goal). But the additional effort may result in a decrease in national per capita consumption, and thus may violate somewhat our vision of progress. In such a situation we would have to ask whether the additional foreign exchange (and the benefits derived from it) is worth a step backward in meeting beef-related consumer wants and desires.

Another example may assist in understanding the terms progress and improvement in relation to the world's beef business. The projections that we have developed demonstrate clearly that more than ample production capacity exists in the beef exporting countries to meet potential needs of the beef importing countries. Thus, apart from specialized situations such as the United States in terms of fed beef or Tanzania in terms of live cattle, few beef producing countries will be able to orient themselves toward the world beef market because the competition is too keen from countries already exporting beef. In addition, the rapid supply response related to increased beef prices (taking into account the necessary lag for planning and biological factors) indicates that increased production, simply for the sake of increasing production, is seldom a justifiable goal. Rather, production-oriented goals should be revised to focus on certain aspects of progress such as increased efficiency (Weber and Gregersen 1977).

Governments can use their fiscal and legislative powers to stimulate the adoption of efficient techniques or practices that will, in turn, reduce costs and thus lead to greater profits, to additional investment, and finally to increased production (Schumacher 1974). There will then be greater per capita consumption of beef because it will be cheaper. In other words, increased production stems from greater producer profits rather than from reduced profits, provided that the mechanisms are available to permit increased private sector investment.

The task of improving the livestock business, especially in countries that have low output levels and inefficient marketing channels, is a complex undertaking that requires multidimensional planning. If government decides to set cattle prices at artificially high levels to stimulate production, then attention must be given to all aspects of the beef chain to insure that the highest possible benefits are achieved. For example, there must be provision for easy transfer of land titles, credit for land purchase and improvements as well as operating expenses, a well-planned and reliable system of semen or bull imports, long-term availability of inputs such as fertilizer, improvements in marketing facilities, and so forth. In other words, a total plan for a long-term period, one that has the commitment of all parties, is required. This means there must be an accurate diagnosis of the entire livestock subsector and related activities, as well as realistic projections of world markets, political changes, and market distortions (Valdez and Nores 1979). We now concentrate on summarizing some policy considerations and on suggesting ways in which the world's beef business can be improved.

POLICY CONSIDERATIONS AND SUGGESTIONS

The complexity of the world's beef business at all levels—domestic and international—has been well documented in this book. Because diverse problems exist in both developed and developing countries and among the various types of systems, the following suggestions are presented as a means of synthesizing the materials discussed in this book and are not directed to specific situations.

Even though specific recommendations or prescriptions for improving production or marketing practices cannot be set forth, several particular problem areas merit attention. In Africa, for example, it is becoming apparent that an efficient, commercially oriented system of beef production will not be possible unless beef producers shift from a nomadic, subsistence existence to one in which cattle are raised in a system of private ownership on private property. In Latin America, beef productivity increased only a total of 3 percent from 1950 to 1970 (in contrast to 58 percent in Europe), again because of institutional constraints, in this case mainly because of inappropriate government policy. Increased beef production in Asia will hinge on resource availability and changes in customs to a greater degree than in Europe and Latin America and vast improvements in productivity can be made despite rapid population growth. With these various considerations in mind we now turn to several broad policy proposals for improving the world's beef business.

FACTOR SUPPLY AND MARKETING. Perhaps most important is the need to recognize that world trade may not grow rapidly over the foreseeable future, and consequently that developing countries should exercise caution in attempts to expand a cattle industry aimed at export trade unless they have a particular, secure market. There is a supreme paradox here: a large number of people in the world are starving and a large number of people who want more beef have the money to pay for it if prices are set at reasonable international rates but government restrictions make it unavailable.

Integral studies are needed to relate alternative production techniques to pricing policies and to determine therefrom optimal development strategies. One failing of past studies is that many have taken both price and production coefficients as given rather than adopting a positive, innovative stance. In other words, we recommend that the economists, animal scientists, and policy planners reorient their thinking so as to determine to a greater extent "what could be" rather than "what is."

Pricing policies of feedstuffs and other inputs must be carefully considered in relation to cattle and beef prices. The key is careful weighing of demand elasticities and supply elasticities to devise optimum pricing policies when there is concern about using the free market system. It has been estimated that with more favorable farm prices, agricultural output in twenty-seven developing countries could have been 40 to 60 percent higher than it was during the late 1960s and early 1970s (Peterson 1979).

Although each situation will be different, we can say that increased economic efficiency is generally tied to greater use of capital and less use of labor. In Europe that means increased use of feedlots and increased size of production units. U.S. cattle operations are becoming increasingly mechanized, so that trucks and airplanes in some cases are replacing horses and riders. Increased productivity per man-hour is also a product of greater efficiency. The next 10 years promise the generalized use of embryo transplants on larger and more sophisticated operations. Computerized record keeping for production and economic decision making will become more common as ranches continue to become more businesslike. Without doubt, good management techniques—for example, determining optimal selling time for calves—will be essential.

Particular attention needs to be given to viewing supply and marketing as part of a livestock system, especially in developing countries or centrally planned economies. Failure to account for any one part, such as credit and appropriate repayment schedules, timely availability of fertilizers, or veterinary supplies, can lead to a serious reduction in productivity.

RESEARCH AND EDUCATION. Although management potential is one limiting factor always mentioned in discussions of or plans for livestock development programs, it seldom receives attention in final implementation. This oversight is unfortunate because the potential benefits of management programs are high. Training programs start at the university, at both the graduate and undergraduate levels. The crying need in the developing countries is for training that covers all aspects of ranch or farm management, from economics to forage production to animal health. In general, however, degrees are awarded only in agronomy or veterinary medicine, and students are not encouraged to cross discipline lines, and sometimes are even prohibited from doing so. We strongly recommend that college administrators in developing countries try to make the curriculums as flexible as possible to encourage student enrollment in courses that the students, themselves, think will be most beneficial to them.

More emphasis needs to be placed on economic and technical research, on a commercial basis, into the use of nontraditional feedstuffs in feed manufacture. Local research is needed in almost every developing country on combining pasture feeding of cattle with short-term feeding of forages and/or concentrates in feedlots.

Increased productive efficiency per animal can be maximized by combining improved nutritional levels with improved management, animal health, and work on animal breeding. Much more work is needed on determining the optimal type of cattle for particular socioeconomic, geographic, and climatic conditions.

Livestock research, especially in developing countries, should be multidisciplinary in nature, incorporating agronomists, economists, and forage specialists as well as professionals from animal and veterinary science. The more complete and precise the "package" of practices described and delivered to the producers, the greater the potential for correct application by producers (McGrann 1976). Producer education through extension programs has a high economic return and must be considered as part of any livestock improvement project.

LIVESTOCK AND FEED PRODUCTION. The constraints in developing countries resulting in shortages of feed resources need to be removed or alleviated. Adequate supplies of feedstuffs, including forage crops, grassland production, feed grains, and other concentrate feeds, are an essential prerequisite for a rapidly growing livestock sector.

Progress in increasing beef production implies that the systems will be more intensive, and therefore that efforts must be made to increase carrying capacities and replace or upgrade native breeds with improved stock. Often the introduced breeds suffer greatly from various health disturbances. New diseases such as salmonellosis or leptospirosis may develop and worm parasitism may increase with more intensive use of grazing facilities. Preventive measures thus become all-important in meeting health expectations, and cattle owners will have to assume greater responsibility as beef production intensifies. As cattle owners' outlooks change so that they no longer regard livestock as chattel or prestige items but see them as commodities for business transactions, the government's role in disease control will increasingly shift from that of being active in actually carrying out health control to providing regulation and information.

Health patterns vary greatly among herds, regions, countries, and continents according to intensity of the production system. Nomadic herd structure and traditional management practices require that health programs be administered largely through government programs, as individuals are powerless to improve their own animals' health without the consensus of the whole group, which is usually difficult to obtain. As cattle owners become settled, disease composition changes. In central Africa the incidence of trypanosomiasis would probably decrease, for example, as production became more intensive, and health problems like brucellosis would increase. On the other hand, if communal grazing were replaced by grazing on enclosed pastures, exposure to disease would be much reduced and more effective control would be possible. Under settled conditions it also becomes possible to isolate some herds from epidemics of infectious diseases. In these situations more emphasis should be placed on preventive medicine.

Experiments have shown that numerous production practices are economically beneficial. For example, experimental evidence demonstrates that some form of rotational grazing is preferable to continuous grazing, even on open ranges where cattle must be herded to other areas. Also, productivity can be improved by reducing stocking rates to allow forage recovery. But stocking rates should be intensified with higher fertilization levels or on improved pastures. It is true that production per animal may decline under these intensified stocking rates, but the important factor—economic return per hectare—may be increased. Thus the important question is to determine the stocking rate that provides the maximum net return in terms of dollars and not the maximum beef production per animal (Simpson 1980).

The need to evaluate agronomic practices in economic terms also holds for fertilizer. Levels suggested by agronomists frequently differ greatly from optimal levels in economic terms. The reason is that with higher beef prices, more fertilizer can be applied. Conversely, higher fertilizer prices mean that this input should be reduced. Too often, recommendations are made without seriously considering the alternatives of different prices on inputs and output. The same holds true at the national level. Widespread fertilization of pastures in Uruguay, for example, may seem to be the way to increase its livestock production, but the input-output price ratios and effects on the balance of payments probably are not favorable.

INSTITUTIONS AND GOVERNMENTS. Livestock policies should be reviewed and integrated with national development policies to insure that each policy stimulates and aids the others' objectives. Furthermore, sound policies depend also on adequate information. Data collection systems in many countries are still poor, however, and frequently the results cannot be used to answer policy questions because the data are of the wrong kind. Data to be collected should be identified on the basis of the types of analyses to be carried out. This precept is particularly applicable to economic analysis of problems concerning livestock production.

It is unrealistic to expect that cattle owners grazing communal lands, whether they be nomadic, seminomadic, or settled, will no longer desire to hold cattle as a store of wealth without a viable substitute. If we acknowledge that two related difficulties first must be overcome, the only long-term solution to a desire for greater output may be a system of land titles which will offer holdings large enough to provide the rancher with an adequate living. There is little expectation, however, that beef production will significantly increase in Africa and the Near East in the 1980s. Of course, some significant changes will take place within individual countries but, despite a great potential, the overall picture is not one for great optimism.

The international community could do much more to promote livestock sector expansion in developing countries by encouraging the transfer of technology. There must be a recognition that such a transfer will not hurt beef producers in developed countries, primarily because they have well-established systems of import quotas.

More attention should be given to improving the world beef market structure by stabilizing volume and price. Action is needed in four priority areas (United Nations 1978). First, exporting countries need to achieve greater reliability of access in return for export restraints in times of surpluses. Second, a multilateral approach to export subsidies and safeguards in trade is needed. Third, the collection and dissemination of market information should be improved so as to assist in decision making. Fourth, a positive approach to health and sanitary regulations is needed.

Apart from the above four recommendations by the FAO committee, major emphasis should be given to understanding causes of beef cycles in the various exporting countries and to seeking ways to mitigate them. A step forward has been taken in the 1979 Meat Import Law of the United States. In addition, continual effort needs to be made by both national and international agencies to reduce tariff and nontariff trade barriers (Kafel 1975). International research indicates that planners associated with development strategies are well advised to recall that a beef cattle operation, like a processing business, must be large enough to yield an adequate income for a family. This is called having an economic unit. Furthermore the planning must provide for expansion of operations so that farm families can continually take advantage of new production and management techniques that are size dependent. In effect, efficiency at the national level implies that each producer will feed a continuously larger number of people above and beyond the rate of population growth. This is why most developed countries have programs that encourage producers to increase the size of their land holdings and increase scale and efficiency of operations.

International cooperation is as important as national policies because of trade implications, credit flows, and technical assistance. In an effort to improve such cooperation, the FAO Intergovernmental Group on Meat set forth guidelines for international cooperation in the livestock and meat sector (see Appendix 17).

SUMMARY AND CONCLUSIONS

Improvement of the world's beef business will depend on a number of hard decisions. Africa, for example, will require especially careful planning because its beef cattle systems will have to undergo radical changes if expectations concerning diet improvement are to be met. Traditional grazing systems in many areas will have to be supplemented with modern fattening enterprises that are highly efficient in the utilization of feed (FAO 1979b). At the same time, a large underutilized grazing potential exists in Africa, since about 7 million km^2 of potentially good grazing land could be brought into cattle production if disease control measures were perfected. Difficult issues will now have to be settled concerning wildlife versus cattle, ecological conditions, land tenure, and social policy. But a wonderful opportunity exists for substantially improving the diets and incomes of people on an entire continent if proper policies are adopted (United Nations 1973; Ferguson 1979).

This book is essentially a case study in industrial organization, and therefore certain principles that it has set forth for the beef business hold for many other industries. Perhaps the most important one in terms of improving the world's beef business is that structure is the major determinant of conduct, which in turn yields a certain level of performance. An industry that has a strong competitive structure containing fair and effective safeguards against predatory actions and abuses of the system will perform well. This performance can be measured by effective use of inputs, avoidance of excessive profit rates, an efficient marketing system, and relatively low beef prices. In other words, a sound structure leads to progress and thus fulfills our multidimensional visions of an effective industry meeting the wants, needs, and desires of both consumers and producers.

REFERENCES

Australia, Bureau of Agricultural Economics. 1979. *Beef Price Stabilization.* Beef Res. Rep. no. 21. Canberra: Australian Government Publishing Service.

Carpenter, G. 1973. *Livestock Industry Trends: Implications for Cooperatives.* USDA, FCS Inf. 92.

FAO. 1979a. Projections of Meat Production, Demand and Trade to 1985. In *Report of the Committee on Commodity Problems, Intergovernmental Group on Meat.* May 7–11. Rome.

______. 1979b. *Agriculture: Toward 2000.* Rome.

______. 1979c. Utilization of Grains in the Livestock Sector: Trends, Factors and Development Issues. In *Report of the Committee on Commodity Problems, Intergovernmental Group on Meat.* Rome.

Farris, D. E., and Dietrich, R. A. 1975. *Opportunities in Cattle Marketing.* Dep. of Agric. Econ. Dir. 77-1. College Station: Texas A & M University.

Ferguson, D. 1979. *A Conceptual Framework for Evaluation of Livestock Production Development Projects and Programs in Sub-Saharan West Africa.* Ann Arbor: University of Michigan. Center for Research on Economic Development (CRED).

Jasiorowski, H. A. 1973. Twenty Years with No Progress? *World Anim. Rev.* 5:1–5.

Johnson, D. 1971. *An Economic Evaluation of Alternative Marketing Methods for Fed Cattle.* Nebraska Agric. Exp. Sta. S.B. 520. Lincoln: University of Nebraska.

Kafel, S. 1975. Exporting Animals and Meat from Developing Countries: The Major Impediments. *World Anim. Rev.* 14:15–19.

McGrann, J. 1976. Technical and Economic Issues in Evaluation of Livestock Development Projects. Coop. Ext. Serv. Unpubl. tech. paper. Ames: Iowa State University.

Mittendorf, H. 1978. *Training in Agricultural and Food Marketing.* Rome: United Nations, Food and Agriculture Organization.

Organization for Economic Cooperation and Development. 1978. *Towards a More Efficient Beef Chain.* Paris.

Peterson, L. 1979. International Farm Prices and the Social Cost of Cheap Food Policies. *Am. J. Agric. Econ.* 61:12–21.

Schumacher, A. 1974. Government Policy and the Latin American Beef Producer. Paper read at CIAT Seminar on Tropical America: Potential to Increase Beef Production. February 18–21, 1974, at Cali, Colombia.

Simpson, J. R. 1979. "Projections of World Beef and Veal Consumption and Production for 1985 and 1990." Unpublished Report. Food Resour. Econ. Dep. Gainesville: University of Florida.

______. 1980. *Determining Optimal Types of Cattle for Tropical and Subtropical Dairy Operations.* Food Resour. Econ. Dep. Staff Pap. 153. Gainesville: University of Florida.

United Nations Economic Commission for Africa. 1973. *African Livestock Development Study.* Addis Ababa.

United Nations Trade and Development Board. 1978. Consideration of International Measures of Meat: Elements of an International Arrangement on Beef and Veal. United Nations Conference on Trade and Development, Geneva, March 20, Rome.

Valdez, Alberto, and Nores, Gustavo. 1979. *Growth Potential of the Beef Sector in Latin America—Survey of Issues and Policies.* Washington, D.C.: International Food Policy Institute.

Weber, A., and Gregersen, M.1977. The Changing Productivity Structure of the World's Cattle Industry in the Course of Economic Development. *Austrian Landwirtsch.*

APPENDIX

Appendix Table A.1. Selected world statistics, by country, 1979

Countries	Total area	Permanent pastures	Population	Population density	Percent in agriculture	Population growth rate	Total cattle	Cattle per person
	(1,000 ha)		*(1,000)*	*(km²)*	*(%)*	*(%)*	*(1,000 head)*	*(head)*
WORLD	13,390,315	3,150,862	4,335,310	32	45.9	2.0	1,212,017	0.28
Africa	3,031,299	806,336	455,873	15	66.1	2.8	170,110	0.37
Algeria	238,174	36,258	17,959	8	50.8	3.3	1,313	0.07
Angola	124,670	29,000	6,901	6	58.3	2.5	3,120	0.45
Benin	11,262	442	3,424	30	46.3		800	0.23
Botswana	60,037	44,000	798	1	81.0	2.8	3,300	4.14
Burundi	2,783	435	4,383	157	83.5	2.6	836	0.19
Cameroon	47,544	8,300	8,248	17	81.1		3,207	0.40
Cape Verde	403	25	319	79	56.8	1.8	12	0.04
Central African Republic	62,298	3,000	2,169	3	87.8	2.3	670	0.31
Chad	128,400	45,000	4,417	3	84.4	2.1	4,070	0.92
Comoros	217	15	328	151	64.0	2.5	77	0.23
Congo	34,200	14,300	1,498	4	35.0	2.6	71	0.05
Djibouti	2,200	244	116	5			32	0.28
Egypt	100,145		40,926	41	50.8	2.3	1,954	0.05
Equatorial Guinea	2,805	104	355	13	75.3	1.8	4	0.01
Ethiopia	122,190	64,500	31,773	26	79.7	2.4	25,900	0.82
Gabon	26,767	4,750	544	2	76.8	0.8	3	0.01
Gambia	1,130	330	587	52	78.4	2.0	280	0.48
Ghana	23,854	10,700	11,317	47	51.5	3.0	930	0.08
Guinea	24,586	3,000	4,887	20	80.6	2.5	1,700	0.35
Guinea Bissau	3,612	1,280	563	16	82.7	1.8	264	0.47
Ivory Coast	32,246	8,000	7,722	24	79.9	2.7	650	0.08
Kenya	58,265	3,770	15,780	27	78.1	3.4	10,470	0.66
Lesotho	3,035	2,000	1,309	43	84.4	2.2	550	0.42
Liberia	11,137	240	1,802	16	70.3	2.5	38	0.02
Libya	175,954	6,730	2,861	2	16.9	3.1	201	0.07
Madagascar	58,704	34,000	8,511	14	84.0	3.0	8,744	1.03

Source: Land area, permanent pastures, percent in agriculture, and total cattle from 1979 *FAO Production Yearbook*, population and population growth rate from *World Population Prospects, 1970–2000, as assessed in 1973*. Pop. Div. Dep. Econ. Social Affairs of the U.N. Secretariat, 1975.

Appendix Table A.1. (*continued*)

Countries	Total area	Permanent pastures	Population	Population density	Percent in agriculture	Population growth rate	Total cattle	Cattle per person
	(1,000 ha)		(1,000)	(km^2)	(%)	(%)	(1,000 head)	(head)
Africa (continued)								
Malawi	11,848	1,840	5,963	50	84.3	2.5	790	0.13
Mali	124,000	30,000	6,465	5	87.4	2.5	4,459	0.69
Mauritania	103,070	39,250	1,588	2	83.3	2.1	1,600	1.01
Mauritius	186	7	976	525	28.7	1.5	56	0.06
Morocco	44,655	12,500	19,642	44	51.7	3.1	3,650	0.19
Mozambique	78,303	44,000	10,199	13	65.4	2.3	1,380	0.14
Namibia	82,429	52,906	980	1	49.2	2.3	3,000	3.06
Niger	126,700	9,300	5,150	4	88.5	2.8	2,995	0.58
Nigeria	92,377	20,850	74,595	81	54.2	2.9	12,000	0.16
Reunion	251	9	517	206	29.0	1.8	21	0.04
Rwanda	2,634	500	4,649	176	90.0	2.9	640	0.14
St. Helena	31	2				0.8	1	
Sao Tome, etc.	96	1	84	88		1.2	3	0.04
Senegal	19,672	5,700	5,518	28	74.9	2.4	2,806	0.51
Seychelles	28					2.2	2	
Sierra Leone	7,174	2,204	3,381	47	65.8	2.6	270	0.08
Somalia	63,766	28,850	3,542	6	80.5	2.8	3,800	1.07
South Africa	122,104	81,200	28,483	23	28.7	2.9	13,200	0.46
Sudan	250,581	24,000	17,865	7	77.4	3.2	17,300	0.97
Swaziland	1,736	1,250	540	31	73.6	3.0	650	1.20
Tanzania	94,509	44,680	17,382	18	81.6	3.3	15,300	0.88
Togo	5,678	200	2,618	46	68.5	2.9	250	0.10
Tunisia	16,361	3,250	6,201	38	41.4	2.7	910	0.15
Uganda	23,604	5,000	12,796	54	81.4	3.1	5,367	0.42
Upper Volta	27,420	13,755	6,728	25	82.0	2.3	2,700	0.40
Western Sahara	26,600	5,000						
Zaire	234,541	24,803	27,519	12	74.8	2.7	1,144	0.04
Zambia	75,261	30,000	5,465	7	67.3	3.1	1,800	0.33
Zimbabwe	39,058	4,856	7,150	18	59.2		5,000	0.70

continued

Appendix Table A.1. *(continued)*

Countries	Total area	Permanent pastures	Population	Population density	Percent in agriculture	Population growth rate	Total cattle	Cattle per person
	(1,000 ha)		*(1,000)*	*(km²)*	*(%)*	*(%)*	*(1,000 head)*	*(head)*
North and Central America	2,241,492	354,693	363,981	16	11.8		174,385	0.48
Antigua	44	3	75	170		0.7	9	0.12
Bahamas	1,394	1	224	16		2.4	4	0.02
Barbados	43	4	251	584	16.8	0.6	19	0.08
Belize	2,296	37	158	7	29.1	2.9	57	0.36
Bermuda	5		59	1,180		1.4	1	0.02
Canada	997,614	23,650	23,691	2	5.3	1.5	12,328	0.52
Cayman Islands	26	2				0.7	5	
Costa Rica	5,070	1,558	2,162	43	35.8	2.7	2,071	0.96
Cuba	11,452	2,200	9,852	86	24.0	2.1	5,844	0.59
Dominica	75	2	79	105		1.4	4	0.05
Dominican Republic	4,873	1,490	5,800	119	56.6	3.4	2,150	0.37
El Salvador	2,104	610	4,663	222	51.0	3.2	1,368	0.29
Greenland	34,170	5				1.8		
Grenada	34	2	98	288		0.4	7	0.07
Guadeloupe	178	21	332	187	17.7	1.5	85	0.26
Guatemala	10,889	880	7,048	65	55.5	1.7	1,575	0.22
Haiti	2,775	515	5,677	205	67.4	1.7	1,000	0.18
Honduras	11,209	2,000	3,565	32	63.0	3.4	1,800	0.50
Jamaica	1,099	210	2,162	197	21.5	1.4	290	0.13
Martinique	110	25	326	296	16.1	1.4	51	.16
Mexico	197,255	74,499	67,676	34	36.9	3.3	29,920	0.44
Montserrat	10	1				0.6	8	
Netherlands Antilles	96		260	271		2.0	8	0.03
Nicaragua	13,000	3,384	2,649	20	43.7	3.3	2,846	1.07
Panama	7,708	1,161	1,899	25	35.2	2.8	1,423	0.75
Puerto Rico	890	332	3,378	380	3.8	1.2	532	0.16
St. Kitts, etc.	36	1	67	186		0.3	8	0.12
St. Lucia	62	3	113	182		1.2	16	0.14
St. Pier, etc.	24					0.0		
St. Vincent	34	2	97	285		1.0	7	0.07

Appendix Table A.1. (continued)

Countries	Total area	Permanent pastures	Population	Population density	Percent in agriculture	Population growth rate	Total cattle	Cattle per person
	(1,000 ha)		(1,000)	(km^2)	(%)	(%)	(1,000 head)	(head)
North and Central America (continued)								
Trinidad, etc.	513	11	1,127	220	16.2	1.0	77	0.07
Turks, Caicos	43					0.0		
United States	936,312	242,0709	220,286	24	2.3	0.9	110,864	0.50
Virgin Islands, U.K.	15	5					2	
Virgin Islands, U.S.	34	9				0.8	7	
South America	1,781,851	444,704	238,912	13	32.4		216,119	0.90
Argentina	276,689	143,400	26,723	10	13.3	1.3	60,174	2.25
Bolivia	109,858	27,100	5,430	5	50.5	2.6	3,990	0.73
Brazil	851,197	167,000	122,879	14	38.9	2.8	90,000	0.73
Chile	75,695	11,850	10,919	14	18.9	1.8	3,607	0.33
Colombia	113,891	17,600	26,253	23	28.3	3.1	26,137	1.00
Ecuador	28,356	2,559	7,779	27	45.1	3.2	2,532	0.33
Falkland Islands	1,217	1,200				0.0	9	
French Guiana	9,100	6	69	1		3.3	3	0.04
Guyana	21,497	999	865	4	22.4	2.2	280	0.32
Paraguay	40,675	15,100	2,979	7	49.3	2.9	5,203	1.75
Peru	128,522	27,120	17,291	13	38.1	2.9	4,187	0.24
Surinam	16,327	10	381	2	18.2	3.1	27	0.07
Uruguay	17,622	13,910	2,905	16	12.2	1.0	10,007	3.44
Venezuela	91,205	16,850	14,437	16	18.7	2.9	9,963	0.69
Asia	2,757,442	623,041	2,509,010	91	58.6		366,579	0.15
Afghanistan	64,750	50,000	21,452	33	78.2	2.7	3,980	0.19
Bahrain	62	4	292	471		3.2	5	0.02
Bangladesh	14,400	600	86,062	598	84.1	2.8	31,741	0.37
Bhutan	4,700	20	1,269	27	93.5	2.5	207	0.16
Brunei	577	6	185	32		1.8	3	0.02
Burma	67,655	361	34,434	51	52.5	2.4	7,560	0.22
China	959,696	220,000	945,018	98	60.6	1.6	63,978	0.07
Cyprus	925	93	648	70	34.6	1.2	38	0.06
East Timor	1,493	150	738	49	59.6			

continued

Appendix Table A.1. *(continued)*

Countries	Total area	Permanent pastures	Population	Population density	Percent in agriculture	Population growth rate	Total cattle	Cattle per person
	(1,000 ha)		*(1,000)*	*(km²)*	*(%)*	*(%)*	*(1,000 head)*	*(head)*
Asia (continued)								
Hong Kong	104		4,714	4,533	2.7	1.4	9	0.00
India	328,759	12,450	678,255	206	64.0	2.5	181,849	0.27
Indonesia	190,435	12,046	148,470	78	59.7	2.6	6,453	0.04
Iran	164,800	44,000	36,938	22	39.3	3.1	7,600	0.21
Iraq	43,492	4,000	12,640	29	40.9	3.4	2,742	0.22
Israel	2,077	818	3,813	184	7.1	2.6	280	0.07
Japan	37,231	554	115,870	311	11.8	1.1	4,120	0.04
Jordan	9,774	100	3,085	32	26.6	3.3	33	0.01
Kampuchea, Democratic	18,104	580	8,718	48	74.4		700	0.08
Korea, DPR	12,054	50	17,489	145	46.7	2.5	925	0.05
Korea, Republic of	9,848	41	37,313	379	39.9	2.0	1,651	0.04
Kuwait	1,782	134	1,293	73	1.7	5.6	10	0.01
Laos	23,680	800	3,633	15	74.3	2.4	551	0.15
Lebanon	1,040	10	3,086	297	10.8	3.2	84	0.03
Macau	2		272	13,600		1.5		
Malaysia	32,975	27	13,297	40	48.6		430	0.03
Maldives	30	1	145	483		2.1		
Mongolia	156,500	124,763	1,622	1	50.2	2.9	2,482	1.53
Nepal	14,080	1,700	13,938	99	92.7	2.5	6,850	0.49
Oman	21,246	1,000	864	4	62.3	3.2	137	0.16
Pakistan	80,394	5,000	79,838	99	53.9	3.2	14,992	0.19
Philippines	30,000	980	49,493	165	46.8	3.2	1,910	0.04
Qatar	1,100	50	210	19		3.2	6	0.03
Saudi Arabia	214,969	85,000	8,112	4	60.8	3.0	440	0.05
Singapore	58		2,370	4,086	2.4	1.6	9	0.00
Sri Lanka	6,561	439	14,608	223	53.5	2.0	1,623	0.11
Syria	18,518	8,421	8,368	45	47.8	3.2	705	0.08
Thailand	51,400	308	46,347	90	75.8	3.2	4,850	0.10
Turkey	78,058	27,400	44,241	57	55.8	2.6	14,941	0.34
U. Arab Emirates	8,360	200	753	9		3.2	25	0.03

Appendix Table A.1. (*continued*)

Countries	Total area	Permanent pastures	Population	Population density	Percent in agriculture	Population growth rate	Total cattle	Cattle per person
	(1,000 ha)		*(1,000)*	*(km²)*	*(%)*	*(%)*	*(1,000 head)*	*(head)*
Asia (continued)								
Viet Nam	32,956	4,870	51,082	155	71.2	2.3	1,600	0.03
Yemen, Arab Republic	19,500	7,000	5,785	30	75.4	3.0	950	0.16
Yemen, PDR	33,297	9,065	1,838	6	59.3	3.0	110	0.06
Europe	487,055	87,016	481,726	99	15.8	0.6	134,535	0.28
Albania	2,875	570	2,671	93	60.9	2.6	474	0.18
Andorra	45	25				2.3		
Austria	8,385	2,071	7,504	89	9.6	0.2	2,594	0.35
Belgium-Luxembourg	3,310	723	10,210	308	3.3	0.4	3,085	0.30
Bulgaria	11,091	1,923	8,840	80	34.4	0.6	1,763	0.20
Czechoslovakia	12,787	1,706	15,250	119	10.8	0.6	4,887	0.32
Denmark	4,307	268	5,099	118	7.4	0.3	3,034	0.60
Faeroe Islands	140		41	29		0.7	2	0.05
Finland	33,701	173	4,770	14	13.8	0.2	1,736	0.36
France	54,703	12,978	53,509	98	9.1	0.8	23,510	0.44
Germany, DR (East)	10,818	1,243	16,745	155	9.9	0.1	5,572	0.33
Germany, FR (West)	24,858	5,155	61,200	246	4.3	0.2	15,007	0.25
Gibraltar	1					0.7		
Greece	13,194	5,255	9,460	72	38.0	0.3	973	0.10
Hungary	9,303	1,310	10,710	115	16.5	0.4	1,966	0.18
Iceland	10,300	2,274	228	2	12.2	1.2	63	0.28
Ireland	7,028	4,800	3,271	47	21.5	1.0	7,178	2.19
Italy	30,123	5,145	56,888	189	11.8	0.5	8,556	0.15
Liechtenstein	16	5				0.9	8	
Malta	32		337	1,053	5.2	0.3	15	0.04
Monaco						0.8		
Netherlands	3,719	1,189	14,039	377	5.6	0.8	5,149	0.37
Norway	32,422	94	4,070	13	8.1	0.6	971	0.24
Poland	31,268	4,071	35,225	113	31.2	0.9	13,036	0.37
Portugal	9,208	530	9,770	106	27.0	0.4	1,050	0.11
Romania	23,750	4,423	22,068	93	48.1	0.8	6,283	0.28

continued

Appendix Table A.1. (*continued*)

Countries	Total area	Permanent pastures	Population	Population density	Percent in agriculture	Population growth rate	Total cattle	Cattle per person
	(1,000 ha)		*(1,000)*	*(km²)*	*(%)*	*(%)*	*(1,000 head)*	*(head)*
Europe (continued)								
San Marino	6					0.9		
Spain	50,478	11,000	36,856	73	18.2	1.0	4,650	0.13
Sweden	44,996	728	8,296	18	5.3	0.6	1,911	0.23
Switzerland	4,129	1,625	6,350	154	5.4	0.6	2,038	0.32
United Kingdom	24,482	11,378	56,076	229	2.1	0.4	13,534	0.24
Yugoslavia	25,580	6,354	22,107	86	38.6	0.9	5,491	0.25
Oceania	850,956	460,772	22,318	3	21.8	1.9	36,203	1.62
American Samoa	20					3.2		
Australia	768,685	446,578	14,324	2	6.0	1.8	27,107	1.89
Canton Island	7							
Christmas Island	13							
Cocos Islands	1							
Cook Islands	23					3.2		
Fiji	1,827	65	609	33	40.9	1.9	170	0.28
French Polynesia	400	20	150	38		3.3	7	0.05
Gilbert Islands	71					2.9		
Guam	55	8				2.7	2	
Nauru	2					1.3		
New Caledonia	1,906	150	149	8		2.9	116	0.78
New Hebrides	1,476	25	106	7		2.7	97	0.92
New Zealand	26,868	13,750	3,096	12	9.5	1.5	8,499	2.75
Niue Islands	26	1				1.4	1	
Norfolk Island	4	1				0.0	2	
Pacific Islands	178	24				2.9	8	
Papua New Guinea	46,169	106	3,003	7	82.7	2.5	136	0.05
Samoa	286	1	159	56			25	0.16
Solomon Islands	2,845	39	214	8			26	0.12
Tokelau Islands	1							
Tonga Islands	70	4	95	136		3.3	6	0.06
Tuvalu Island	3							

Appendix Table A.1. (continued)

Countries	Total area	Permanent pastures	Population	Population density	Percent in agriculture	Population growth rate	Total cattle	Cattle per person
	(1,000 ha)		*(1,000)*	*(km²)*	*(%)*	*(%)*	*(1,000 head)*	*(head)*
Oceania (continued)								
Wallis Islands, etc.	20					0.0		
USSR	2,240,220	374,300	263,500	12	17.3		114,086	0.43
Developed Market Economies	3,376,054	880,390	779,780	23	8.9		276,953	0.36
North America	1,933,926	265,720	243,977	13	2.5		123,192	0.50
Western Europe	385,163	71,770	370,217	96	11.0		100,555	0.27
Oceania	795,553	460,328	17,420	2	6.6		35,606	2.04
Other developed	161,412	82,572	148,166	92	14.1		17,600	0.12
Developing Market Economies	6,592,839	1,530,663	2,156,610	33	59.4		717,313	0.33
Africa	2,382,515	694,406	365,738	15	69.9		137,454	0.38
Latin America	2,055,218	533,672	358,799	17	34.7		267,312	0.75
Near East	1,208,353	267,207	211,685	18	54.6		51,262	0.24
Far East	857,151	34,929	1,215,380	142	62.9		260,687	0.21
Other developing	89,602	449	5,015	6	72.1		597	0.12
Centrally Planned Economies	3,521,422	739,809	1,398,920	40	49.4		217,751	0.16
Asian Centrally Planned Economies	1,179,310	350,263	1,023,920	29	61.0		69,685	0.07
Eastern Europe and USSR	2,342,112	389,546	375,009	16	20.7		148,066	0.39
Developed, all	5,618,166	1,269,936	1,154,780	21	13.1		425,019	0.37
Developing, all	7,772,149	1,880,926	3,180,530	41	60.0		786,997	0.25

Appendix Table A.2. Production of beef and veal in 54 selected countries, 1960, 1970, and 1979, by region

	Year			Change
	1960	1970	1979	1960–1979
	(1,000 metric tons)			*(%)*
North America				
Canada	631.1	850.6	955.0	51
United States	7,195.0	10,103.0	9,932.0	38
Total	7,826.1	10,953.6	10,887.0	39
Central America, Caribbean, and Mexico				
Costa Rica	23.4	46.3	81.4	248
Dominican Republic	25.0	31.5	40.0	60
El Salvador	19.2	20.3	37.6	96
Guatemala	34.4	57.1	96.4	180
Honduras	15.4	30.0	53.0	244
Mexico	354.0	590.0	1,037.0	193
Nicaragua	23.6	62.6	98.4	317
Panama	19.0	35.3	41.0	116
Total	514.0	873.1	1,484.8	189
South America				
Argentina	1,892.8	2,624.0	2,990.0	58
Brazil	1,359.0	1,845.0	2,100.0	55
Chile	138.7	182.3	195.0	41
Colombia	292.2	425.4	598.4	105
Ecuador	38.5	44.5	86.8	125
Peru	65.8	109.3	81.0	23
Uruguay	273.8	378.8	268.3	−2
Venezuela	120.8	200.9	291.6	141
Total	4,181.6	5,810.2	6,611.1	58
Europe				
Western, EEC				
Belgium-Luxembourg	218.0	270.0	290.6	33
Denmark	155.7	191.6	245.0	57
France	1,325.0	1,565.2	1,791.0	35
Germany, FR	1,004.8	1,356.5	1,520.0	51
Ireland	112.2	216.3	378.0	237
Italy	536.0	1,077.0	1,065.0	99
Netherlands	236.3	326.2	342.0	45
United Kingdom	833.0	948.1	1,010.0	21
Total	4,421.0	5,950.9	6,641.6	50
Western, other				
Austria	131.3	153.8	198.4	51
Finland	71.6	106.0	105.0	47
Greece	28.9	90.1	98.0	239
Norway	49.4	57.2	70.0	42
Portugal	40.3	87.2	90.0	123
Spain	159.6	308.2	387.0	142
Sweden	128.6	161.1	150.7	17
Switzerland	100.0	136.6	160.1	60
Total	709.7	1,100.2	1,259.2	77

Source: Adapted from USDA, *Foreign Agriculture Circular: Livestock and Meat,* FLM 10-78 and FLM 2-80, 1978 and 1980.

Appendix Table A.2. *(continued)*

	Year			Change
	1960	1970	1979	1960–1979
	(1,000 metric tons)			*(%)*
Europe *(continued)*				
Eastern				
Bulgaria	41.1	93.2	107.0	160
Czechoslovakia	203.3	299.2	350.6	72
Germany, DR	188.7	300.5	397.0	110
Hungary	99.3	117.5	135.3	36
Poland	316.6	599.4	732.5	131
Yugoslavia	138.0	233.0	345.0	150
Total	987.0	1,592.8	2,066.8	109
Total, Europe	6,118.1	8,643.9	9,967.6	63
Soviet Union	3,024.4	5,015.5	6,510.0	115
Africa				
South Africa	401.6	443.8	624.0	55
Morocco	30.8	60.7	77.2	151
Total	432.4	504.5	701.2	62
Asia				
Republic of China (Taiwan)	3.3	9.1	13.4	306
India	155.6	176.1	192.3	24
Iran	36.8	51.6	90.0 (est)	145
Israel	12.6	19.4	23.0	83
Japan	142.5	278.0	405.0	184
Republic of Korea	22.6	54.2	115.2	410
Philippines	82.4	75.9	123.8	50
Turkey	121.9	183.6	158.3	30
Total	577.7	847.9	1,121.0	94
Oceania				
Australia	684.8	1,002.8	1,793.8	162
New Zealand	240.0	392.8	490.6	104
Total	924.8	1,395.6	2,284.4	147
Total, selected countries	23,598.3	34,044.3	39,567.1	68

Appendix Table A.3. Per capita consumption of beef and veal in 50 selected countries, 1961, 1970, and 1979, by region

	1961	1970	1979	Change 1961–1979
		(kg)		*(%)*
North America				
Canada	35.6	40.0	40.7	14
United States	42.8	53.3	49.8	16
Central America, Caribbean, and Mexico				
Costa Rica	11.4	12.1	16.5	45
Dominican Republic	7.0	6.5	6.9	−1
El Salvador	8.0	5.9	7.3	−9
Guatemala	8.3	7.6	10.9	31
Honduras	7.2	4.9	5.8	−19
Mexico	9.1	10.6	14.9	64
Nicaragua	12.9	15.4	12.7	−2
Panama	19.3	22.4	20.7	7
South America				
Argentina	83.2	82.4	89.6	8
Brazil	18.4	18.6	17.5	−5
Chile	19.7	23.1	16.9	−14
Colombia	19.2	19.5	22.3	16
Peru	7.4	9.0	4.5	−39
Uruguay	79.0	77.8	67.2	−15
Venezuela	17.0	18.5	23.1	36
Europe				
Western, EEC				
Belgium-Luxembourg	22.8	27.1	27.8	22
Denmark	16.9	20.9	15.5	−8
France	29.0	29.9	33.2	14
Germany, FR	20.7	25.1	24.2	17
Ireland	16.2	19.2	24.3	50
Italy	15.0	25.5	24.6	64
Netherlands	18.9	19.7	19.5	3
United Kingdom	26.4	25.0	27.0	2
EEC average	22.3	25.8	26.4	18
Western, other				
Austria	18.7	22.5	25.3	35
Finland	17.5	20.9	23.3	33
Greece	5.6	18.1	23.0	311
Norway	14.9	14.4	19.6	32
Portugal	6.1	10.5	10.6	74
Spain	5.9	12.2	12.5	112
Sweden	19.8	18.6	18.0	−9
Switzerland	23.2	26.4	25.9	12
Eastern				
Bulgaria	7.3	11.3	11.0	51
Czechoslovakia	17.9	27.1	22.8	27
Germany, DR	15.4	19.6	22.3	45
Hungary	9.6	9.3	10.9	14

Source: Adapted from USDA, *Foreign Agriculture Circular: Livestock and Meat,* FLM 5-78 and FLM 5-80, 1978 and 1980.

Appendix Table A.3. (*continued*)

	1961	1970	1979	Change 1961–1979
		(kg)		*(%)*
Europe, Eastern (*continued*)				
Poland	10.1	16.1	19.9	97
Yugloslavia	6.2	8.9	14.6	135
Soviet Union	12.2	20.7	27.0	51
Africa				
South Africa	22.5	19.6	23.0	2
Asia				
Republic of China (Taiwan)	0.3	0.6	1.4	367
Iran	1.6	1.7	NA	NA
Israel	7.3	19.1	20.0	174
Japan	1.6	3.0	4.9	206
Philippines	2.4	2.3	2.8	17
Turkey	5.0	5.2	3.6	−28
Oceania				
Australia	42.2	40.8	52.1	23
New Zealand	41.3	40.6	60.5	46

Appendix Table A.4. Consumption of beef and veal in 54 selected countries, 1961, 1970, and 1979, by region

	Year			Change
	1961	1970	1979	1961–1979
	(1,000 metric tons)			*(%)*
North America				
Canada	651.1	871.9	989.0	52
United States	7,867.7	10,915.9	10,996.9	40
Total	8,518.8	11,787.8	11,985.9	41
Central America, Caribbean, and Mexico				
Costa Rica	14.6	20.9	33.6	130
Dominican Republic	22.8	27.2	36.3	59
El Salvador	20.7	20.5	31.1	50
Guatemala	34.0	40.1	73.8	117
Honduras	14.4	13.3	21.4	49
Mexico	339.3	538.8	1,033.1	204
Nicaragua	19.3	29.6	63.5	229
Panama	21.6	32.7	39.6	83
Total	486.7	723.1	1,332.4	174
South America				
Argentina	1,749.0	1,956.0	2,320.0	33
Brazil	1,333.5	1,721.9	2,073.0	55
Chile	155.6	224.3	203.0	30
Colombia	314.7	417.6	579.2	84
Ecuador	36.5	44.6	86.8	138
Peru	76.5	119.7	81.1	6
Uruguay	203.7	211.7	192.3	−6
Venezuela	137.2	200.9	309.7	126
Total	4,006.7	4,896.8	5,845.2	46
Europe				
Western, EEC				
Belgium-Luxembourg	216.9	270.9	285.6	32
Denmark	78.0	103.1	83.5	7
France	1,339.1	1,520.2	1,734.1	29
Germany FR	1,168.8	1,526.8	1,470.0	26
Ireland	45.8	56.5	78.0	70
Italy	758.5	1,366.5	1,410.0	86
Netherlands	220.4	256.9	275.0	25
United Kingdom	1,396.0	1,385.2	1,470.0	5
Total	5,218.6	6,486.1	6,806.2	30
Western, other				
Austria	132.8	166.6	193.8	46
Finland	78.2	96.2	103.9	33
Greece	47.0	158.9	213.0	353
Norway	53.7	56.0	78.2	46
Portugal	54.5	91.8	102.2	88
Spain	182.2	410.7	451.5	148
Sweden	149.1	149.7	150.8	1
Switzerland	127.6	163.4	171.2	34
Total	825.1	1,293.3	1,464.6	78

Source: Adapted from USDA, *Foreign Agriculture Circular: Livestock and Meat,* FLM 10-78 and FLM 2-80, 1978 and 1980.

Appendix Table A.4. (*continued*)

	Year			Change
	1961	1970	1979	1961–1979
	(1,000 metric tons)			*(%)*
Europe (*continued*)				
Eastern				
Bulgaria	57.6	95.6	98.0	70
Czechoslovakia	247.0	387.9	348.0	41
Germany, DR	264.5	335.1	373.5	41
Hungary	96.7	96.1	117.3	21
Poland	303.7	525.0	697.5	130
Yugoslavia	115.0	180.7	318.0	177
Total	1,084.6	1,620.4	1,952.3	80
Total, Europe	7,128.2	9,399.5	10,223.0	43
Soviet Union	2,652.0	5,028.0	6,635.6	150
Africa				
South Africa	394.8	439.6	602.5	53
Morocco	32.2	60.6	79.2	146
Total	427.0	500.3	681.7	60
Asia				
Republic of China (Taiwan)	3.7	9.1	24.4	559
India	156.7	176.1	178.3	14
Iran	37.7	51.8	108.0 (est)	186
Israel	15.9	56.8	60.0	277
Japan	150.3	311.2	564.1	275
Republic of Korea	23.5	54.9	167.3	612
Philippines	69.1	84.9	135.5	96
Turkey	141.9	183.6	157.3	11
Total	599.0	928.3	1,286.9	115
Oceania				
Australia	445.6	510.1	781.0	75
New Zealand	100.0	112.6	200.0	100
Total	545.6	622.7	981.0	80
Total, selected countries	24,364.1	33,886.8	38,971.8	60

Appendix Table A.5. World trade in beef for 47 selected countries, 1961–1979

Region and country	Average exports			Average imports			Net imports or exports[a]		
	1961–1963	1970–1974	1977–1979	1961–1963	1970–1974	1977–1979	1961–1963	1970–1974	1977–1979
					(1,000 metric tons)				
North America									
Canada	14.2	41.9	48.5	26.7	84.9	90.4	(12.5)	(43.0)	(41.9)
United States	15.2	28.0	65.1	628.2	837.7	1,020.3	(613.0)	(809.7)	(955.2)
Total	29.4	69.9	113.6	654.9	922.6	1,110.7	(625.5)	(852.7)	(997.1)
Central America, Caribbean, and Mexico									
Costa Rica	7.7	31.0	46.2	...	0.4	0.1	7.7	30.6	46.1
Dominican Republic	0.9	7.5	2.2	0.2	0.4	...	0.7	7.1	2.2
El Salvador	...	3.8	5.0	0.1	0.1	0.3	(0.1)	3.7	4.7
Guatemala	5.5	19.9	21.4	...	...	...	5.5	19.9	21.4
Honduras	5.4	21.1	28.5	...	0.1	...	5.4	21.0	28.5
Mexico	40.2	43.4	21.6	...	0.6	0.8	40.2	42.8	20.8
Nicaragua	12.7	32.6	39.5	...	...	...	12.7	32.6	39.5
Panama	0.3	2.1	1.7	0.1	0.3	...	0.2	1.8	1.7
Total	72.7	161.4	166.1	0.4	1.9	1.2	72.3	159.5	164.9
South America									
Argentina	557.7	548.8	664.3	...	...	...	557.7	548.8	664.3
Brazil	29.1	156.2	151.1	...	12.6	79.9	29.1	143.6	71.2
Chile	0.2	...	1.9	6.9	37.9	9.1	(6.7)	(37.9)	(7.2)
Colombia	...	22.3	19.4	...	...	...	...	22.3	19.4
Peru	...	...	...	3.8	8.6	1.9	(3.8)	(8.6)	(1.9)
Uruguay	83.3	121.9	107.0	...	...	1.4	83.3	121.9	105.6
Total	670.3	849.2	943.7	10.7	59.1	92.3	659.6	790.1	851.4

Source: Compiled from USDA, *Foreign Agriculture Circular: Livestock and Meat,* various issues.
Note: Carcass weight equivalent basis, excludes fat, offal, and live animals.
[a]Net imports shown within parentheses.

Appendix Table A.5. (*continued*)

Region and country	Average exports			Average imports			Net imports or exports[a]		
	1961–1963	1970–1974	1977–1979	1961–1963	1970–1974	1977–1979	1961–1963	1970–1974	1977–1979
					(1,000 metric tons)				
Europe, Western, EEC									
Belgium-Luxembourg	4.6	28.3	48.0	16.6	30.2	53.5	(12.0)	(1.9)	(5.5)
Denmark	87.3	100.7	178.8	0.2	1.3	1.8	87.1	99.4	177.0
France	136.6	184.3	247.5	14.8	127.8	244.3	121.8	56.5	3.2
Germany, FR	11.9	74.0	240.0	93.4	266.8	251.7	(81.5)	(192.8)	(11.7)
Ireland	86.9	175.1	303.4	0.1	0.6	2.8	86.8	174.5	300.6
Italy	7.5	8.1	18.1	137.2	348.0	344.9	(129.7)	(339.9)	(326.8)
Netherlands	48.6	131.1	171.5	20.2	73.5	97.6	28.4	57.6	73.9
United Kingdom	0.8	43.3	117.0	503.9	432.8	581.3	(503.1)	(389.5)	(464.3)
Total	384.2	744.9	1,324.3	786.4	1,281.0	1,577.9	(402.2)	(536.1)	(253.6)
Europe, Western, other									
Austria	1.6	5.3	11.9	2.9	11.9	15.9	(1.3)	(6.6)	(4.0)
Finland	...	7.6	1.1	5.4	1.7	1.2	(5.4)	5.9	(0.1)
Greece	0.2	...	...	20.5	51.4	109.4	(20.3)	(51.4)	(109.4)
Norway	3.2	0.9	0.1	0.8	3.5	9.9	2.4	(2.6)	(9.8)
Portugal	0.3	0.1	...	9.2	23.2	28.9	(8.9)	(23.1)	(28.9)
Spain	0.3	0.7	0.4	45.0	64.5	62.8	(44.7)	(63.8)	(62.4)
Sweden	9.9	12.2	11.7	10.8	10.7	13.6	(0.9)	1.5	(1.9)
Switzerland	0.4	0.9	0.4	21.4	35.9	15.9	(21.0)	(35.0)	(15.5)
Total	15.9	27.7	25.6	116.0	202.8	257.6	(100.1)	(175.1)	(232.0)
Europe, Eastern									
Bulgaria	6.7	9.8	24.7	3.3	6.1	16.5	3.4	3.7	8.2
Czechoslovakia	1.5	13.3	16.0	54.3	61.8	14.2	(52.8)	(48.5)	1.8
Germany, DR	0.4	...	30.7	90.3	21.4	9.3	(89.9)	(21.4)	21.4
Hungary	26.2	33.9	45.0	15.5	10.1	30.3	10.7	23.8	14.7
Poland	32.6	37.8	38.0	9.0	9.5	32.9	23.6	28.3	5.1
Yugoslavia	63.1	56.1	68.5	0.4	4.4	42.2	62.7	51.7	26.3
Total	130.5	150.9	222.9	172.8	113.3	145.4	(42.3)	37.6	77.5

continued

Appendix Table A.5. (*continued*)

Region and country	Average exports			Average imports			Net imports or exports[a]		
	1961–1963	1970–1974	1977–1979	1961–1963	1970–1974	1977–1979	1961–1963	1970–1974	1977–1979
					(1,000 metric tons)				
Total, Europe	530.6	923.5	1,572.8	1,075.2	1,597.1	1,980.9	(544.6)	(673.6)	(408.1)
Soviet Union	56.9	32.3	19.3	44.8	88.9	181.0	12.1	(56.6)	(161.7)
Oceania									
Australia	361.0	623.7	1,080.6	0.1	...	...	360.9	623.7	1,080.6
New Zealand	165.4	283.4	352.8	0.1	...	...	165.3	283.4	352.8
Total	526.4	907.1	1,433.4	0.2	...	...	526.2	907.1	1,433.4
Other									
Iran	...	...	...	0.5	1.2	25.0[b]	(0.5)	(1.2)	(25.0)[b]
Israel	...	...	...	10.3	24.0	36.9	(10.3)	(24.0)	(36.9)
Japan	...	...	...	5.1	86.6	146.1	(5.1)	(86.6)	(146.1)
Philippines	...	...	...	14.9	5.3	15.3	(14.9)	(5.3)	(15.3)
South Africa	21.0	46.8	33.3	0.3	33.5	33.2	20.7	13.3	0.1[b]
Total	21.0	46.8	33.3	31.1	150.6	265.5[b]	(10.1)	(103.8)	(223.2)
Total, world	1,907.3	2,990.2	4,282.2	1,817.3	2,820.2	3,622.6	90.0	170.0	659.6
Total net imports							1,638.4	2,249.4	2,431.5
Total net exports							1,728.4	2,419.4	3,091.1

[b]Estimation.

Appendix Table A.6. U.S. trade in cattle, beef, and veal products compared with all livestock and meat products, 1979

Commodity	Imports	Exports	Net[a]
	(millions of dollars)		
Beef and veal			
Fresh or frozen	1,757.8	227.2	(1,530.6)
Prepared, pickled, canned, and variety meats	209.0[b]	213.2[c]	4.2
Tallow and greases	0.6	697.1	696.5
Meat extract	2.9	...	2.9
Hides and skins			
Cattle parts	0.7	16.9	16.2
Cattle	31.1	905.9	874.8
Live cattle and calves	247.0	68.0	(179.0)
Total	2,249.1	2,128.3	(120.8)
Other meats and animals[d]	1,171.0	1,108.8	(62.2)
Total	3,420.1	3,237.1	(183.0)

Source: USDA, *Foreign Agriculture Circular: Livestock and Meat,* FLM-MT 8-80, April 1980.

[a]Net imports shown in parentheses.

[b]Includes canned, and prepared and preserved beef products.

[c]Includes pickled or cured beef and veal, and beef variety meats.

[d]Also includes items like wild animals and fur skins.

Appendix Table A.7. Bovine hide production by selected countries, 1960-1978

Country	Annual average 1960-1964	Annual average 1965-1969	Annual average 1970-1974	1978
		(metric tons)		
Argentina	248,298	275,810	246,701	370,842
Australia	92,073	97,346	129,290	202,075
Brazil	172,144	201,603	331,406	336,000
Canada	72,733	88,489	89,669	106,075
France	146,281	146,650	143,615	162,831
Germany, FR	125,985	129,041	145,783	164,516
Mexico	52,136	62,072	80,032	124,000
United States	809,685	979,593	1,011,608	1,133,509
Others	1,309,321	1,510,008	1,760,485	1,939,405
Total, 49 selected countries	3,028,656	3,490,612	3,938,589	4,539,253
		(%)		
Argentina	8	8	6	8
Australia	3	3	3	4
Brazil	6	6	8	7
Canada	2	2	2	2
France	5	4	4	4
Germany, FR	4	4	4	4
Mexico	2	2	2	3
United States	27	28	26	25
Others	43	43	45	43
Total, 49 selected countries	100	100	100	100

Source: Adapted from USDA, *Foreign Agriculture Circular: Livestock and Meat,* FLM-5-79, October 1979.

Appendix Table A.8. Bovine hide exports by selected countries, 1959–1978

Country	Annual average 1959–1963	1963–1967	1967–1971	1971–1975	1978
	(metric tons)				
Argentina	147,000	163,500	166,300	52,900	69,200
Australia	28,400	52,500	64,300	119,600	151,000
Belgium-Luxembourg	14,300	23,900	25,500	29,300	40,000
Canada	26,300	51,400	57,000	62,500	85,600
France	42,100	59,200	83,400	101,400	120,000
Germany, FR	33,700	50,300	60,300	72,800	102,000
South Africa	31,000	39,100	43,400	43,600	40,800
United States	168,000	335,400	407,500	514,300	700,800
Others	146,600	213,600	246,100	260,200	263,000
Total, 49 selected countries	637,400	988,900	1,153,800	1,256,600	1,572,400
	(%)				
Argentina	23	17	14	4	4
Australia	4	5	6	10	10
Belgium-Luxembourg	2	2	2	2	3
Canada	4	5	5	5	5
France	7	6	7	8	7
Germany, FR	6	5	5	6	6
South Africa	5	4	4	4	3
United States	26	34	36	41	45
Others	23	22	21	20	17
Total, 49 selected countries	100	100	100	100	100

Source: Adapted from USDA, *Foreign Agriculture Circular: Livestock and Meat*, FLM-5-79, October 1979.

Appendix Table A.9. Bovine hide imports by selected countries, 1959–1978

Country	Annual average 1959–1963	1963–1967	1967–1971	1971–1975	1978
	(metric tons)				
Czechoslovakia	32,200	41,000	41,100	46,900	46,900
Germany, FR	73,300	93,100	80,800	51,300	54,400
Italy	76,600	123,500	178,100	199,500	330,000
Japan	97,500	150,800	198,600	225,900	271,300
Mexico	6,600	23,900	46,500	56,500	46,900
Poland	29,200	39,500	43,100	51,600	41,600
Romania	5,100	15,100	28,100	39,900	43,000
Spain	16,500	34,700	51,300	72,600	80,000
Others	333,800	466,100	455,700	441,100	560,200
Total, 49 selected countries	670,800	987,700	1,123,300	1,184,100	1,474,300
	(%)				
Czechoslovakia	5	4	3	4	3
Germany, FR	11	9	7	4	4
Italy	10	13	16	17	22
Japan	15	15	18	19	19
Mexico	1	2	4	5	3
Poland	4	4	4	5	3
Romania	1	2	2	3	3
Spain	3	4	5	6	5
Others	50	47	41	37	38
Total, 49 selected countries	100	100	100	100	100

Source: Adapted from USDA, *Foreign Agriculture Circular: Livestock and Meat*, FLM-5-79, October 1979.

Appendix Table A.10. Cattle breeds native to the Western Hemisphere

Type of cattle or breed	Principal distribution	Purpose	Mature weight Females	Mature weight Males	Remarks
			(kg)		
Humpless cattle					
Caracú	Brazil	Beef, milk			Most important native breed of Brazil.
Criollo	Central America, Caribbean, and South America	Beef, milk	350–450	400–500	
Blanco Orejinegro	Colombia	Milk, beef	350–450	450–550	
Criollo Rio Limon	Venezuela	Milk			
Romo-Sinuano	Colombia	Beef	500–700	600–800	Believed to have originated from crossbreeding between Red Poll and/or Aberdeen Angus and Horned Sinú in the latter 1800s.
San Martinera	Colombia	Beef			
Texas Longhorn	United States	Beef			Originated from Spanish cattle and other cattle imported into the Americas; considerable expansion of numbers during the 1970s.

Source: Adapted from W. J. A. Payne, *Cattle Production in the Tropics. Vol. I: Breeds and Breeding* (London: Longman Group Ltd. 1970); and J. Rouse, *World Cattle*, vol. 3 (Stillwater: Oklahoma State University Press, 1970); and consultation with Tim Olson, University of Florida, and other breed specialists.

Appendix Table A.11. Cattle breeds recently developed, and crossbred types

Type of cattle or breed	Principal distribution	Purpose	Mature weight		Remarks
			Females	Males	
			(kg)		
Humpless cattle					
Cattalo (Beefalo)	United States	Beef	variable		Not breed; cross of buffalo (American Bison) on any type of cattle.
Hays Converter	Canada, United States	Beef	590	up to 1,270	Developed in Canada from Brown Swiss, Holstein, and Hereford crosses.
Holstein-Friesian	United States with worldwide distribution	Milk	450–700	925–1,200	Most important U.S.-developed breed; distinctly different from Dutch Friesian, which is dual-purpose breed.
Luing	United Kingdom	Beef	. . .	. . .	Developed from cross of Shorthorn and Scottish Highland cattle
Polled Hereford	United States, Canada, scattered distribution in rest of Western Hemisphere	Beef	390–600	450–1,100	Developed in United States form Horned Herefords; about one-third of all Herefords in United States are Polled.
Senepol(Nelthropp)	St. Croix Island in Virgin Islands	Beef	500	790	Cross of Red Poll on N'Dama.
Humpless × humped cattle					
Basuto	South Africa	Beef, milk, draft	. . .	. . .	Sanga type developed using Africander and Holstein breeds.
Barzona	United States	Beef	500–570	up to 850	Developed in Arizona from crosses of Hereford, Africander, Angus, and Santa Gertrudis.
Beefmaster	United States with more limited numbers in other Western Hemisphere countries	Beef	640 up to 730	1,000 up to 1,180	Initially composed of approximately ½ Brahman, ¼ Hereford and ¼ Shorthorn, breeding developed by Lasater Ranch in Texas.
Bonsmara	South Africa, scattered distribution in much of world	Beef, milk	250–390	540–770	Developed by J. C. Bonsmara; from base of about 5/8 Africander, 3/16 Shorthorn, and 3/16 Hereford.

Source: Adapted from W. J. A. Payne, *Cattle Production in the Tropics. Vol I: Breeds and Breeding* (London: Longman Group Ltd., 1970); and J. Rouse, *World Cattle,* vol. 3 (Stillwater: Oklahoma State University Press, 1970); and consultation with Tim Olson, University of Florida, and other breed specialists.

Appendix Table A.11. (continued)

Type of cattle or breed	Principal distribution	Purpose	Mature weight Females	Mature weight Males	Remarks
			(kg)		
Humpless cattle × humped cattle (continued)					
Braford	United States	Beef	550–590	800–820	About 5/8 Hereford and 3/8 Brahman; developed in Florida in the 1950s and 1960s on Adams Ranch.
Brahmousin	United States	Beef	...	...	Cross of Limousin and Brahman.
Brangus	United States with more limited numbers in other Western Hemisphere countries	Beef	550	890–930	Developed from foundation of 3/8 Brahman and 5/8 Angus; except for coloration, Red Brangus is very similar.
Charbray	United States with more limited numbers in other Western Hemisphere countries	Beef	...	...	Crossbred of 3/4–7/8 Charolais and 1/4–1/8 Brahman; developed in United States.
Droughtmaster	Australia	Beef	...	...	Developed from base of 3/8 to 5/8 Zebu mainly derived from Red Brahman imported from Texas and some Santa Gertrudis; Humpless breeds are Shorthorn, Hereford, and Red Poll.
Holmonger	Southwest Africa	Beef	...	...	1/2 Africander and 1/2 Brown Swiss.
Jamaica Black	Jamaica	Beef	...	...	Developed from cross of Angus and Zebu. Similar to Brangus.
Jamaica Hope	Jamaica	Milk	...	...	Crossbred type of Jersey and Sahiwal.
Jamaica Red Poll	Jamaica	Beef	...	...	Developed by crossing Red Poll cows with Zebu × Devon crossbred bulls.
Santa Gertrudis	United States, but widely dispersed all over Western Hemisphere and in many parts of world	Beef	610 up to 850	910–1,090	Developed from base of 3/8 Zebu and 5/8 Shorthorn developed by King Ranch in Texas.
Simbrah (Brahmental)	United States	Beef	...	...	Cross of Simmental and Brahman.
Humped cattle					
Brahman (American Brahman)	United States, but widely dispersed all over the Western Hemisphere	Beef	500–600	850–1,000	Developed in Gulf Coast area of United States between 1834 and 1926; Brahman is Zebu developed from various Indian breeds; Red Brahman is similar except red in color.

continued

Appendix Table A.11. (*continued*)

Type of cattle or breed	Principal distribution	Purpose	Mature weight		Remarks
			Females	Males	
			(*kg*)		
Humped cattle (*continued*)					
Gyr (Zebu)	Brazil, much of Latin America	Beef	similar to Brahman		Developed in Brazil from seedstock of Indian Gir.
Guzerat (Zebu)	Brazil, much of Latin America	Beef	similar to Brahman		Developed in Brazil from seedstock of Indian Kankrej.
Indo-Brazil	Brazil, Mexico, United States	Beef	550–700	900–1,050	Originated primarily from crosses of Gyr and Guzerat breeds, noted for its extremely long ears.
Jamaica Brahman	Jamaica	Beef	580–600	850–1,000	Essentially same as American Brahman.
Nellore (Zebu)	Brazil, much of Latin America	Beef	similar to Brahman		Developed in Brazil from seedstock on Ongole breed in India.
Romana Red	Dominican Republic	Beef	450–550	850–1,000	Developed from crosses of Zebu and Criollo cattle.
Senepol	Virgin Islands	Beef	400–500	700–900	Developed from Red Poll crosses with African N'Dama.

Appendix Table A.12. Cattle breeds originating in Europe, the USSR, and the United Kingdom, by country of origin

Type of cattle or breed	Principal distribution	Purpose	Mature weight		Remarks
			Females	Males	
			(kg)		
Austria					
Fleckvieh	Austria, Germany	Beef, milk	680–770	up to 1,180	Essentially Simmental cattle bred in Austria and Germany.
Grauvich (Gray cattle on Gray Mountain)	Austria	Milk, beef	450–550	680–820	
Pinzgauer	Austria, Romania	Beef, milk	590–680	up to 1,090	
Belgium					
Moyenne et Haute Belgique (Blue Spotted)	Belgium	Beef, milk	up to 860	up to 1,270	Cross of Charolais and Shorthorn on native cattle.
Pie Rouge de la Campine (Campine Red Spotted)	Belgium	Beef, milk	660	1,090	Related to Meuse-Rhine-Yessel.
Pie Rouge de la Flandre	Belgium	Beef, milk	700	up to 1,320	Related to Meuse-Rhine-Yessel.
Rouge de la Flandre Occidentale (Red cattle)	Belgium	Beef, milk	640–730	1,180 up to 1,450	Consolidation of Veune Ambaelt and Polders.
Bulgaria					
Iskur	Bulgaria	Draft, milk, beef	450	640	Derived from Gray Steepe cattle. Distinctive large upswept horns.
Czechoslovakia					
Czech Red and White	Czechoslovakia	Milk, beef	550	up to 1,090	Derived from European Red and White with some crossing of Simmental.

Source: Adapted from W. J. A. Payne, *Cattle Production in the Tropics. Vol. I: Breeds and Breeding* (London: Longman Group Ltd., 1970); and consultation with Tim Olson, University of Florida, and other breed specialists.

Appendix Table A.12. *(continued)*

Type of cattle or breed	Principal distribution	Purpose	Mature weight Females	Mature weight Males	Remarks
			(kg)		
Denmark					
Danish Black and White	Denmark	Milk, beef	up to 680	1,090–1,270	Derived from Dutch Friesian.
Danish Red and White	Denmark	Milk, beef			Cross of Shorthorn with German Red and White and Meuse-Rhine-Yessel.
Red Danish	Denmark, much of Europe, United States	Milk, beef	660	1,090–1,270	Cross of Angler with native cattle.
France					
Abondance	France, United States	Milk, beef			Derived from Swiss Simmental.
Armoricaine	France	Milk, beef	640–730	1,090	Derived from Shorthorn crossed with native cattle.
Charolais	France, scattered in Europe, and quite widely distributed in Western Hemisphere	Beef	640–730 or more	1,140–1,360 or more	Very widely distributed breed; first continental breed to be widely distributed.
Flamande (Flemish)	France	Milk, beef	590–700	up to 1,140	
Limousin	France, Canada, United States	Beef	610	1,090	One of few breeds developed exclusively for beef after originally being draft animal.
Maine-Anjou	France, United States	Beef, milk	910	1,360	Largest of French breeds.
Marchigiana	France, United States				
Montbéliard	France, United States	Milk, beef			Derived form Swiss Simmental.
Normande	France, United States, and Colombia	Milk, beef	590–680	up to 1,180	Derived from Shorthorn cattle crossed with native cattle.
Pie Rouge de l'Est (Red Spotted of the East)	France	Milk, beef			Derived from Swiss Simmental.
Salers	France, United States	Milk, beef	590 up to 770	1,000 up to 1,300	
Tarentaise	France, United States	Milk, beef	590	up to 1,090	Selected for milk production.

continued

Appendix Table A.12. (*continued*)

Type of cattle or breed	Principal distribution	Purpose	Mature weight Females	Males	Remarks
			(kg)		
Germany (East and West)					
Angler (Angeln)	West Germany	Milk, beef	590–640	1,000	
Polish Black and White	Poland	Milk, beef	590		Basically Friesian developed in Germany.
Polish Red	Poland	Beef, milk	450	up to 820	Derived from German Red.
Polish Red and White	Poland	Milk, beef	680		Basically German Red and White.
Rotvieh (Red cattle)	West Germany	Milk, beef	500–640	up to 950	
Wäldervich (Forest cattle)	West Germany	Milk, beef	up to 590	820	
Hungary					
Hungarian Spotted	Hungary	Milk, beef	450	up to 820	Simmental crossed on native cattle.
Italy					
Chianina	Italy, United States	Beef	820	up to 1,820	Started as draft animal; first Italian breed to reach United States.
Marchigiana	Italy, United States	Beef, milk, draft	590–680	up to 910	
Modenese	Italy	Beef	660–770	1,140	Similar to Romagnola.
Modicana	Italy	Milk, draft, beef	410–590	450–730	Sicilian breed.
Piemontese	Italy	Beef, milk, draft	640	820	Similar to Romagnola, but smaller.
Prete	Italy	Draft, beef	320–360	up to 450	Sicilian breed.
Reggiana	Italy	Milk, beef	550–640	820	
Romagnola	Italy, United States	Beef	660–800	1,140	Started as draft animal.
Netherlands					
Friesian	Worldwide	Milk, beef	640–700	1,040	
Groningen	Netherlands	Milk, beef	moderate		
Meuse-Rhine-Yessel	Netherlands	Milk, beef	730	1,140	

continued

Appendix Table A.12. (*continued*)

Type of cattle or breed	Principal distribution	Purpose	Mature weight		Remarks
			Females	Males	
			(kg)		
Norway					
Norwegian Red	Norway, United States	Milk			
Portugal					
Alentejana	Portugal	Draft, beef	550	320	
Romania					
Brown Maramures	Romania	Milk, beef			Swiss Brown cattle upgraded from Gray Steppe cows.
Gray Steppe	Romania	Milk, draft			Practically identical to Iskur of Austria.
Red Lativia	Romania	Milk, beef			Developed from cross of Red Danish on Gray Steppe cattle.
Simmental (Romania Spotted)	Romania	Beef, milk			Developed from cross of Swiss Simmental on native cattle.

Appendix Table A.13. Cattle breeds indigenous to Africa

Type of cattle or breed	Principal distribution	Purpose	Mature weight Females	Males	Related types	Remarks
			(kg)			
Humpless cattle						
Brown Atlas	Tunisia, Algeria, Morocco	Milk, beef, draft,	300–350	350–450	Moroccan, Tunisian	Indigenous at least since Roman times.
Dwarf Shorthorn	Gambia, Cameroon	Beef, milk, social, and some draft	120–350	120–450	West African Shorthorn, Nigerian Shorthorn, Lagoon, Gambian Dwarf, Bakasi, Somba, Lagone, Manjaca, Gold Coast Shorthorn	Tolerant to trypanosomiasis.
Kuri	Chad, Nigeria, Niger, Central African Republic	Milk, beef	300–800	300–800	None	Largest breed in West Africa; has gigantic bulbous horns.
Libyan	Libya	Milk, beef, draft	250–300	300–400	Brown Atlas	More of dairy than beef conformation.
N'Dama	West Africa	Beef, milk	210–350	220–420	Senegal N'Dama, Sierra Leone	Inherently tolerant to trypanosomiasis.
N'Dama × Dwarf Shorthorn Crossbreds	Ghana, Togo, Ivory Coast	Beef			Baoule', Gold Coast Shorthorn	Tolerant to trypanosomiasis.
Mauritius Creole	Mauritius	Milk	340–410	340–410		Often kept in small, dark, sheds at all times.
Humpless × humped cattle						
Africander	Southern Africa	Beef	360–540	450–900		Developed in 17th and 18th centuries from cattle owned by Hottentots.

Source: Adapted from W. J. A. Payne, *Cattle Production in the Tropics. Vol. I: Breeds and Breeding* (London: Longman Ltd., 1970); and J. Rouse, *World Cattle,* vol. 3 (Stillwater: Oklahoma State University Press, 1970); and consultation with Tim Olson, University of Florida, and other breed specialists.

continued

Appendix Table A.13. (*continued*)

Type of cattle or breed	Principal distribution	Purpose	Mature weight Females	Males	Related types	Remarks
			(kg)			
Humpless × humped cattle (continued)						
Alur	Uganda, Zaire	Beef, milk	300–350	300–350		
Angolan	Angola	Milk, beef				Longhorned Sanga. Similar to Ovampo.
Ankole	Uganda, Rwanda, Burundi, Tanzania	Social, milk, beef	200–400	350–500	Bahima, Watusi, Kigezi, Bashi	Being replaced by Zebu.
Baila	Zambia	Draft, beef, milk	large			
Bambara	Mali	Beef, milk	Small			Stabilized breed from cross of N'Dama and Zebu.
Barotse	Zambia, Rhodesia, Angola	Draft, beef, milk	Large		Ila	Longhorned Sanga type.
Bechuana	Botswana, Rhodesia, South Africa	Beef, milk, draft	400–480	580–710	Batawana, Southern, Sengologa, Seshaga, Damara	Longhorned Sanga; triple purpose.
Borgu	Nigeria, Dahomey, Ivory Coast	Draft	Small			Stabilized breed from cross of N'Dama, Dwarf Shorthorn, and White Fulani.
Danakil	Ethiopia	Milk, beef	200–300	250–370	Aliab Dinka, Nuer, Awail Dinka, Eastern Nuer, Anuak	Only longhorned Sanga in Ethiopia.
Egyptian	Egypt	Draft, milk, beef			Similar to cattle all over Egypt.	Essentially shorthorn humpless type.
Kuri × Zebu Crossbreds	Chad, Nigeria	Milk, beef			Kanem, Jotko, Kilara, Toubou	
Mashona	South Africa, Rhodesia	Beef, draft	160–400	360–630		Medium horned Sanga.
Matabele	Rhodesia	Milk, beef	Wide variation in size		Inkone, Govuvu	Intermediate type between medium and longhorned Sanga.
N'Dama × Dwarf Shorthorn × Zebu Crossbreds	West Africa	Milk, beef			N'Dama, Sanga, Gambia	

Appendix Table A.13. (*continued*)

Type of cattle or breed	Principal distribution	Purpose	Mature weight Females	Males	Related types	Remarks
			(kg)			
Humpless × humped cattle (continued)						
N'Dama × Zebu Crossbreds	Senegal, Gambia, Guinea	Milk, beef			Djakore', Hamitic, N'Gabou	
Nganda	Uganda	Milk, beef	300–340	280–420	Kyaro, Kyago, Serere	Fairly good milk production, but more of beef type.
Nguni	Swaziland, South Africa, Mozambique, Rhodesia	Beef, milk, draft	220–420	320–680	Bapedi, Bavenda	Medium-horned Sanga.
Nilotic	Ethiopia, Sudan, Egypt	Milk, beef, social	180–260	280–380		Good potential for beef production.
Ovampo	Southwest Africa	Milk, beef, draft				Longhorned Sanga.
Tonga	Zambia	Draft, milk, beef	450–680	450–680		Medium-horned Sanga.
Tuli	Rhodesia	Beef, milk, draft	up to 450	up to 680		Longhorned Sanga type; a useful, triple-purpose breed.
Humped cattle						
Adamawa	Cameroon, Nigeria	Milk, draft, beef	250–480	350–650	Banyo, Yola, Prewakwa	Triple-purpose breed; medium horned.
Abyssinian	Ethiopia	Draft, milk, beef	230–250	290–310	Ingessana, Murle, Begait	Quiet disposition.
Angoni	Zambia, Malawi, Mozambique	Milk, beef, draft	180–470	270–720	Zambia Angoni, Malawi Angoni, Mozambique Angoni, Malawi Zebu	Shorthorned Zebu cattle, wide variation in utility depending on region and size.
Azaouak	Upper Volta, Central Niger, Nigeria, Cameroon, Chad	Milk, draft, beef	300–400	350–500	Somewhat similar to Shuwa	Especially drought resistant; shorthorned.
Boran	Ethiopia, Somali, Kenya	Milk, beef	260–450	320–680	Somali Boran, Tanaland Boran, Kenya Boran	Polled animals are common.

continued

Appendix Table A.13. (*continued*)

Type of cattle or breed	Principal distribution	Purpose	Mature weight		Related types	Remarks
			Females	Males		
			(kg)			
Humpled cattle (continued)						
Diali	Niger	Beef	250–450	250–450		Medium horned.
Karamajong	Sudan, Uganda, Kenya	Milk, beef	320–350	400	Toposa, Jie, Turkana, Suk	Often considered subtype of Boran.
Madagascar Zebu	Madagascar	Beef, draft	320–440	320–450		Make up most of cattle on Madagascar.
Maure	Mauritania, Mali, Upper Volta	Milk, draft, beef	250–300	250–450	Tuareg	Completely dependent on grazing; shorthorned.
Red Bororo	Nigeria, Cameroon, Chad	Beef, milk, draft	300–450	350–500	Peul-Mber	Wild and intractable; poor milkers; long lyre-shaped horns.
Senegal Fulani	Senegal, Mauritania	Beef, milk, draft	250–300	300–350		Triple-purpose breed; lyre horned.
Shuwa	Chad, Nigeria, Cameroon	Milk, beef, draft	250–300	350–400	Fellata, Chad	Triple-purpose breed; shorthorned.
Small East African Zebu	East Africa, Sudan, Zaire					Major type of cattle found in East Africa.
Bukedi	Uganda	Beef, milk, draft	270–320	250–450		Frequently polled.
Lugware	Sudan, Zaire, Uganda	Beef, milk, draft	230–250	300–350		Polls are known; poor milkers.
Masai	Kenya, Tanzania	Milk, beef, social	260–360	350–400		Seldom slaughtered.
Mongalla	Sudan	Beef, milk	150	150		Poor milkers, but excellent beef animals; to some extent tolerant of trypanosomiasis.
Nandi	Kenya, Tanzania	Draft, milk, beef	200–320	310–410		Frequently polled; hump sometimes not noticeable in female.

Appendix Table A.13. (*continued*)

Type of cattle or breed	Principal distribution	Purpose	Mature weight		Related types	Remarks
			Females	Males		
			(*kg*)			
Humped cattle, Small East African (*continued*)						
Tanzania Zebu	Tanzania	Milk, beef	200–220	180–330		
Zanzibar Zebu	Island of Zanzibar	Milk, beef	250	320		
Small Somali Zebu	Somali, Ethiopia	Milk, draft beef			North Somali, Gasara, Garre	Small hump.
Sokoto	Nigeria	Milk, draft, beef	300–340	500–540	Similar to grey-white breeds of Indian cattle.	Essentially a dairy-draft breed; shorthorned.
Sudanese	Sudan	Milk, draft, beef	250–350	300–500	Kenana, Butana, White Nile, Western, Northern Province	Majority of these cattle kept by nomads.
Sudanese Fulani	Mali	Beef, milk	250–400	250–400	Toronke'	Fairly wild; lyre horned.
White Fulani	Nigeria, Cameroon	Milk, beef, draft	250–350	250–350	Wodabe	Triple-purpose breeds; lyre horned.

Appendix Table A.14. Cattle breeds indigenous to Asia

Type of cattle or breed	Principal distribution	Purpose	Mature weight (kg) Females	Males	Related types	Remarks
Humpless cattle						
Chinese Yellow	China		275	275		All cattle in China are referred to as "Chinese Yellow," which distinguishes them from the water buffalo and yak.
Bantangas	Philippines	Beef			Large Ilcos, Small Ilcos, Iloilo	Closely related to Chinese Yellow.
Manchurian	China	Beef, milk and draft				
Mongolian	China	Beef, milk, and draft				
Peking Black Pied	China	Milk				Cross of Pinchow and Holstein.
Pien niu	China, Tibet	Work, milk				Cross of Tibetan with yak.
Pinchow	China	Milk				Cross of Manchurian with Simmental.
Sanho	China	Milk				Cross of Mongolian with Siberian and Holstein.
Tibetan	China, Tibet	Draft, milk				Dwarf cattle.
Oksh	Western Asia	Milk, draft, beef	220–270		Jaulon, Kundi, Grey Steepe, Anatolian Black, and Eastern Anatolian Red	Principal use depends on area.
Humpless × humped cattle						
Burmese	Burma	Work	200–230	180–360	Chaubauk, Racing cattle	
Chinese Yellow	China					Result of interbreeding between humpless cattle in north and Zebu type from south.

Source: Adapted from W. J. A. Payne, *Cattle Production in the Tropics. Vol. I: Breeds and Breeding* (London: Longman Group Ltd., 1970).

Appendix Table A.14. (continued)

Type of cattle or breed	Principal distribution	Purpose	Mature weight		Related types	Remarks
			Females	Males		
			(kg)			
Humpless × humped cattle, Chinese Yellow (*continued*)						
Chinchwan	China	Draft				Perhaps best draft cattle in China.
Chowpei	China	Draft, beef				
Hwangpei	China	Draft				
Nanyang	China	Draft, milk				
Shantung	China	Beef, draft				Best beef cattle in China.
Szechwan	China	Draft				
Damascus	Syria, Turkey, Iran, Cyprus, Egypt	Milk	140–270	140–320		Usually tethered and not allowed to graze.
Indo-Chinese	Indo-China	Work	200–250	300–350		
Lebanese	Lebanon, Syria	Milk				Milking breed.
Siri	Bhutan, Darjeeling, Bengal, India, Nepal, Sikkim	Draft, milk	320–410	320–540	Kachcha Siri, Tarai, Subalpine cattle of Nepal	Well adapted to mountainous conditions.
Taiwan	Taiwan	Draft				Used for draft, but otherwise not too useful.
Thai	Thailand	Draft	340–500	340–500	Two types, Northern and Southern	Northern cattle true work cattle while Southern ones also used for fighting.
Humped cattle						
Alambadi	India	Draft				Longhorned Zebu; active and hardy draft breed.
Amrit Mahal	India	Work				Longhorned Zebu; wild disposition.
Bachaur	India	Draft			Hariana	Shorthorned Zebu; survive under poor feeding conditions.
Bhagnari	India	Draft, milk	290–400	400–590		Shorthorned Zebu; widely used for field operations.
Burgur	India	Draft				Longhorned Zebu; unsurpassed in speed and endurance in trotting.

continued

Appendix Table A.14. *(continued)*

Type of cattle or breed	Principal distribution	Purpose	Mature weight Females	Males	Related types	Remarks
			(kg)			
Humped cattle (continued)						
Dangi	India	Draft			Deoni	Lateral-horned Zebu; semi-nomadic in grazing.
Deoni	India	Draft, milk	290	450–680	Red Khandari	Lateral-horned Zebu; grazed in forest.
Dhanni	India	Draft	270–400	360–590		Lateral-horned Zebu; possess marked stamina.
Gaolao	India	Transport	340	430	Ongole	Shorthorned Zebu; developed as fast trotting breed for army transport.
Gir	India	Draft, milk	380	540	Mewati, Red Sindhi, Krisha Valley, Nimari	Lateral-horned Zebu; one of more widely exported breeds for beef purposes.
Hallikar	India	Draft	230	340	Gujamaua	Longhorned Zebu; one of the best all-around draft breeds in Southern India.
Hariana	India	Milk, work	350	360–540	Bhagnari, Gaolao, Ongole	Shorthorned Zebu; compact animal of graceful appearance.
Kangayam	India	Draft	110–270	140–410	Umblachery	Longhorned Zebu; has been exported to Ceylon.
Kankrej	India	Milk, draft	340–450	450–680		Lyre-horned Zebu; has excellent potential for beef production.
Khillari	India	Draft	320–360	450–630		Longhorned Zebu; fast, powerful draft animals; have been exported to Ceylon.
Krishna Valley	India	Draft	320	500–720	Ongole, Mysore, Gir, Kankrej	Shorthorned Zebu; have been exported to Brazil and United States.

Appendix Table A.14. *(continued)*

Type of cattle or breed	Principal distribution	Purpose	Mature weight Females	Males	Related types	Remarks
			(kg)			
Humped cattle (continued)						
Kumauni	India	Work				Small shorthorned or lyre-horned Zebu.
Lohoni	India	Milk, draft, pack	180–270	230–360	Lohoni, Afghan	Small shorthorned or lyre-horned Zebu; usually stall fed.
Malvi	India	Draft	300	350	Kankrej, Kenwariya, Kheri-garh	Lyre-horned Zebu; size and type dictate type of work.
Mewati	India	Draft, milk			Harina, Gir, Kankrej, Malvi	Shorthorned Zebu; this breed evolved from several breeds.
Nagori	India	Transport, draft, milk	320	320–360	Hariana and lyre-horned types	Shorthorned Zebu; fast trotting breed used for fast road work.
Nimari	India	Work, milk	300	400	Khamala, Khamgaon	Lateral-horned Zebu; has special ability to work in rough areas.
Ongole (Nellore)	India	Work, milk	430–450	540–610	Gaolao	Shorthorned Zebu; one of more widely exported breeds for beef purposes.
Ponwar	India	Draft	270–300	320–360	Morang (in Nepal)	Small shorthorned or lyre-horned Zebu; rarely handled so relatively intractable.
Rath	India	Work, milk			Hariana, Mewati, and Nagori	Shorthorned Zebu; considered a breed of the poor.
Red Sindhi	India	Milk, work	320–450	250–340	La Bela	Lateral-horned Zebu; majority of owners of these cattle are nomadic.
Sahiwal	Pakistan	Milk, work	270–410	450–590		Lateral-horned Zebu; one of most productive tropical dairy breeds; has been widely exported.

continued

Appendix Table A.14. *(continued)*

Type of cattle or breed	Principal distribution	Purpose	Mature weight		Related types	Remarks
			Females	Males		
			(kg)			
Humped cattle (*continued*)						
Shahabadi	India	Draft, milk				Small shorthorned or lyre-horned Zebu; could be good dual-purpose breed.
Sinhala	Ceylon	Draft, milk	180–200	230–320		Small shorthorned or lyre-horned Zebu.
South Chinese Zebu	China	Draft			Kwantung, Hainan, Kwang-si	
Taiwan Zebu	Taiwan	Draft				Cross of Kwantung cattle with Zebu breeds.
Bos Bibos						
Bali (Bos Bibos Banteng)	Indonesia	Work, beef	250–300	350–400		Has high dressing percentage; especially agile in work; domesticated Banteng; crosses with cattle produce sterile F_1 males but fertile females.
Banteng (Bos Bibos Banteng)	Indonesia	Wild	250–300	350–400	Malay, Siamese, and Burmese Bantengs.	Wild relative of cattle.
Gaur (Bos Bibos Gaurus)	India, Southeast Asia	Wild			Malaysian Gaur, Readi	Wild relative of cattle, not true cattle; conformation similar to that of some types of *Bos Indicus* cattle; domesticated Gauri called Mitbrans; could possibly be of some importance as beef producers; Male F_1 progeny of crosses with cattle are sterile.
Kouprey (Bos Bibos Sauveli)	Cambodia	Wild	250–300	350–400	In some respects an intermediate between Gaur and Banteng	Wild; small numbers, but could be of economic importance.
Madura	Indonesia	Work	230–350	450–500	Javanese	Thought to have originated from cross between Zebu and Banteng at least 1,500 years ago.

Appendix Table A.15. Level and prevalence of 20 bovine diseases, by region and country, temperate countries, 1977

Region, Country, and FAO Country Number	Foot-and-Mouth Disease	Rinderpest	Bovine Rhinotracheitis (IBR)	Contagious Bovine Pleuropneumonia	Rabies	"Heartwater"	Leptospirosis	Anthrax	Blackleg	Intestinal Salmonella Infection	Bovine tuberculosis	Johne's Disease	Actinomycosis (Lumpy Jaw)	Brucella abortus (Bang's Disease)	Anaplasmosis	Trypanosomiasis (insect borne)	Mange and Scabies	Warble Infestation	Distomatosis (Liver Fluke)	Echinococcosis-Hydatidosis
	1	2	3	4	5	6	7	8	9	10	11	12	13	14	15	16	17	18	19	20
North and Central America																				
4.02 Canada			xx		x		x	x	xx	x	x	x	x	x			x	xx	x	
4.03 United States			xx		xx		xx	x	xx	xx	x	x	x	x	x		x	xx	x	
South America																				
3.10 Argentina	xx				x		x	x	xx	x	xx	x	x	x	x		xxx		xxx	xxx
3.13 Chile	x				x		xx	x	xx	x	xxx	x	x	xxx			x	x	xxx	xxx
3.08 Falkland Islands																				x
3.11 Uruguay	x						xx	x	xx	x	x	x	xx	x	x		xx	x	xxx	xxx
Asia																				
6.19 Afghanistan	xx				x			x	x		xx		x	x	x		x	x	xx	x
7.19 China				x				x												
6.11 Cyprus			x							x			x				x			x
6.18 Iran	x		x		x		x	x	xx	xx	x	x	x	xx	x	x	x	xx	x	xx
6.17 Iraq	xxx				x			x	x	x	x		x	x	x		x	x	x	xx
6.15 Israel	x		x		x		x	x		xx		x	x	x	xx			x	x	x
8.02 Japan			xx					x	x	x	x	x	x	x				x	xxx	

Source: Adapted from FAO, *Animal Health Yearbook*, 1970.
Note: x = low, xx = moderate, xxx = high.

continued

Appendix Table A.15. (*continued*)

Region, Country, and FAO Country Number	Foot-and-Mouth Disease 1	Rinderpest 2	Bovine Rhinotracheitis (IBR) 3	Contagious Bovine Pleuropneumonia 4	Rabies 5	"Heartwater" 6	Leptospirosis 7	Anthrax 8	Blackleg 9	Intestinal Salmonella Infection 10	Bovine tuberculosis 11	Johne's Disease 12	Actinomycosis (Lumpy Jaw) 13	Brucella abortus (Bang's Disease) 14	Anaplasmosis 15	Trypanosomiasis (insect borne) 16	Mange and Scabies 17	Warble Infestation 18	Distomatosis (Liver Fluke) 19	Echinococcosis-Hydatidosis 20
Asia (continued)																				
6.16 Jordan	x										x	x	x				x	x	x	x
8.01 Korea			x				x	x	x	x	x								xx	
6.14 Lebanon	xx	x						x	x	x	x		x	x	x		xx		xx	xx
7.20 Mongolia	x			xx	x			x	x			x		xxx			xx	xxx	x	x
6.13 Syria	x				x			x	xx		x	x	x	x	xx		x	x	xx	xx
6.12 Turkey					x		x	x	x		x	x	x	xx	x			xx	x	xx
Europe																				
6.09 Albania							x	x			x	x	x	x				x	x	x
5.16 Austria			x		x		x		xx	x	x		x	x			x	x	xxx	x
5.11 Belgium			x		x		x	x	x	x	x	x	x	x			xx	x	xx	
5.12 Luxembourg					x					x		x		x				x	xx	x
6.07 Bulgaria								x			x	x	x		x			x	xx	xx
6.04 Czechoslovakia			x		x		x	x	x	x	x		x				x	x	x	x
5.06 Denmark			x		x			x		x	x	x	x				x		xx	
5.09 Finland								x		x	x		x					x	x	
5.13 France			xx		x		x	x	x	x	x	x	x	xx			x	xx	xx	x
6.03 Germany, DR	x				x		x	x		x	x		x	x			x	x	x	x

Appendix Table A.15. *(continued)*

Region, Country, and FAO Country Number	Foot-and-Mouth Disease 1	Rinderpest 2	Bovine Rhinotracheitis (IBR) 3	Contagious Bovine Pleuropneumonia 4	Rabies 5	"Heartwater" 6	Leptospirosis 7	Anthrax 8	Blackleg 9	Intestinal Salmonella Infection 10	Bovine tuberculosis 11	Johne's Disease 12	Actinomycosis (Lumpy Jaw) 13	Brucella abortus (Bang's Disease) 14	Anaplasmosis 15	Trypanosomiasis (insect borne) 16	Mange and Scabies 17	Warble Infestation 18	Distomatosis (Liver Fluke) 19	Echinococcosis-Hydatidosis 20
Europe (continued)																				
5.14 Germany, FR	x		x		x		x	x	x	x	x	x	x	x			x	x	x	x
6.10 Greece			x		x			x	x	x	x		x	x	x		x	xx	x	xx
6.05 Hungary			x		x		x	x			x		x	x				x	x	x
4.01 Iceland												x	x							
5.05 Ireland			x				x		x	x	x	x	x	xx				x	xxx	x
5.17 Italy	xx		x				x	x	x	x	x	x	x	x			x	x	xx	x
5.18 Malta	x						x			xx	x			xx					x	x
5.10 Netherlands			xx				x	x	x	x	x	x	x	x			x	x	x	x
5.07 Norway			x					x	x	x			x						x	
6.02 Poland			x		x		x	x	x		x		x	x			x	x	xx	x
5.20 Portugal										x	xx		x	xx	x		x	x	xx	xx
6.06 Romania			x		x		x	x	x	x	xx	x	x					x	xx	x
5.19 Spain								x	x	x	x			x			x	xx	xx	x
5.08 Sweden									x	x	x		x						x	
5.15 Switzerland			x		x		x	x	x	x		x	x				xx	x	x	x
5.01 United Kingdom			xx				x	x	x	xx	x	x	x	x			x	x	xx	x
6.08 Yugoslavia	x		x		x		x	x	x	x	xx	x		x			x	x	x	x
6.01 USSR	x		x		x		x	x	x	x	x	x	x	x	x		x	x	x	x

Appendix Table A.16. Level and prevalence of 20 bovine diseases, by region and country, warm countries, 1978

Region, Country, and FAO Country Number	Foot-and-Mouth Disease 1	Rinderpest 2	Bovine Rhinotracheitis (IBR) 3	Contagious Bovine Pleuropneumonia 4	Rabies 5	"Heartwater" 6	Leptospirosis 7	Anthrax 8	Blackleg 9	Intestinal Salmonella Infection 10	Bovine Tuberculosis 11	Johne's Disease 12	Actinomycosis (Lumpy Jaw) 13	Brucella abortus (Bang's Disease) 14	Anaplasmosis 15	Trypanosomiasis (insect borne) 16	Mange and Scabies 17	Warble Infestation 18	Distomatosis (Liver Fluke) 19	Echinococcosis-Hydatidosis 20
Africa																				
1.04 Algeria	x				x			x	x		xx	x		x		x	x	x	x	x
3.07 Angola	x			xxx		xx		xx	xx		xx		x	xx	xx	xx	xx	x	xx	xx
2.15 Benin	x			x	x			x			x					xxx	x			
3.03 Botswana	x				x	xx		x	xx	xx	x		x	xxx	xxx	x	x		x	x
1.18 Burundi	xx				x	x		x	xx	xx	xx	xx	x	xxx	x	xx	x		xxx	x
2.17 Cameroon	x			x		x		x	xx		x			x		xxx	x		xxx	x
2.07 Cape Verde								x		x	x		x	x			x		xx	x
2.20 Central African Republic	xx					x		x	x		x			x		xxx	x		xx	x
2.01 Chad	x			xx		x		x	x		x			x	x	xxx	x		xx	x
1.20 Congo						x					xx		x	xx		xxx			x	xx
1.10 Djibouti											x				x		x		xxx	xx
1.07 Egypt	xx							x	x	x	x			x	x		xx	x	xxx	x
2.18 Equatorial Guinea	x										x						x			x
1.09 Ethiopia	xxx	x	x	x	x	xxx	x	x	xxx	x	x		x	x	x	xxx	xx	xx	xxx	xx

Source: Adapted from FAO, *Animal Health Yearbook,* 1978.
Note: x = low, xx = moderate, xxx = high.

Appendix Table A.16. (continued)

Region, Country, and FAO Country Number	Foot-and-Mouth Disease 1	Rinderpest 2	Bovine Rhinotracheitis (IBR) 3	Contagious Bovine Pleuropneumonia 4	Rabies 5	"Heartwater" 6	Leptospirosis 7	Anthrax 8	Blackleg 9	Intestinal Salmonella Infection 10	Bovine Tuberculosis 11	Johne's Disease 12	Actinomycosis (Lumpy Jaw) 13	Brucella abortus (Bang's Disease) 14	Anaplasmosis 15	Trypanosomiasis (insect borne) 16	Mange and Scabies 17	Warble Infestation 18	Distomatosis (Liver Fluke) 19	Echinococcosis-Hydatidosis 20
Africa (continued)																				
2.19 Gabon											x			x		xx	xx			
2.06 Gambia								x	x					x	x	xx	x			
2.13 Ghana	x	x		x	x	x		xxx	x	xx	x	x	x	x	xxx	xxx	xxx		x	
2.09 Guinea				x	x			xx	xx	xx				x		xxx	x		xx	xx
2.08 Guinea-Bissau		x				x		xxx	xxx		x			xx	xx	xx	x		xx	x
2.12 Ivory Coast	x			xxx	x	x		xx	xx		x			xxx	x	xxx	x		xxx	x
1.12 Kenya	xxx	x	x	x	x	x	x	xx	xx	x		x	x	xx	xxx	x	x	x	xx	xxx
3.04 Lesotho						xx		x	xx				x	x	xx		xx		xx	x
2.11 Liberia				x				x	x		x			x	xx	xxx	x		xxx	
1.06 Libya	x							x			xx		x	x	x		x	x		
8.21 Madagascar					x	x		x	xx	x	xx				x				xxx	
1.15 Malawi					x	x			xx	x	xx		x	x	xx	x	x		xx	x
2.04 Mali		x		xx	x	x	x	xx	xx		xx			xx	x	xxx	xx		xx	xx
1.01 Mauritania	x	x	x	xx		x		x	x		x	x	x	x	x	x	x		x	x
8.20 Mauritius											x			x	x		x			
1.03 Morocco	x				x			xx	xx	x	x		x	x	xx		x	x	xx	xxx
3.01 Mozambique	xx				x	xxx	x	x	x	xx	xx		x	xx	xxx	xxx	x		xxx	xx
3.08 Namibia	x		x	x	xx	x		x	xx	x			x	x	xxx	x	xx	x	x	x

continued

Appendix Table A.16. (continued)

Region, Country, and FAO Country Number	Foot-and-Mouth Disease	Rinderpest	Bovine Rhinotracheitis (IBR)	Contagious Bovine Pleuropneumonia	Rabies	"Heartwater"	Leptospirosis	Anthrax	Blackleg	Intestinal Salmonella Infection	Bovine Tuberculosis	Johne's Disease	Actinomycosis (Lumpy Jaw)	Brucella abortus (Bang's Disease)	Anaplasmosis	Trypanosomiasis (insect borne)	Mange and Scabies	Warble Infestation	Distomatosis (Liver Fluke)	Echinococcosis-Hydatidosis
	1	2	3	4	5	6	7	8	9	10	11	12	13	14	15	16	17	18	19	20
Africa (*continued*)																				
2.02 Niger	x	x		x				xx	xx		xx		x	x	x	xxx	xx		xxx	xx
2.16 Nigeria	xx		x	xx	x	xx	x	xx	xx	x	x	x	x	xx	xx	xxx	xx		xxx	x
8.22 Reunion						x	x			x	x			x					x	x
3.02 Zimbabwe	x		xxx		x	x		x	x	x	x	x	x	xxx	xxx	x	x		xxx	x
1.17 Rwanda	x				x	x		x	x	x	x		x	x	x	xxx	x	x	xxx	x
2.05 Senegal	x			x				x	x	x	x			x	x	xx	x		xx	
2.10 Sierra Leone				x	x			x	x	x				xx	x	x	x		xx	x
1.11 Somalia	x			xx		x	x	x	x	x	x		x	x	x	x	x		x	x
3.06 South Africa	x		x		x	xxx	x	x	x	xxx	x	x	x	xx	xxx		x	x	xxx	x
1.08 Sudan	xx	x		x	x		x	xx	x		x	x	x	x	x	x	x	x	x	x
3.05 Swaziland					x	xx		x	xxx	x	x	x	x	xx	x		x		xxx	xxx
1.14 Tanzania	xxx				x	x		xx	xxx		x		x	xx	xx	xxx			xxx	x
2.14 Togo	x	x		x	xx			x	x	x	x		x	x		xx	x	x	xxx	xx
1.05 Tunisia					xx		x	x	xx		x			xx	xx		x	x	xxx	xxx
1.13 Uganda	xxx			x	x	xxx		x	x	x	xxx	x	x	xxx	xxx	xxx	xxx	x	xxx	xxx
2.03 Upper Volta	x	x		xx	x		x	x	x	x	x				x	xx	xx	x	xx	x

Appendix Table A.16. *(continued)*

Region, Country, and FAO Country Number	Foot-and-Mouth Disease 1	Rinderpest 2	Bovine Rhinotracheitis (IBR) 3	Contagious Bovine Pleuropneumonia 4	Rabies 5	"Heartwater" 6	Leptospirosis 7	Anthrax 8	Blackleg 9	Intestinal Salmonella Infection 10	Bovine Tuberculosis 11	Johne's Disease 12	Actinomycosis (Lumpy Jaw) 13	Brucella abortus (Bang's Disease) 14	Anaplasmosis 15	Trypanosomiasis (insect borne) 16	Mange and Scabies 17	Warble Infestation 18	Distomatosis (Liver Fluke) 19	Echinococcosis-Hydatidosis 20
Africa (continued)																				
1.19 Zaire						x		xx	xxx	xx	xx	x	x	xx	xxx	xxx	xxx		xxx	xx
1.16 Zambia					x	xx		x	x	x	xx	x	x	xx	xx	xxx	x		xxx	x
North and Central America																				
4.13 Bahamas													x	x						
4.05 Belize					x		xxx		xx					x	x					
4.12 Bermuda														x						
4.10 Costa Rica					x		x	x	xx	x	x	x	x	xxx	x		x		x	
4.14 Cuba			x		x		x		x	xx	x	x	x	x	xx		x		x	
4.17 Dominican Republic			x		x		xxx				xx		x	xxx	xx		x		x	
4.07 El Salvador			x		xx		x	xxx	xxx	x	xx		x	x	xxx		xx			
4.06 Guatemala					xx		x	xxx	xxx	x	xxx	x	x	xxx	xxx		xx		xx	xx
4.16 Haiti							x	x		x			x	x			x	x	xxx	x
4.08 Honduras					x		x		xxx		x		x	xx	x	x	x	x		
4.15 Jamaica							x		xx		xx	x		x	xx				xx	

continued

Appendix Table A.16. *(continued)*

Region, Country, and FAO Country Number	Foot-and-Mouth Disease	Rinderpest	Bovine Rhinotracheitis (IBR)	Contagious Bovine Pleuropneumonia	Rabies	"Heartwater"	Leptospirosis	Anthrax	Blackleg	Intestinal Salmonella Infection	Bovine Tuberculosis	Johne's Disease	Actinomycosis (Lumpy Jaw)	Brucella abortus (Bang's Disease)	Anaplasmosis	Trypanosomiasis (insect borne)	Mange and Scabies	Warble Infestation	Distomatosis (Liver Fluke)	Echinococcosis-Hydatidosis
	1	2	3	4	5	6	7	8	9	10	11	12	13	14	15	16	17	18	19	20
North and Central America (continued)																				
4.04 Mexico			xx		xxx		xx	xx	xxx	x	xx	x	x	xxx	xxx		x	x	xxx	x
4.09 Nicaragua					x		xx	xx	xx	x	x		xx	xx	x	x	x		xx	
4.11 Panama					x		x		xx	x	x		x	xx	x	x		xx		
4.19 Trinidad, etc.					x		x			x	x	x			xxx		x			
4.20 Virgin Islands, U.S.							x			x			xx		xx					
South America																				
3.14 Bolivia	xx				xxx		xxx	xx	xxx	x	xx	x	x	xx	xxx	x	x		xxx	xx
3.09 Brazil	xx		x		xx		xx	x	xx	x	xx		x	xx	xx		x		x	xx
3.17 Colombia	xx		x		x		x	x	xx	xx	x	x	x	xx	xx	x	x	x	xx	
3.16 Ecuador	xx				x		xx	x	xx	xx	x	x	x	xx	xx	x		xxx	xxx	
3.19 Guyana	x				xx		x	x	x	x	x		x	x	xx	xx	x			
3.12 Paraguay	x				x		x	xx	x	x	xx		x	xxx	x		xx			x
3.15 Peru	x		x		x		x	x	x	x	xx	x	x	x	x		x		xx	x
3.20 Surinam					x					x	x				xx		x			
3.18 Venezuela	xx				x		xx	x	xx	xx	x	x	x	xx	xx	xx	x		x	x

Appendix Table A.16. (continued)

Region, Country, and FAO Country Number	Foot-and-Mouth Disease 1	Rinderpest 2	Bovine Rhinotracheitis (IBR) 3	Contagious Bovine Pleuropneumonia 4	Rabies 5	"Heartwater" 6	Leptospirosis 7	Anthrax 8	Blackleg 9	Intestinal Salmonella Infection 10	Bovine Tuberculosis 11	Johne's Disease 12	Actinomycosis (Lumpy Jaw) 13	Brucella abortus (Bang's Disease) 14	Anaplasmosis 15	Trypanosomiasis (insect borne) 16	Mange and Scabies 17	Warble Infestation 18	Distomatosis (Liver Fluke) 19	Echinococcosis-Hydatidosis 20
Asia																				
7.07 Bahrain				x									x				x		x	x
7.13 Bangladesh	xx				x			x	x					x		x		x	xx	
7.12 Bhutan	xxx			x	x		x	xx	xx		xx		x				x	x	xxx	x
8.10 Brunei							x								x	x	x		xxx	
7.14 Burma	xxx			x	x			x	xx	x	x	x	x	x	x	xx	x		xx	
8.05 East Timor								xxx	xxx	xx					x		xx		xxx	xx
7.22 Hong Kong	x								x	x	x	x		xxx	x				xx	x
7.10 India	xx	x		x	x		x	x	xx	xx	xx	xx	x	xxx	xx	xx	xx	xx	xxx	xxx
8.04 Indonesia	x						x	x	x		x	x	x	x	x	xx	xxx	x	xxx	
7.16 Kampuchea, Dem.	x				x			x	x	xx	x					x			xx	x
7.08 Kuwait	x	x		x				x		x	x		x	x			xx	x	x	x
7.17 Laos	x				x			x	x	xx	x			xx		x	x		xx	xx
7.21 Macao	x									x	x			x					x	x
8.07 Malaysian Peninsula	x						x			x	x	x		x	x	x	x		xx	

continued

Appendix Table A.16. (continued)

Region, Country, and FAO Country Number	Foot-and-Mouth Disease 1	Rinderpest 2	Bovine Rhinotracheitis (IBR) 3	Contagious Bovine Pleuropneumonia 4	Rabies 5	"Heartwater" 6	Leptospirosis 7	Anthrax 8	Blackleg 9	Intestinal Salmonella Infection 10	Bovine Tuberculosis 11	Johne's Disease 12	Actinomycosis (Lumpy Jaw) 13	Brucella abortus (Bang's Disease) 14	Anaplasmosis 15	Trypanosomiasis (insect borne) 16	Mange and Scabies 17	Warble Infestation 18	Distomatosis (Liver Fluke) 19	Echinococcosis-Hydatidosis 20
Asia (*continued*)																				
8.08 Malaysian Sabah											x				x					
8.09 Malaysian Sarawak									x	x									xx	
7.11 Nepal	xxx	xxx		x	xx					x	xx	x		xx	x	x	xx	x	xxx	xx
7.04 Oman	xx							x	x					x	x		x	x	x	x
6.20 Pakistan	xxx	xx						x	xxx		xx				x		x	xx	xxx	x
8.03 Philippines	xx				x		xx	x	x	x	x		x	x	x	x	x	x	xxx	x
7.06 Qatar	xx														x		xx		x	
7.01 Saudi Arabia	xx	xx			x						x		x	x	x		x		x	
8.06 Singapore											x								x	
7.09 Sri Lanka	xx				x			x	x	x	x	x	x	x	x		x	x	x	
7.15 Thailand	xx							x	x		x			x	xx	x	xx		xxx	x
7.05 U. Arab Emirates	xx																x		x	x
7.18 Viet Nam	x	x					x	x	x		x			x		x			xx	
7.02 Yemen, Arab Republic	xx	xx																	xx	x

Appendix Table A.16. (*continued*)

Region, Country, and FAO Country Number	Foot-and-Mouth Disease	Rinderpest	Bovine Rhinotracheitis (IBR)	Contagious Bovine Pleuropneumonia	Rabies	"Heartwater"	Leptospirosis	Anthrax	Blackleg	Intestinal Salmonella Infection	Bovine Tuberculosis	Johne's Disease	Actinomycosis (Lumpy Jaw)	Brucella abortus (Bang's Disease)	Anaplasmosis	Trypanosomiasis (insect borne)	Mange and Scabies	Warble Infestation	Distomatosis (Liver Fluke)	Echinococcosis-Hydatidosis
	1	2	3	4	5	6	7	8	9	10	11	12	13	14	15	16	17	18	19	20
Asia (continued)																				
7.03 Yemen, PDR	x						x				x			x	x		x			x
Oceania																				
8.12 Australia			x				x	x	x	x	x	x	x	xx	x		x		xx	xx
8.15 Fiji										x	x	x	x	x						
8.18 French Polynesia													x	x						
8.17 New Caledonia										x							xx			
8.16 New Hebrides			x				x			x	x		x	x			x			
8.11 New Zealand			x				xxx		x	x	x	x	x	x			x		x	x
8.13 Papau New Guinea							x			x			x		x	x			xx	x
8.19 Samoa							x				x		xx	xx						x
8.14 Solomon Islands			x				x				x		x	x						

Appendix 17

GUIDELINES FOR INTERNATIONAL COOPERATION IN THE LIVESTOCK AND MEAT SECTOR

(as adopted by the United Nations, Food and Agriculture Organization, Intergovernmental Group on Meat, at its Sixth Session, Rome, October 1976)

A. General objectives of international cooperation

The broad objective of international cooperation in the livestock and meat sector should be to secure a balanced expansion in meat production and consumption—particularly in countries where animal protein deficiency exists—and trade. The attainment of this objective should be beneficial to both producers and consumers and should create equitable conditions for sustaining the development efforts of developing countries. In particular, measures taken should:

(i) support the efforts of developing countries to develop their livestock and meat industry through integrated technical assistance and aid and investment programs (including genetic improvement, research, training and extension), so as to help develop fully their production potential to satisfy the growing domestic demand for meat. Such development efforts should pay particular attention to promoting livestock production at the small farmers' level and to improving their standard of living, taking into consideration the development of indigenous technology and the utilization of local resources;

(ii) improve consumption and nutritional levels and promote the efficiency of meat production and marketing, and thereby farm incomes, as well as the overall conditions of international trade in meat;

(iii) take into account the interests of both exporting and importing countries and the special contribution which the livestock and meat sector makes to the development process of developing countries;

(iv) aim at mitigating the impact of market instability on the incomes and foreign exchange earnings of countries engaged in international trade in livestock and meat, and in particular of the developing countries;

(v) promote greater participation of developing countries in the international trade of meat.

B. National measures

(i) Since policies affecting the profitability of cereals production may have important influences on the livestock sector, governments should endeavour to ensure that such policies and policy instruments avoid any destabilizing effects on domestic and external livestock and meat economies, and are without prejudice to the meat imports especially from developing countries.

(ii) In the event of domestic oversupply, measures for increasing the demand and consumption should be given priority before resorting to measures designed to stimulate exports.

(iii) Improvements in the processing and marketing of meat should be encouraged as a means of facilitating a continuing adjustment of meat supply and demand and of reducing market instability and of expanding overall production and consumption.

(iv) In order to promote greater harmonization among national meat policies, the Intergovernmental Group on Meat should periodically review national policies affecting production, consumption, and international trade of meat.

C. Trade policies

(i) Governments should endeavour to ensure that the consequences of instabilities arising in national livestock and meat industries do not harm the livestock sectors of other countries and in particular those of developing countries.

(ii) To the extent that an overall world imbalance between demand and supply of meat is due to developments within the livestock and meat industries of countries engaged in international trade in these products, an exchange of views should take place among governments of the countries concerned with a view to assuring, under satisfactory conditions, both outlets for the production of exporters and continuity of supplies to meet requirements of importers. Such exchange of views should take full account of the need for developing countries with production potential to expand output and exports at remunerative prices as part of their development efforts.

(iii) In order to safeguard the interests of meat exporting and importing countries, consultations should take place in the appropriate manner and fora and in particular within the General Agreement on Tariffs and Trade (GATT), among governments of the countries concerned whenever either side intends to take action which could cause harmful interference with the normal patterns of international trade or which could adversely affect the development efforts of developing exporting countries.

(iv) When trade restrictions and other measures of a temporary and exceptional nature are introduced by importing countries, they should pay particular attention to safeguarding the development interests of meat exporting developing countries; and to this end, when necessary, special and preferential arrangements should be made by developed countries in favour of imports from developing countries.

(v) When accumulated stocks of meat are disposed of on concessional terms in foreign markets, such disposals should be carried out in accordance with the FAO Principles of Surplus Disposal and Consultative Obligations.

(vi) Where feasible and economically advisable, governments should consider entering into longer-term contracts for exports and imports of livestock and meat.

(vii) Importing countries should provide for the uniform and consistent application over time of their animal health and meat hygiene regulations to imports from all sources.

(viii) Governments should use the opportunity offered in the Intergovernmental Group on Meat for the regular exchange of information on national measures affecting international trade and for consultations on possible remedial action when any special difficulties arise.

INDEX

import quotas, 13
leader in price making, 182, 185